METABOLISM AND ENZYMOLOGY OF NUCLEIC ACIDS
Including Gene Manipulations

METABOLISM AND ENZYMOLOGY OF NUCLEIC ACIDS
Including Gene Manipulations

Edited by
Ján Zelinka and Jozef Balan
Slovak Academy of Sciences
Bratislava, Czechoslovakia

PLENUM PRESS • NEW YORK AND LONDON

Library of Congress Cataloging in Publication Data

International Symposium on Metabolism and Enzymology of Nucleic Acids Including Gene Manipulations (6th: 1987: Smolenice, Czechoslovakia)
  Metabolism and enzymology of nucleic acids, including gene manipulations /
edited by Ján Zelinka and Jozef Balan.
    p.   cm.
  "Proceedings of the Sixth International Symposium on Metabolism and Enzymology of Nucleic Acids Including Gene Manipulations, held June 7–11, 1987,
at Smolenice Castle, Czechoslovakia" — T.p. verso.
  Sponsored by the Slovak Academy of Sciences.
  Includes bibliographical references and indexes.

  ISBN-13: 978-1-4612-8063-7   e-ISBN: 978-1-4613-0749-5
  DOI: 10.1007/978-1-4613-0749-5

  1. Nucleic acids—Congresses. 2. Nucleases—Congresses. 3. Methylation—
Congresses. 4. Molecular cloning—Congresses. 5. Gene expression—Congresses. I. Zelinka, J. (Ján) II. Balan, J. (Jozef) III. Slovenská akadémia vied. IV. Title.
QP620.I56 1987   88-39215
574.87'3282—dc19   CIP

Proceedings of the Sixth International Symposium on Metabolism and
Enzymology of Nucleic Acids, Including Gene Manipulations,
held June 7–11, 1987, at Smolenice Castle, Czechoslovakia.
This symposium was held under the auspices of the President of the
Slovak Academy of Sciences, Academician Vladimír Hajko.

© 1988 Plenum Press, New York
Softcover reprint of the hardcover 1st edition 1988
A Division of Plenum Publishing Corporation
233 Spring Street, New York, N.Y. 10013

OPENING OF THE SYMPOSIUM

I would like to open the Sixth International Symposium on Metabolism and Enzymology of Nucleic Acids Including Gene Manipulations, organized by the Institute of Molecular Biology of the Slovak Academy of Sciences under the auspices of Academician Vladimír Hajko, President of the Slovak Academy of Sciences.

This symposium, in a similar way as our previous meetings, is part of the multilateral cooperation of the Academies of Sciences of the socialist countries in the field of molecular biology and biochemistry.

We are glad to be able to welcome here, at Smolenice Castle, numerous outstanding scientists from many countries and our friends whom we traditionally meet here.

Fourteen years have passed since our first symposium in 1973, and tremendous advances have been achieved in the field of the metabolism and enzymology of nucleic acids which have resulted in gene manipulations and have stimulated the formation of modern biotechnologies. We are happy that our Institute has contributed to this development in a modest way. Our work was helped and supported by the organization of these symposia, giving our younger co-workers the opportunity to find new stimulation for their work and to get acquainted with scientists from other countries, thus fostering creative international cooperation. It is especially valuable that this cooperation takes place between countries without regard to their socio-economic setup. The scientists who meet here hold different political opinions, but in spite of this, we are long-time friends, we respect and support each other, and each of us contributes to the development of science which serves all of mankind. This is our contribution to world peace and mutual understanding.

Let me use this opportunity to thank all our foreign friends who have contributed in the past or are at present contributing to our mutual scientific successes or are helping us to master new methodological approaches.

I would like to wish you a successful meeting and to express my sincere hope that you will feel at home at Smolenice Castle.

J. Zelinka

Corresponding Member of the Czechoslovak and
    Slovak Academy of Sciences
Director, Institute of Molecular Biology,
    Slovak Academy of Sciences

CONTENTS

DNA METHYLATION AND DNA METHYLTRANSFERASE IN WHEAT

Hartmut Follmann, Roland Schleicher
and Jörg Balzer

Fachbereich Chemie, Abteilung Biochemie
der Philipps Universität
D-3550 Marburg, Fed. Rep. Germany

Plant DNA is distinguished from the deoxyribonucleic acid of all other eukaryotic organisms by its substantial amount of the modified base, 5-methylcytosine (mCyt). mCyt is not detectable in most fungi and in invertebrates, it constitutes only up to 1% of the bases in vertebrate DNA, but it contributes 5-8 mol% of the total base composition in average plant DNA. Thus, one fifth to one third (20-35%) of the cytosine residues are modified in all plants, and nearly all cytosines are replaced by the methylated analog in some special plant DNA fractions.

This unique base composition of plant DNA was recognized more than three decades ago [1]; however details of the distribution and formation of mCyt in the plant kingdom have received little attention until recently and specific functions are still unknown today. It is almost impossible to conceive that all these modified cytosine residues could serve as built-in regulatory signals that determine gene inactivity or activity, the function which has been demonstrated in recent years for the few mCyt residues present in mammalian genes [2].

DNA methylation is a postreplicational event in which methyl groups are transferred from S-adenosylmethionine onto specific cytosine residues (mostly those in C-G sequences) of DNA; no cytosine modification occurs on the mononucleotide level. The reaction is catalyzed by DNA methyltransferases (EC 2.1.1.37) which have thus far been characterized almost exclusively from mammalian sources.

We reasoned that the high amount of mCyt in plants requires an efficient enzyme system which might differ from the DNA methylase found in animal cells in activity and in specificity, and could thus provide insight into the origin and functions of DNA methylation in plants. With the exception of methylase activity in extracts from pea seedlings [3] and the enzyme purified from the unicellular algae, Chlamydomonas reinhardii [4] DNA methyltransferase has not been demonstrated in plants. Because the DNA of the green algae does not exhibit plant-specific mCyt level (0.7%) it was even uncertain whether the Chlamydomonas enzyme could be compared with that of higher plants at all.

It is our aim to develop an in vitro system in which plant DNA methylation can be simulated qualitatively and quantitatively. To this aim we have purified to near homogeneity a highly active DNA methyltransferase from wheat embryo and initiated an analysis of the distribution of mCyt during wheat seed germination and in wheat DNA fractions of different sequence complexity.

WHEAT DNA METHYLTRANSFERASE

Germinating seeds are a better source than green tissues for the characterization of the enzymes of nucleic acid metabolism in plants. DNA methylase activity in germinating wheat embryo had been shown to increase two- to threefold during and following the first round of DNA replication (i.e., 15-20 hours after the uptake of water) but the enzyme was also present in dry, ungerminated seeds [5]. Thus, unlike the enzyme of deoxy-ribonucleotide biosynthesis [6], DNA methylase is not confined to the S (or G2) phase of a cell cycle. The preparation of intact cell nuclei from wheat embryo by Percoll density gradient centrifugation provided a basis for the purification of wheat DNA methyltransferase [7]. After separation of tightly bound endogenous DNA on a column of DEAE cellulose the enzyme was purified further by chromatography on Sephadex G-75, Blue Sepharose, and DNA cellulose. This enzyme preparation had a specific activity of 2 nmol CH_3 incorporation$\cdot h^{-1} \cdot mg^{-1}$ and was enriched 300-fold over the cell-free tissue homogenate. It exhibited two protein bands on SDS poly-acrylamide electrophoresis gels, corresponding to M_r = 50,000 and 35,000. Under native conditions, a molecular weight of 50,000 - 55,000 was established by centrifugation in sucrose gradients. DNA methylase from maize seedlings could be obtained in a similar way.

The properties of wheat DNA methyltransferase are compared with those of the enzyme from Chlamydomonas and of some mammalian proteins in Table 1. It is obvious that the enzymes isolated from a unicellular and a higher plant are similar, but differ strongly in molecular weight and in specificity from the mammalian enzymes.

DNA methylation in plants vs. animals is not only characterized by a much higher level of methylcytosine but also by lower sequence specificity

Table 1. DNA Methyltransferases purified from mammalian and plant sources.

Source of enzyme	M_r	specificity for DNA	spec.activity $pmol \cdot h^{-1} \cdot mg^{-1}$	ref.
calf thymus	130 000	ss > ds, hetero > homologous	600	15
rat liver	115 000	" "	5 000	16
mouse ascites cells	180 000	" "	1 500	17
human placenta	135 000	" "	2 000	18
Chlamydomonas reinhardii	58 000	ds > ss, homo = heterologous	180	4
wheat embryo	55 000	ds > ss, homo > heterologous	2 000	7

of the modification; besides the preferred C-G methylation sites the other C-N dinucleotides and in particular C-N-G trinucleotides are methylated to considerable degree [8]. We have therefore studied the specificity of purified wheat DNA methylase in vitro in great detail. The enzyme has a strong preference for doublestranded, native DNA; singlestranded DNAs are modified only at 3- to 5-fold lower rates. In contrast, the enzyme is unspecific with respect to the origin of and the type of methylation already present in a substrate DNA: it will introduce methyl groups into native, methylated plant DNA, little methylated mammalian or bacterial DNA, an unmethylated linear plasmid (pBR 322), and (although with low efficiency) into poly[d(C-G)] and other doublestranded polydeoxyribonucleotides of alternating base sequence. Plant DNA is always a better substrate than mammalian or bacterial DNA although it contains already many more methylated cytosines (Figure 1). On the other hand, highest reaction

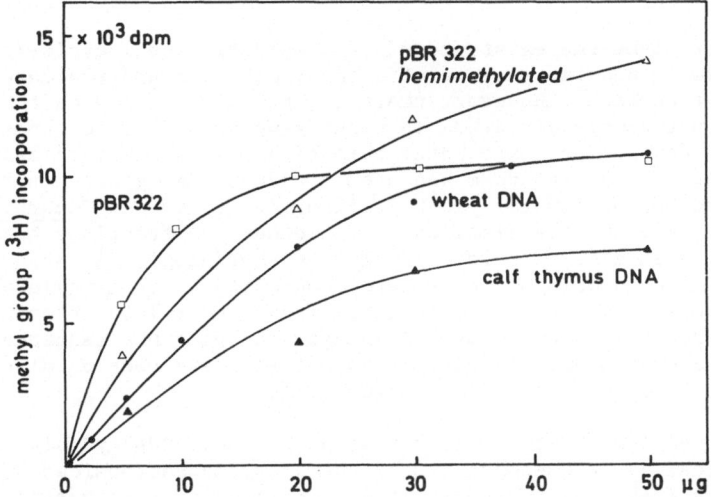

Fig. 1. Substrate specificity of wheat DNA methylase. Identical standard assays (1 h at 25°C, pH 7.5, 40 μM AdoMet) were performed with different substrate DNA concentrations. pBR322 was linearized with EcoRI, and the hemimethylated derivative prepared in vitro as described[19].

Table 2. Sequence specificity of wheat DNA methylation by purified DNA methylase in vitro. Data for the original methylation are from [8]. Nearest-neighbor analysis was done by nick-translation of in vitro ^{3}H-CH$_3$-labeled DNA with ^{32}P-labeled dNTPs as described in [15].

Sequence	^{32}P radioactivity in		original methylation	methylation in vitro	increase in methylation
	m^{5}dC-3'-p	dC-3'-p			
	counts/min		%	%	
d(C-G)	10 062	2 407	82	81	none
d(C-A)	14 665	8 996	19	62	3.3
d(C-T)	2 982	2 226	19	57	3.0
d(C-C)	3 891	11 690	7	25	3.5

rates are observed with plasmid DNA and with a hemimethylated derivative of the plasmid. These experiments demonstrate that the plant enzyme has the capacity to carry out both de novo and maintenance methylation of DNA strands.

In view of the remarkable (over)methylation of plant DNA in vitro we have determined the site specificity of the newly added methyl groups by nearest-neighbor-analysis (Table 2). It was found that the high proportion of methylated C-G sequences remains unchanged (as expected) and that cytosines adjacent to A, C, or T residues are increasingly modified, again confirming the low in vitro specificity of wheat DNA methylase. Undermethylated wheat DNA, in which C-G sequences were still unmethylated, would be a more interesting substrate but such a plant DNA has not been available until recently (see table 3).

METHYLCYTOSINE DISTRIBUTION IN WHEAT DNA

At first sight the existence of an unspecific, cell cycle-independent DNA methylase in plants seems to offer a straightforward explanation for the high mCyt content. However, there is increasing evidence that methylcytosine is not evenly dispersed in plant genomes and is not constant during plant development, and hence this view may be an oversimplification. An enrichment of mCyt residues has been noted in rRNA genes, in repetitive and satellite DNA of several plant families (e.g., in the Cucurbitaceae and Liliaceae [9,10]) but the distribution in other DNA fractions has not been systematically studied. It is probably not physiologically meaningful to use total DNA (of homologous or heterologous origin) as methylase substrate in vitro. We felt it necessary to more closely correlate cytosine methylation in wheat with developmental changes and with DNA sequence complexity. The only available data in this field have been obtained previously by B.F. Vanyushin and co-workers [11].

The proportion of methyldeoxycytidine in wheat DNA hydrolysates has been determined by HPLC separation of the deoxyribonucleosides on reverse phase columns (Table 3). We share the experience that enzymatic DNA digestion to the nucleoside level is the only reliable procedure for accurate mCyt determination because acid hydrolysis in 88% HCOOH at 175°C will inadvertently deaminate a small, variable fraction of mCyt to thymine and thus obscure subtle changes in mCyt content. A substantial decrease (about 30%) in cytosine methylation was observed during the early germination of wheat embryos. This decrease must be contrasted with the increase in incorporation of methionine-derived methyl groups into DNA during the same period of time [5], suggesting turnover and demethylation reactions; the wheat embryo might provide a suitable experimental system for the study of these little-known processes.

Germination of wheat embryos for 15 hours in the presence of 0.5 mM 5-azacytidine, which is a potent inhibitor of mammalian DNA methylation, resulted in a decrease of mCyt in wheat DNA (down to 1 mol%) and opens a practical way for the preparation of undermethylated plant DNA. No change in DNA methylation was observed in aged wheat seeds (comparing 1973-1985 crops of the same winter wheat, "Kormoran" variety) where DNA precursor metabolism undergoes major changes [12].

In another series of experiments we have fragmented wheat DNA into 400-600 bp long pieces, denatured and separated them into fractions of different reassociation kinetics by thermal hydroxylapatite chromatography. Three fractions of highly repetitive ($c_o t$ 0 - 0.04), moderately repetitive ($c_o t \sim 0.2$), and unique sequences ($c_o t \geqslant 100$), single copy DNA) were produced and were compared in mCyt content and as DNA methylase substrates

Table 3. 5-Methyldeoxycytidine content in wheat DNA during germination.
The data represent mean values from 2-3 analyses. Wheat embryos
were germinated on agar plates [5] and the DNA hydrolyzed and
analyzed as in [20].

Germination (hours)	mol% m^5dC	mol% dC	$\dfrac{m^5dC}{m^5dC + dC}$ (%)
0	5.48 ± 0.05	17.66	23.7
5	4.49 ± 0.15	18.55	19.5
10	4.20 ± 0.07	19.21	18.0
15	4.14 ± 0.04	19.50	17.5
20	4.24 ± 0.05	20.69	17.0
25	3.94 ± 0.04	20.02	16.6
30	3.84 ± 0.12	21.34	15.2
seedlings (8 d)	5.58	17.23	24.9

(Table 4). Single copy-DNA has much lower (less than half the average)
mCyt content; at the same time it exhibits proportionately lower substrate
activity for DNA methylation in vitro and in vivo (cf. the italicized
numbers in Table 4) although it must contain more methylatable sites.
However when total DNA is used as enzyme substrate in vitro these mCyt-
poorer DNA fractions are indeed modified more strongly. The above observations confirm the notion that functional studies of DNA methylation in
vitro require defined DNA substrates. The systematic analysis of wheat DNA
methyltransferase specificity with different DNA fractions, derived from
different stages of germination and resolved more precisely, has yet to be
undertaken.

Table 4. Differential methylcytosine distribution and formation in wheat
DNA. Total DNA was methylated and then fractionated, or DNA was
fractionated first and the fractions then methylated by DNA
methylase in vitro. In the last experiment, purified wheat
nuclei were labeled with $^3H-CH_3$-methionine "in vivo" and the DNA
then isolated and fractionated.

DNA fraction	total DNA	unique sequences	moderately repetitive	highly repetitive
C_ot		>100	0.2	0-0.04
% of total	100	22	46	32
mCyt/mCyt+Cyt (%)	22	*10*	21	24
in vitro methylation of total DNA, fractionation cpm [^3H]/ µg DNA	640	1 090	520	660
in vitro methylation of separated fractions		*110*	270	330
DNA methylation in nuclei (in vivo) cpm [^3H]/ mg DNA	61 400	*38 300*	54 400	72 500

5

There are still more questions than answers about methylcytosine in plant DNA. Taking together all currently available information, we advance the hypothesis that the mCyt content of plant DNA may be divided into a small number of specific cytosine residues, localized in the unique sequences, and a bulk quantity of methylation in repetitive DNA which is abundant (50-95% of total DNA) in most plants. The first fraction should correspond to the regulatory, gene activity-controlling mCyt residues of mammalian DNA; the mCyt level in this DNA fraction from wheat (which was separated from the rest with only a low degree of resolution in our present work) was not more than twice that found in animal cells. Whether the activity of individual plant genes correlates with reduced methylation is not known with certainty [13,14] but the decrease in methylation during germination, a period of high general gene activity would clearly be consistent with that correlation.

If this model is true, the question arises how individual methylation patterns can be controlled in different DNA regions without gross changes in base composition. We have not observed additional or minor enzyme fractions in the chromatographic purification of wheat DNA methylase; nevertheless it cannot be excluded that plants contain more than one methyltransferase activity to differentiate between specific methylation at some sites and random methylation of others. Other discriminating factors may be sought in DNA conformations (Z-DNA?) and chromatin components.

It is still reasonable to assume that an unspecific DNA methylase like the one characterized in wheat mainly serves for the efficient and not very specific modification of repetitive DNA. As long as a selective advantage of highly methylated over little methylated DNA in plants is not known it may even be speculated that high mCyt content is a consequence of the presence of the high number of repetitive genes and an active, unspecific methylase in plant nuclei. Wheat has a genome of average size and repetitive DNA content (83%). We plan to extend the determination of DNA methylase activity to other plant species with low (as in Cruciferae) and very high percentage of repetitive DNA (as in Liliaceae) to probe for such a relationship.

Acknowledgement

Our work on DNA methylation has been supported by grants from Deutsche Forschungsgemeinschaft and from Fritz-Thyssen-Stiftung, Köln.

REFERENCES

1. G. R. Wyatt, Biochem. J., 48:584 (1951); G. Brawerman and E. Chargaff, J. Amer. Chem. Soc., 73:4052 (1951).
2. Review: A. Razin, A. D. Riggs, Science, 210:604-610 (1980); W. Doerfler, Angew. Chemie Intl. Ed., 23:919-931 (1984); R. L. P. Adams and R. H. Burdon, "Molecular Biology of DNA Methylation", Springer-Verlag, New York, Berlin, Heidelberg, Tokyo (1985).
3. F. Kalousek and N. R. Morris, Science, 164:721-722 (1969).
4. H. Sano and R. Sager, Eur. J. Biochem., 105:471-480 (1980).
5. G. Theiss and H. Follmann, Biochem. Biophys. Res. Commun., 94:291-297 (1980).
6. B. Bachmann, R. Hofmann, B. Klein, M. Lammers and H. Follmann, in: "Metabolism and Enzymology of Nuclei Acids", J. Zelinka and J. Balan, eds., 5:59-68, Slovak Academy of Sciences, Bratislava
7. G. Theiss, R. Schleicher, G. Schimpff-Weiland and H. Follmann, Eur. J. Biochem., in press (1987).

8. Y. Gruenbaum, T. Naveh-Many, H. Cedar and A. Razin, _Nature_, 292:860-862 (1981).

9. R. Shmookler Reis, J. N. Timmis and J. Ingle, _Biochem. J._, 195:723-734 (1981).

10. B. Deumling, _Proc. Natl. Acad. Sci._, USA., 78:338-342 (1981).

11. G. E. Sulimova, A. P. Drozhdenyuk and B. F. Vanyushin, _Molekulyarnaya Biologiya_, 12:496-504 (1978); I. B. Kudryashova, S. A. Volkova, M. D. Kirnos and B. F. Vanyushin, _Biokhimiya_, 48:1031-1034 (1983).

12. G. Schimpff, H. Müller and H. Follmann, _Biochim. Biophys. Acta_, 520: 70-81 (1978).

13. A. Spena, A. Viotti and V. Pirrotta, _J. Mol. Biol._, 169:799-811 (1983).

14. H. Nick, B. Bowen, R. J. Ferl and W. Gilbert, _Nature_, 319:243-246 (1986).

15. H. Sano, H. Noguchi and R. Sager, _Eur. J. Biochem._, 135:181-185 (1983).

16. D. Simon, F. Grunert, U. v. Acken, H. P. Döring and H. Kröger, _Nuclei Acids Res._, 5:2153-2167 (1978).

17. R. L. P. Adams, E. L. McKay, L. M. Craig and R. H. Burdon, _Biochem. Biophys. Acta_, 561:345-357 (1979).

18. G. P. Pfeifer, S. Grönwald, T. L. J. Boehm and D. Drahovsky, Biochem. Biophys. Acta, 740:323-330 (1983).

19. Y. Gruenbaum, H. Cedar and A. Razin, _Nucl. Acids Res._, 9:2509-2515 (1981).

20. C. W. Gehrke, R. A. McCune, M. A. Gama-Sosa, M. Ehrlich and K. C. Kuo, _J. Chromatography_, 301:199-219.

THE X-RAY ANALYSIS OF RIBONUCLEASE Sa

J. Ševčik, *E. J. Dodson,
*G. G. Dodson and J. Zelinka

Institute of Molecular Biology of the Slovak Academy
of Sciences, 84251 Bratislava, Czechoslovakia

*University of York, Chemistry Department
Heslington, York YO1 5DD, Great Britain

INTRODUCTION

The tertiary structure of ribonuclease Sa from Streptomyces aureofaciens as well as the structure of its complex with 3'guanylic acid
(3'GMP) have been determined by X-ray analysis. The electron density that
was finally interpreted was calculated with phases obtained by multiple
isomorphous replacement technique and solvent flattening at a resolution of
2.5 A. Refinement against 1.8 A data yields R factor of 0.172 and 0.184
for the native and complex structures, respectively. The three dimensional
structure of the enzyme is similar to that of other microbial ribonucleases
and, as expected from the amino acid sequences, is closest to RNase St.
For comparison sequences of ribonucleases T1 (Takahashi, 1980), C2
(Bezborodova, 1983), Pch1 (Shlyapnikov, 1986), Pb1 (Shlyapnikov, 1984), Ba
(Hartley, 1972), Bi (Aphanasenko, 1980), St (Yoshida, 1976), and Sa
(Shlyapnikov, 1986) are given in Table 1.

EXPERIMENTAL

Ribonuclease Sa (E.C.3.1.4.8) isolated from Streptomyces aureofaciens
(Gasperik, 1982) is a guanylate specific endoribonuclease. The enzyme
specifically hydrolyses phosphodiester bonds of RNA at the 3' side of
guanosine nucleotides (Zelinkova, 1971). The amino acid sequence of the
enzyme has been determined (Slyapnikov, 1986). The molecule contains 96
aminoacid residues.

As described earlier (Sevcik, 1982), the enzyme is crystallized by a
vapor equilibrium technique from phosphate buffer at pH 7.2. As a pre-
cipitant the saturated ammonium sulphate solution is used at a concentration about 25%. The crystals are orthorhombic, the space group is
P212121 with cell parameters a = 64.90 A, b = 78.32 A, c = 38.79 A. There
are two molecules in the asymmetric unit, which will be referred to as
molecule A and B, respectively, and solvent constitutes 48% of the cell by
weight.

Three heavy atom derivative crystals were prepared, containing mercury
(mersalyl), platinum (potassium tetranitro platinum) and iodine. The

Table 1.   Alignment of Microbial Ribonuclease Amino Acid Sequences.

```
T1     A C D Y T C G S N C Y S S S D V S T A Q A A G Y - - - - - Q L H E D G E T V G S
C2     D C D Y T C G S H C Y S A S A V S D A Q S A G Y - - - - - Q L E S A G Q S V G R
Pchl   A C A A T C C G S V C Y T S S A I S S A A Q E A G Y - - - - - D L Y S A N D D V - S
Pbl    A C A A T C C G T V C Y T S S A I S S A A Q A A G Y - - - - - N L Y S T N D D V - S

Ba     A Q V I N T F D G V A D Y L Q T Y H K L P N D Y I T K S Q E Q A L G W V A S K G N L A D V A - P - G K
Bi     A V I N T F D G V A D Y L I R Y K R L P N D Y I T K S Q A S A L G W V A S K G D L A E V A - P - G K
St     Q A P C G D T S G F E Q V R L A D L P P E A T D T Y - - - E L I E K G G P Y P Y
Sa     D D V S G T - V C - L S A L P P E A T D T - - L - - N L I A S D G P F F P Y *

T1     N S Y P H K Y N N Y E G F D F S V S S - P Y Y E W P I L S S G D V Y S G P G S G A D R V V F N E N - N
C2     S R Y P H Q Y R N Y E G F N F P V S G - - N Y Y E W P I L S S G S T Y N G G G P G A D R V V F N D N - D
Pchl   N - Y P H E Y R N Y E G F D F P V S G - - T Y Y E F P I L R S G A V Y S G N S P G A D R V V F N G N - D
Pbl    N - Y P H E Y H N Y E G F D F P V S G - I T Y Y E F P I L K S G K V T G S S P G A N R V I F N D D - D

Ba     S I G G D I F S N R E G K L P G K S G R T W R E A D I N Y T - - - I - S G F - R N S D R I L Y S S D - W
Bi     S I G G D V F S N R E G R L P S A G S R T W R E A D I N Y V - - - I - S G F - R N A D R I V Y S S D - W
St     P E D G T V F E N R E G I L P D C A E G Y Y H E Y T V K T P I - - - I - S G D D R G A R R F V V G D G - G
Sa     S Q D G V V F Q N R E S V L P T Q S Y G Y Y H E Y T V I T P G - - - - - A R T R G T R R I I C G E A T Q

T1     Q L A G V I T H T G A S G N N F V E C T
C2     E L A G L I T H T G A S G D G F V A C Y
Pchl   Q L A G V I T H T G A S G N N F V A C D
Pbl    E L A G V I T H T G A S G N N F V A C T

Ba     L I Y K T T D H Y Q T F T K I R
Bi     L I Y K T T D H Y A T F T R I R
St     E Y F Y T E D H Y E S F R L T I V N
Sa     E D Y Y T G D H Y A T F S L I D Q T C *
```

method for preparing heavy atom derivative crystals was to add heavy atom reagents in small quantities to the crystals which were floating in drops of mother liquor. Mersalyl and potassium tetranitro platinum were added until a state of saturation was reached. While iodinating the crystals, the saturated solution of I_2, KI and $C_2H_2O_2$ was added to the mother liquor until the crystals became a yellow-brown color.

The procedure for preparing crystals of the RNase Sa-3'GMP complex was analogous to that of preparing heavy atom derivatives: 3'GMP was added in small quantities at a time to the crystals until the solution was saturated. The diffraction pattern recorded from crystals after having been soaked with 3'GMP for about a week showed small but clearly recognizable changes in intensities.

X-RAY ANALYSIS

Diffraction data were collected by diffractometric and photographic methods. Three sets of data with Friedel equivalents were collected with a Hilger-Watts diffractometer (ω scan mode). Another four sets of data were collected by photographic methods using the Arndt-Wonacott rotation camera. For platinum and iodine containing derivative crystals a laboratory X-ray source was used. High resolution data were collected from native and 3'GMP soaked crystals at the Synchrotron Radiation Source at Daresbury.

Intensities from the films were measured by a Joyce-Loebel Scandig 3 densitometer.

The data sets collected by diffractometer were used for initial calculations and for filling in the reflections missing from the photographic data sets.

The details of the collected data sets are given in Table 2, where crystals are named according to their substitution; native means the free enzyme. In this Table the resolution limit to which the data were collected is given, method of collecting the data, source of radiation used, total number of measurements, number of independent reflections, number of anomalous measurements for heavy atom derivative crystals and merging R factor.

The wavelength of the synchrotron radiation used was 1.488 A. The data sets all proved to be of good quality (see Table 2). However, there were problems with merging the blind region data collected from 3'GMP crystals, which reduced the internal agreement. The R factor for this set of data without the blind region was 0.05, for blind region data 0.06, however, after merging 0.117.

Two mercury atoms, five platinum atoms and six iodine atoms were located in the asymmetric unit. Positions of mercury and platinum atoms were identified from the difference Patterson function using isomorphous data with coefficients ($|FPH|-|FP|$), where FP are observed amplitudes of the protein and FPH are those for the derivatives. Positions of I atoms were subsequently determined using phases calculated from Pt isomorphous and anomalous differences. The positions of Pt and Hg atoms were in good agreement with those determined by direct methods from MULTAN (Wilson, 1978). Parameters of the heavy atoms were refined by least squares minimization to the experimental heavy atom contribution FHLE (Dodson, 1975). Phases were calculated from the native and all three derivative data sets to 2.5 A resolution using Blow and Crick's combination procedure (Blow, 1959).

11

Table 2. Details of Data Collection and Processing.

Crystal	Res. A	Method	Source	Total measur.	Indep. refl.	Anomal. measur.	Merging R
native	2.5	diffr.	CuKα	9626	7122		0.055
mercury	3.15	diffr.	CuKα	12642	3623	3297	0.124
platinum	6.0	diffr.	CuKα	3558	604	321	0.057
iodine	2.5	rot.cam.	CuKα	26846	6923	5636	0.075
platinum	2.5	rot.cam.	CuKα	29631	6923	5636	0.057
native	1.8	rot.cam.	SRS	85946	17202		0.056
3'GMP	1.8	rot.cam.	SRS	85834	17105		0.117

To improve phasing in electron density maps solvent flattening procedures were used (Wang, 1985). Several maps were calculated at different resolutions and plotted on transparent plastic sheets. The most useful map was that calculated with a combined 2.5 A resolution data set.

In Table 3 a list of heavy atoms is given with their occupancies, binding sites and coordinates. It is worth noticing, that the binding sites are not the same for both molecules which can only partly be explained by steric hindrance.

From published ribonuclease structures it was expected that there would be an α-helix and 3 strands of β-sheet structure in the molecule. Once these were located, the map was clearly interpretable in all parts except for several residues at the chain termini and a big loop connecting β-sheet strands where the connectivity was not easily disentangled. These segments became clear, however, in the first stages of refinement.

Table 3. Heavy Atom Statistics.

Heavy atom	occupancy	Binding site		Coordinates x	y	z
mercury 1	0.235		His 85B	0.110	0.900	0.200
mercury 2	0.235	His 53A		0.158	0.445	0.033
platinum 1	0.489	His 85A		0.567	0.829	0.018
platinum 2	0.394	His 53A		0.842	0.945	0.467
platinum 3	0.365		His 85B	0.103	0.940	0.194
platinum 4	0.332		His 53B	0.190	0.006	0.744
platinum 5	0.209			0.047	0.852	0.277
iodine 1	0.946		Asp 84B	0.081	0.773	0.139
iodine 2	0.547	His 85A		0.068	0.739	-0.001
iodine 3	0.170	Glu 54A		0.595	0.904	0.034
iodine 4	0.286	His 85A		0.132	0.709	0.065
iodine 5	0.238	Tyr 49A		0.713	0.983	0.090
iodine 6	0.450		Tyr 59B	0.897	0.807	0.101

Molecules in the asymmetric unit were built using the Evans and Sutherland PS 300 graphic system and the program FRODO (Jones, 1982). In the refinement process all data were used with a cut-off at 10 A. All atoms are now located with positional average errors ranging from 0.15 A to 0.2 A; the difference Fourier maps are essentially featureless.

DESCRIPTION OF THE ENZYME AND 3'GMP COMPLEX STRUCTURES

The Native Enzyme - Structure 1

The RNase Sa molecule contains one regular α-helix (residues 13-25) and an antiparallel, twisted β-sheet (res. 51-57, 68-74, 78-82). There are two large loops in the molecule; the first connecting the α-helix with the β-sheet (res. 26-50) and the second connecting the two successive β-sheet strands (res. 58-67). The C terminus of the molecule (Cys 96) is linked to Cys 7 by a S-S bridge. The N-terminus (Asp 1) is fixed by a hydrogen bond to Val 43. The secondary structure of the enzyme is illustrated in Figure 1. Glu 54, Arg 69 and His 85 are active site residues. One of the 6 proline residues - Pro 27 - adopts the CIS conformation. As compared to RNases T1, C2, Pb1, Pch1, Bi, Ba and St there is one insertion - Thr 76 - which was clearly identified in the electron density maps. The overall dimensions of the molecule vary from 26 to 35 A. There are about 1100 water molecules in the asymmetric unit, 264 of them have been positioned and refined. The average B value of the protein and water molecules is 20.90 A^2 ranging from 5.2 to 68 A^2. The structure was refined in 16 cycles by a restrained least square procedure (Konnert, 1980) alternating with rebuilding on the graphics. The starting crystallographic index R was 0.53, the final 0.172 (R = $\Sigma||Fo|-|Fc||/\Sigma|Fo|$, where Fo and Fc are observed and calculated structure factors, respectively).

Both molecules in the asymmetric unit are identical in structure except for the loop (res. 60-64) in which the main chain difference is more than 1 A. One sulphate ion was located in the A molecule forming several contacts with Arg 63, and Arg 68B of a neighboring molecule. An overlap of molecules A and B with clearly visible differences at sites of residues 60-64 is shown in Figure 2.

Structure of the RNase Sa-3'GMP Complex - Structure 2

The coordinates obtained from the crystal of the native enzyme were refined against the enzyme-substrate complex data. The crystallographic R index for the native enzyme at this stage was 0.225 which gave a value of

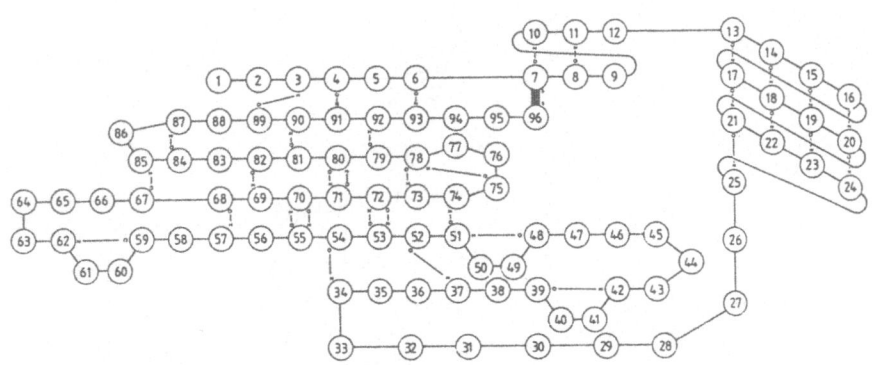

Fig. 1. The secondary structure of RNase Sa.

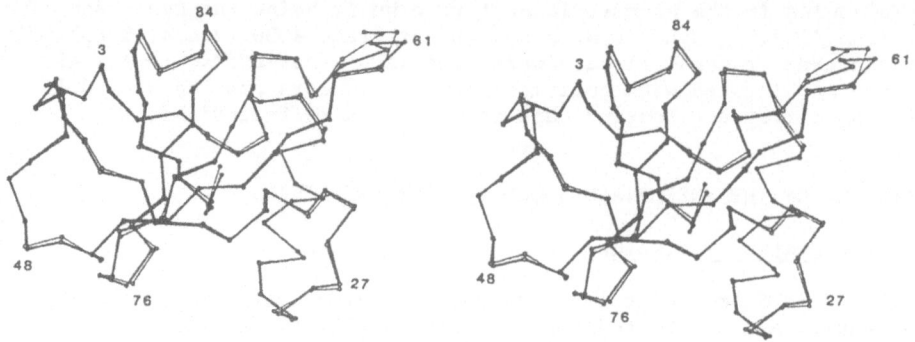

Fig. 2. Overlap of Bl molecule (●—●) on to Al (o—o).

0.296 in the structure 2. The structure of the complex was then refined independently in 7 cycles to a final R value of 0.184. 3'GMP coordinates were included in phasing only in the last three refinement cycles.

The density for 3'GMP was found only at the active site of the molecule A. The density was very clear and it was possible to determine the positions of all the substrate analogue non-hydrogen atoms accurately (see Figure 5). In molecule B access to the active site is blocked by a crystal contact. In Figure 3 the phosphate group of the substrate analogue is shown in the equivalent site of molecule B. As can be seen binding of the substrate analogue at this site is prevented by the close approach of Glu 20 from a symmetry related molecule, which makes a hydrogen bond to Arg 65B.

The protein structure in the RNase Sa-3'GMP complex is essentially identical to the structure of unsubstituted RNase Sa. In the structure 2, 210 water molecules were placed and refined in the asymmetric unit. The sulphate anion was identified in both molecules at the Arg 63 site. The

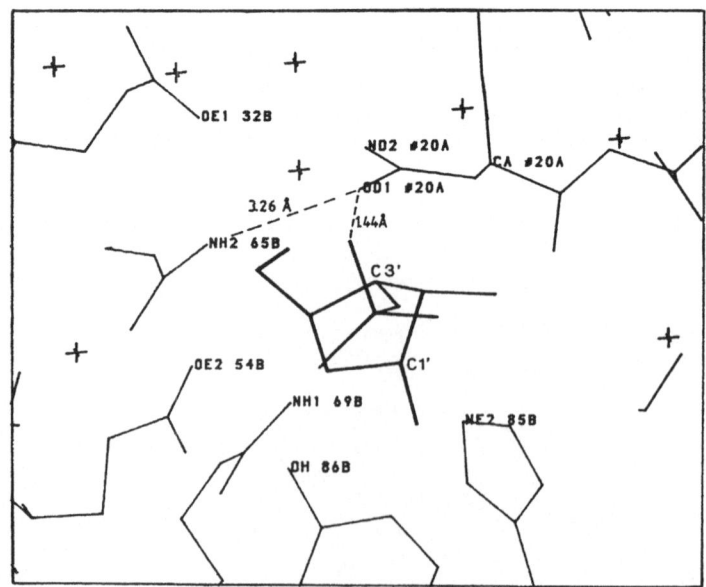

Fig. 3. 3'GMP shown at the active site of molecule B.

average B value for all protein atoms, water molecules and sulphate anions is 21.6 A². The average B value for 3'GMP atoms is 35.6 A².

The base of 3'GMP lies flat on the surface of the molecule. In Figure 4 there is a view on the A molecule parallel to the base with all residues involved in binding shown (except Glu 54). The phosphate group is inserted into the molecule's active site forming salt bridges and hydrogen bonds with active site residues Arg 69 and His 85, and with the residues Gln 32 and Arg 65. The base contacts the main chain nitrogen atoms of Gln 38, Asn 39 and Arg 40, and the side chain of Glu 41 through good hydrogen bonds. The hydrogen bonding contents and salt bridges made by 3'GMP to RNase Sa are illustrated in Figure 5 and the list of bonds between the substrate and the enzyme is given in Table 4.

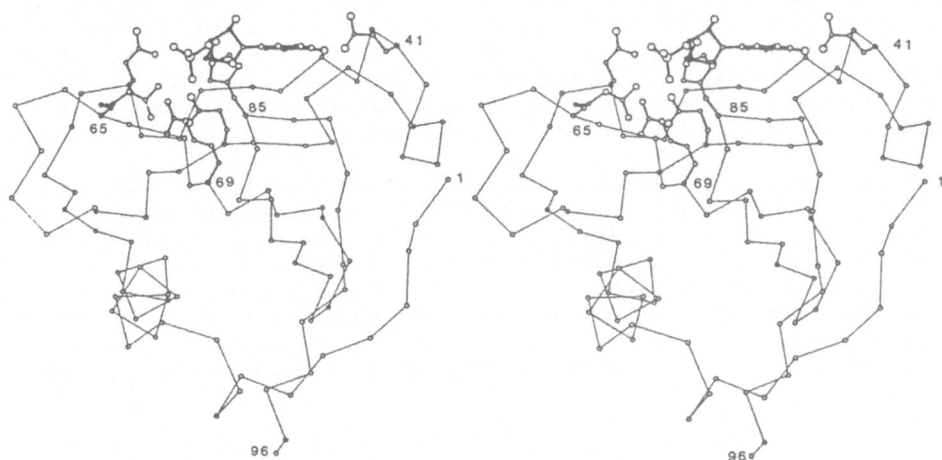

Fig. 4. Alpha-carbon backbone with 3'GMP.

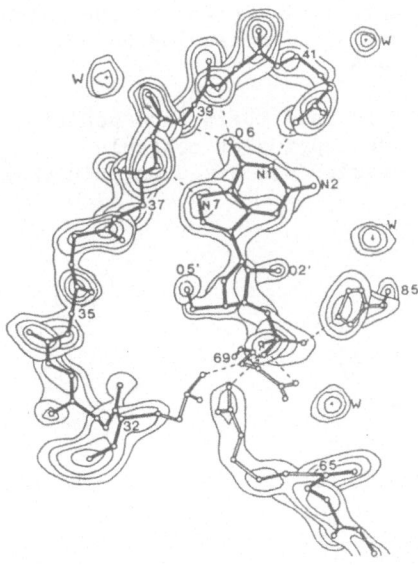

Fig. 5. The hydrogen bonding contents and salt bridges made by 3'GMP to RNase Sa.

Table 4. List of Bonds between the Substrate and the Enzyme.

3'GMP	RNase Sa	Distance	3'GMP	RNase Sa	Distance
N 7	N 38	2.77	O 31	NH2 65	2.95
O 6	N 39	2.83	O 31	NE2 32	2.91
O 6	N 40	2.60	O 32	NE2 85	2.52
N 1	OE1 41	2.62	O 33	NH2 69	2.95
			O 33	NE 69	2.92

In the crystal complex 3'GMP ribose moiety adopts C(2')-endo type puckering. The torsion angle around the glycosyl link is in the anti range. The ribose ring does not make any hydrogen bonds with the enzyme. The orientation of the ribose ring is such that O2', the nucleophilic center of catalysis, points towards His 85. In the most likely mechanism, however, it is Glu 54, that abstracts the proton from 2'hydroxyl group. It may be that following reaction to produce 3'GMP the ribose ring rotates bringing 2'hydroxyl to the surface and near His 85.

The difference in position of residues 60-64 between the A and B molecules was unaffected by the presence of the substrate in structure 2. This loop protrudes substantially above the molecule surface and is not tied down by any side chain contacts to the molecule. The crystal packing for each molecule is different: for example, whilst Arg 63A makes contacts with Glu 14B, Arg 63B makes contacts with the main chain oxygen O 75A of the symmetry related molecule. This difference illustrates the nature of the changes introduced by crystal packing to the structure of a protein molecule.

The largest apparent change in the side chains following bonding of 3'GMP is at Ser 31. The adjacent residue, Gln 32 is involved in the phosphate group binding; its interaction might have caused a small movement in the main chain position away from the substrate. This perhaps might become even more pronounced after Ser 31 side chain rotates, creating a hydrogen bond to the main chain oxygen of Asp 25B of the neighboring symmetry related molecule. An overlap of molecules 2A and 1A with differences at Ser 31 is shown in Figure 6.

The maximum changes in main chain atom positions, as well as side chain positions, after overlapping of molecules from structure 1 and 2 are given in Table 5. The overlap was based on main chain atoms (MAIN) and all

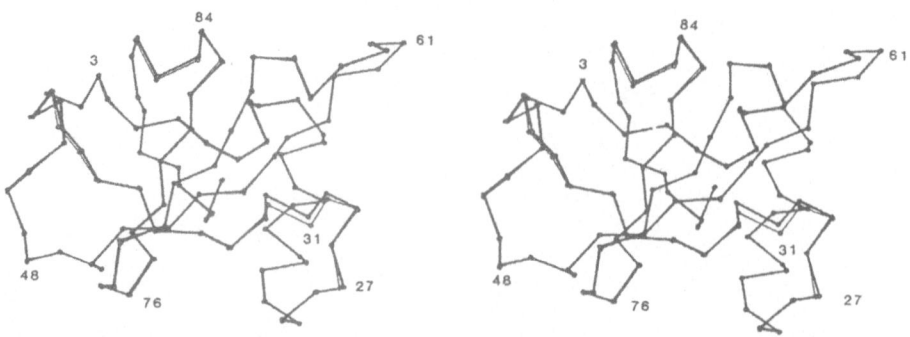

Fig. 6. Overlap of A2 molecule (●—●) on to A1 (○—○).

Table 5. A Survey of Maximum Changes in the Main Chain and Side Chain Positions.

Overlap			Average displacement	Maximum displacement	Site	(A)
1 A 1 B		Main	0.351	1.615	610	(1.6)
					630	(1.3)
					62C	(1.1)
1 A 1 B		All	0.525	5.707	40NH2	(5.7)
					70OE1	(5.4)
					63NH2	(4.6)
2 A 2 B		Main	0.363	1.747	610	(1.7)
					630	(1.2)
					310	(1.1)
2 A 2 B		All	0.502	6.126	40NH2	(6.1)
					63NH2	(4.8)
					32OE1	(3.7)
2 A 1 A		Main	0.170	1.148	310	(1.1)
					10	(0.4)
2 B 1 B		Main	0.136	0.712	4CA	(0.7)
					270	(0.4)
2 A 1 A		All	0.245	4.977	74OE1	(4.9)
					54OE1	(3.6)
					310G	(3.4)
2 B 1 B		All	0.197	3.759	250D2	(3.7)
					76CG2	(3.1)
					74OE2	(1.5)

atoms (ALL), respectively. The largest changes in side chain positions are due to the contacts between molecules in the crystal.

A comparison of residues which are involved in the substrate analogue binding shows only slight changes in their positions with respect to the B molecule, except Glu 54, the side chain of which moves apparently to avoid a steric clash with the substrate. Glu 54 makes no hydrogen bonds with the substrate analogue.

A comparison of A molecules from structure 1 and 2 shows the same difference in Glu 54 side chain as that between molecule A and B in structure 2. His 85 in molecule A, structure 1, is disordered. In the structure 2, however, after binding to the substrate analogue it became highly ordered. There are no other significant differences.

CONCLUSION

RNase Sa belongs to a homologous family of microbial ribonucleases. The knowledge of its three-dimensional structure will contribute to the picture of the evolutionary pathways of the RNases.

The classification of microbial ribonucleases according to their tertiary structures showed subsets of more closely related enzymes, Ba and Bi, Tl, Ms and C2 (Hill, 1983). This grouping can now be extended to another subset, the closely related enzymes, St and Sa. It may be that grouping of RNases according to their tertiary structures will allow their evolutionary relationship to be better established.

Finally, the structure of RNase Sa-3'GMP complex has identified clearly the contacts that determine specificity for the substrate and has provided a firm structural basis for analysis of the enzymatic mechanism.

REFERENCES

Aphanasenko, G. A., Dudkin, S. M., Kaminir, L. B., Leshchinskaya, I. A., and Severin, E. S., 1980, FEBS Lett., 97:77-80.
Bezborodova, S. I., Khodova, O. M., and Stepanov, V. M., 1983, FEBS Lett., 159:256-257.
Blow, D. M., and Crick, F. H., 1959, Acta Cryst., 12:794-802.
Dodson, E. J., 1975, in: "Crystallographic Computing Techniques", F. R. Ahmed, ed., Munskgaard, Copenhagen, 260-268.
Gasperik, J., Prescakova, S., and Zelinka, J., 1982, Biologia, Bratislava, 37:377-381.
Hartley, R. W., and Baker, E. A., 1972, Nature, 235:15-16.
Hill, C., Dodson, G., Heineman, V., Seanger, W., Mitsui, Y., Nakamura, K., Borisov, S., Tischenko, G., and Polyakov, S., 1983, TIBS, 8:366-369.
Jones, T. A., 1978, J. App. Crystallogr., 11:268-272.
Konnert, J. H., and Hendricson, W. A., 1980, Acta Cryst., A36:344-350.
Sevcik, J., Gasperik, J., and Zelinka, J., 1982, Gen. Phisiol. Biophys., 1:255-259.
Shlyapnikov, S. V., Kulikov, V. A., and Yakovlev, G. I., 1984, FEBS Lett., 177:246-248.
Shlyapnikov, S. V., Bezborodova, S. I., Kulikov, V. A., and Yakovlev, G. I., 1986, FEBS Lett., 196:29-33.
Shlyapnikov, S.V., Both, V., Kulikov, V.A., Dementiev, A.A., Sevcik, J., Zelinka, J. 1986 FEBS Lett., 209:335-339.
Takahashi, K. J., 1971, Biochem., Tokyo, 70:945-960.
Wang, B. C., 1985, in: "Methods in Enzymology", H. W. Wyckoff, C. H. Hirs, and S. N. Timasheff, eds., 114:90-112.
Wilson, K. S., 1978, Acta Cryst., B34:1599-1608,
Yoshida, N., Sasaki, A., Rashid, M. A., and Otsuka, M., 1976, FEBS Lett., 64:122-125.
Zelinkova, E., Bacova, M., and Zelinka, J., 1971, Biochem. Biophys. Acta, 235:343-352.

THE NATURE OF DISTRIBUTION OF RIBONUCLEOTIDE REDUCTASES

H. Follmann, J. Harder and H. P. C. Hogenkamp*

Fachbereich Chemie der Philipps-Universität
D-3550 Marburg, Germany

*Department of Biochemistry, University of Minnesota
Minneapolis, Minnesota 55455, USA

Ribonucleotide reductases catalyze the irreversible reduction of the 2'-hydroxyl group of ribonucleoside 5'-phosphates to the corresponding 2'-deoxyribonucleoside 5'-phosphates. This reaction is a prerequisite for DNA replication and thus controls the S phase of a cell cycle and cell proliferation. It seems reasonable that <u>all</u> prokaryotic and eukaryotic organisms utilize the same pathway for DNA precursor biosynthesis. It is still not known, however, how many different enzyme systems occur in the different phyla, and the exact mechanism of ribonucleotide reduction is not yet understood.

Three different enzyme systems have been described thus far:

1. Ribonucleotide reductases which require deoxyadenosylcobalamin as cofactor are found in many bacteria and cyanobacteria.
2. Ribonucleotide reductases which contain a protein-bound binuclear iron center and an organic free radical (localized on a tyrosine residue) are found in animals and plants and in some viruses. This type of enzyme has also been found and characterized in great detail in <u>Escherichia coli</u>.
3. Coryneform bacteria possess manganese-containing ribonucleotide reductases which are apparently structural analogs of the iron proteins.

All these enzymes have in common that catalysis of the hydrogen transfer reaction from an SH group to the ribonucleotide substrate requires a transition metal center together with a stable or transient organic radical.

Several previous reviews have dealt with the properties of the various ribonucleotide reductases [1-5]. As the physiological importance of the reaction becomes more widely acknowledged, new organisms and new components are continuously added to the list through the efforts of research laboratories in Sweden, Germany, England, Italy, Czechoslovakia, the Soviet Union, the United States of America, and Canada. In the present review we have compiled the ribonucleotide reductases that have recently been characterized with respect to cofactor requirement, protein structure, and/or reaction mechanism. We will briefly comment on current developments in the field.

Previous studies have suggested that deoxyadenosylcobalamin-dependent ribonucleotide reduction predominates among aerobic and anaerobic bacteria, and the blue-green algae (cyanobacteria). This type of reaction can be detected in cell extracts by the characteristic tritium exchange reaction between $[5'-^3H_2]$deoxyadenosylcobalamin and water [6], or more indirectly by the fact that DNA synthesis in these organisms is insensitive to hydroxyurea.

The enzymes of Lactobacillus leichmannii [7], Thermus aquaticus [8], and Anabaena 7119 [9] (monmeric proteins of M_r 72,000-80,000), Corynebacterium nephridii (a dimer composed of two M_r = 100,000 subunits) [10], and of Rhizobium meliloti (an aggregated protein of M_r = 110,000-130,000 subunits) [11] have been purified and characterized.

A similar ribonucleotide reductase was isolated from the mycelium of the antibiotic producer, Streptomyces aureofaciens [12]. This enzyme (M_r 200,000 under native conditions) appears to resemble that of C. nephridii. A rapid activity increase is observed when cultures are infected with actinophage μ 1/6 [13]. The thioredoxin of this organism was also characterized (cf. this volume).

Deoxyadenosylcobalamin-requiring ribonucleotide reductase from Pseudomonas stutzeri [14] may represent yet another structure. Enzyme activity was obtained from cell extracts only in particulate form tightly associated with other components. This finding is of interest in the light of the "replitase" multienzyme aggregate of dNTP and DNA biosynthesis.

The ribonucleotide reductase of E.coli consists of two non-identical subunits, one of which (B1, M_r = 160,000) binds substrate and effector nucleotides while the other (B2, M_r = 80,000) contains a μ-oxo-bridged di-iron(III) center and a tyrosine free radical [15]. The radical is eliminated by hydroxyurea and consequently the enzyme is inactivated by this inhibitor. The protein of E.coli is the most intensely studied enzyme of its kind (see below). It is surprising, however, that the iron-dependent ribonucleotide reductase has not been studied in any other prokaryote, probably because there is no straightforward and specific test for this type of ribonucleotide reductase in cell extracts.

A third class of bacterial ribonucleotide reductases has been discovered in organisms which have an absolute requirement for manganese ions in DNA replication [16]. Ribonucleotide reductase of Brevibacterium ammoniagenes was purified to homogeneity, and a manganoprotein structure of subunit B2 (M_r = 100,000) was demonstrated by the incorporation of ^{54}Mn [17]. The absorption spectrum with bands at 455 and 485 nm suggests the presence of a μ-oxo-di-manganese(III) complex. Enzyme activity is strongly inhibited by hydroxyurea, and an (as yet unidentified) organic radical could be observed by ESR spectroscopy at 4 K. Thus, the B. ammoniagenes enzyme is analogous to the iron-enzyme of E.coli in many properties. Manganese-dependent ribonucleotide reductases were also found in the coryneform strains Micrococcus luteus, Arthrobacter citreus, A. globiformis, and A. oxidans, and the Mn^{++} requirement of DNA synthesis suggests their presence in even more gram-positive bacteria (Brevibacterium lactofermentum, Micrococcus lylae, M. roseus, M. varians, Nocardia opaca) [18].

RIBONUCLEOTIDE REDUCTION IN AN ARCHAEBACTERIUM

Recently a ribonucleotide reductase has been detected in cell-free extracts of the strict anerobe, Methanobacterium thermoautotrophicum [19].

This enzyme is very unstable in the presence of oxygen. Active extracts can only be obtained from cells harvested and disrupted under strictly anaerobic conditions. This reductase does not require adenosylcobalamin for activity and indeed cell-free extracts do not catalyze the tritium exchange reaction between $[5'-^3H_2]$deoxyadenosylcobalamin and water. The reduction of CDP requires ATP and that of GDP requires dTTP as allosteric effectors. At low concentrations dATP acts as a positive effector for CDP reduction, however at higher concentrations dATP is a powerful inhibitor. These effects are similar to those reported for the ribonucleotide reductases isolated from eukaryotic organisms.

EUKARYOTIC RIBONUCLEOTIDE REDUCTASES

Like in bacteria, ribonucleotide reduction shows variability in the unicellular eukaryotes whereas its components appear to be very uniform in higher animals and plants.

The alga Euglena gracilis is the only known eukaryotic organism with an adenosylcobalamin-dependent ribonucleotide reductase. This enzyme has been studied in several laboratories [20-22]. Its properties, in particular allosteric regulation, resemble those of the bacterial adenosylcolamin-requiring enzymes (e.g., of Lactobacillus leichmannii).

In contrast, typical green algae like Chlorella or Scenedesmus, which we have chosen as model organisms for deoxyribonucleotide biosynthesis in the plant kingdom, possess the iron-organic radical type of ribonucleotide reductase. The enzyme of Scenedesmus obliquus was purified from over-producing cells and shown to consist of two non-identical subunits U1 (M_r = 170,000, nucleotide-binding) and U2 (M_r = 75,000) [23]. The latter contains an organic free radical and exhibits an ESR doublet signal with additional hyperfine structure (g = 2.005) [24]. This signal is virtually identical with that observed in mouse cell ribonucleotide reductase.

Surprisingly little is known about ribonucleotide reduction in fungi. The reported deoxyadenosylcobalamin requirement for CDP reduction by extracts of Pithomyces chartarum [25] remains uncertain because these extracts failed to catalyze the tritium exchange reaction from radioactive adenosylcobalamin to water (unpublished results).

Permeabilized cells or extracts of baker's yeast (Saccharomyces cerevisiae) contain higher initial ribonucleotide reductase activity than most other organisms [26-28]. However the enzyme purified from yeast extracts is unstable and decays with a half-life of about 12 hours. Some of its properties (two non-identical subunits, M_r = 250,000 of holoenzyme) are similar to the algal and mammalian enzyme systems but thus far no cofactor, metal ion, or radical involvement has been detected. Hydroxyurea inactivates yeast ribonucleotide reductase slowly and only at high concentrations; 20 mM hydroxyurea does not completely eliminate enzyme activity in vitro, and cultures continue to grow at even higher concentrations.

Ribonucleotide reduction is frequently studied in mammalian tissues, because it is the key enzyme in cell proliferation and thus an obvious target for the development of chemotherapeutic agents. Ribonucleotide reductases were isolated from calf thymus, regenerating rat liver, many mouse cell lines, human lymphoblasts, Ehrlich ascites tumor, and Yoshida sarcoma cells [5,29]. The mammalian enzyme consists of non-identical subunits M1 and M2 of which the latter carries a non-heme iron center and a tyrosyl free radical [30]. Thus, these proteins are usually considered very similar to the E.coli ribonucleotide reductase. However there are conflicting reports on the structure and number of nucleotide-binding

subunits M1 which may be composed of different, pyrimidine or purine ribo-
nucleotide-specific polypeptide chains. The question has been thoroughly
reviewed and it has been suggested that in addition to the mature holo-
enzyme incomplete precursor forms may exist in mammalian cells [29]. Other
differences with E.coli ribonucleotide reductase involve the stability and
generation of the tyrosyl radical (see below). Ribonucleotide reductase is
a cytosolic, and not a nuclear enzyme [31].

Ribonucleotide reduction in invertebrates has only recently received
more attention when it was recognized that one of the most abundant
proteins synthesized after the fertilization of sea urchin (Arbacia
punctulata) or clam (Spisula solidissima) oocytes is the small subunit of
ribonucleotide reductase (M2, M_r = 2 x 41,000) [32]. In contrast the large
subunit (M1, M_r = 2 x 86,000) is already present in unfertilized eggs [33].
M2 can be easily obtained by immunoaffinity chromatography because it
reacts with a monoclonal antibody (YL 1/2) directed against tubulin.

VIRAL RIBONUCLEOTIDE REDUCTASES

A phage-specific ribonucleotide reductase is synthesized in E.coli
after infection with bacteriophage T4. Its structure is similar to but not
identical with that of E.coli reductase. Physical and chemical studies
indicate that the tyrosyl radical geometry and accessibility are slightly
different [34].

New virus-encoded ribonucleotide reductases are also induced in
mammalian cells infected with Herpes simplex virus [35], Epstein-Barr virus
[36], and pseudorabies virus [37], as well as after infection of primate
cells with vaccinia virus [38]. In contrast to the eukaryotic host cell
enzymes which are highly regulated by effector nucleotides, the herpes
virus ribonucleotide reductases are not. They are also more resistant to
hydroxyurea inhibition and can be distinguished immunologically. HSV
ribonucleotide reductase is composed of two subunits, H1 and H2 [39].
These viral proteins do not form functional hybrids with the host cell
enzymes but nevertheless there are highly conserved regions in the genes or
in amino acid sequences of the entire enzyme family comprising E.coli,
animal, T4, and herpes virus ribonucleotide reductases [40].

REACTION MECHANISM

It has been difficult to unravel the mechanism of ribonucleotide
reduction because no obvious model system in organic chemistry is
available. An impressive number of ^{2}H-, ^{3}H-, ^{18}O-labeled, or halogen and
azido-substituted ribonucleotide derivatives has been synthesized and used
as substrates or inactivators of the Lactobacillus leichmannii, Escherichia
coli, and Herpes simplex virus ribonucleotide reductases. These experiments have lad to the conclusion that enzymes from different sources
utilize similar mechanisms [3,41,42]. Furthermore the presence of a free
radical in the iron enzymes and a transient cob(II)alamin 5'-deoxyadenosyl
radical pair in the case of L. leichmannii reductase make a radical
mechanism very likely. The experimental evidence for a chain of hydrogen
transfers has been discussed in a comprehensive review [42]. Salient
features are the initial abstraction of hydrogen from the 3' carbon of a
substrate (I; cf. reaction scheme) by the enzyme radical, loss of the
2'-hydroxyl group (II) to form a nucleotide cation radical (III) which is
then reduced to the 2'-deoxyribonucleotide product (IV) with regeneration
of the enzyme radical. However, the proposed cation radical (III) has not
yet been detected, and stopped-flow experiments have so far failed to
demonstrate intermediate reduction of the tyrosyl radical.

1 e to tyrosine radical   2 e to leaving OH⁻   1 e from active site tyrosine, 2 e from thiols

$$
\text{I substrate} \longrightarrow \text{II} \longrightarrow [\text{III cation radical}] \longrightarrow \text{IV product}
$$

I
substrate

II

III
cation radical

IV
product

The only other substituent that can be catalytically removed from C-2' with the formation of a deoxyribonucleotide is fluorine, as was recently shown for 2'-deoxy-2'-fluoro-UTP and -CTP.

PROTEIN AND ACTIVE SITE STRUCTURES

The complete three-dimensional structure of a ribonucleotide reductase is still unknown but progress has been made in characterizing ribonucleotide reductase sequences, nucleotide sites, and the catalytic center of the iron-tyrosyl radical type enzyme.

The E.coli ribonucleotide reductase subunit B1 gene (nrdA) and the genes of mouse and EBV exhibit domains of about 30% homology [44]. Three conserved cysteine residues at positions 225, 439, and 462 (revised numbering) may represent thiol groups engaged in hydrogen transfer. Photochemical affinity labeling of an allosteric site with ³H-dTTp has identified the amino acid region 288-295, with the nucleotide most likely attached to Cys-289 in as yet unknown covalent linkage [45].

The sequences of ribonucleotide reductase small subunits (nrdB gene) have been established for E.coli (B2, 375 amino acids), clam, HSV, EBV, and phage T4 [40,46]. The most important conserved residue is tyrosine 122. This tyrosine (out of 16) was identified as the radical site in a protein engineering experiment in which Tyr-122 was replaced by phenylalanine [47]: the modified protein was without enzyme activity, its light absorption and ESR spectrum showed the absence of a radical, but it did contain the complete iron center.

Recently, EXAFS studies [48] have defined the μ-oxo-bridged iron center, and these observations are consistent with the conserved amino acids (other than Tyr-122) found in subunit B2 [40]. It is likely that the binuclear complex is ligated to the protein by two histidines (His-118 and 241) and two μ-carboxylato ligands (Glu-115 and Asp-237 or Glu-238).

RADICAL GENERATION

In deoxyadenosylcobalamin-dependent ribonucleotide reductases the coenzyme undergoes a homolytic cleavage of the carbon-cobalt bond [51]. The radical-bearing subunits of other ribonucleotide reductases have to be modified post-translationally in order to become catalytically active species. The incorporation of metal ions conceivably proceeds by "spontaneous self-assembly" like in inorganic model complexes [52]. In vitro the apoenzyme B2 of E.coli is converted to the active subunit in the presence of ferrous ions and oxygen [53]. Hydroxyurea-treated or aged enzyme which still contains the iron center requires oxygen and dithiothreitol, magnesium or iron ions, and an unphysiological pH for reactivation [54]. In vivo, radical formation is enzyme-catalyzed. The activating system characterized in extracts of E.coli [49,50] includes superoxide dismutase, two unknown proteins, FMN, and NADPH, and it generates the radical in the presence of oxygen, dithiothreitol, and Mg^{++}.

Such a radical generating system has not yet been studied in eukaryotes. It is probably not essential as in E.coli because inactivated ribonucleotide reductase samples from mammalian or algal cells can undergo spontaneous reactivation in the presence of iron and oxygen [23,55,56]. In eukaryotes it is the de novo synthesis of one enzyme subunit which limits enzyme activity in vivo [57,58].

The tyrosyl radical of ribonucleotide reductases is formed aerobically by one-electron oxidation. The study of anaerobic organisms may in the future reveal yet unknown radicals in the ever exciting field of ribonucleotide reduction research.

REFERENCES

1. P. Reichard and A. Ehrenberg, Science, 221:514-519 (1983).
2. B. M. Sjöberg and A. Gräslund, Adv. in Inorg. Biochem., 5:87-110 (1983).
3. M. Lammers and H. Follmann, "Structure and Bonding", 54, Springer-Verlag, Berlin, Heidelberg (1983).
4. H. P. C. Hogenkamp, Pharmac. Ther., 23:393-405 (1984).
5. L. M. Nutter and Y. C. Cheng, Pharmac. Ther., 26:191-207 (1984).
6. F. K. Gleason and H. P. C. Hogenkamp, Biochim. Biophys. Acta, 277:466-470 (1972).
7. D. Panagou, M. D. Orr, J. R. Dunstone and R. L. Blakley, Biochemistry, 11:2378-2388 (1972).
8. G. N. Sando and H. P. C. Hogenkamp, Biochemistry, 12:3316-3322 (1973).
9. F. K. Gleason and T. D. Frick, J. Biol. Chem., 255:7728-7733 (1980).
10. P. K. Tsai and H. P. C. Hogenkamp, J. Biol. Chem., 255:1273-1278 (1980).
11. S. Inukai, K. Sato and S. Shimizu, Agric. Biol. Chem., 43:637-646 (1979).
12. M. Kolarova, P. Halicky, D. Perecko, G. Bukovska and J. Zelinka, Biologia (Bratislava), 37:777-785 (1982).
    M. Kolarova, P. Halicky, G. Bukovska and J. Zelinka, Biologia (Bratislava), 38:1189-1195 (1983).
13. M. Kolarova, K. Szuttorova, P. Halicky, O. Kudela and J. Zelinka, Biologia (Bratislava), 38:1197-1201 (1983).
14. R. G. Holt, B. D. Mehrotra and F. D. Hamilton, Biochem. Biophys. Res. Commun., 128:1044-1050 (1985).
15. L. Petersson, A. Gräslund, A. Ehrenberg, B. -M. Sjöberg and P. Reichard, J. Biol. Chem.., 6706-6712 (1980).
16. G. Schimpff-Weiland, H. Follmann and G. Auling, Biochem. Biophys. Res. Commun., 102:1276-1282 (1981).
17. A. Willing, H. Follmann and G. Auling, Eur. J. Biochem., in press (1987).
18. J. Plönzig and G. Auling, Arch. Microbiol., 146:396-401 (1987).
19. H. P. C. Hogenkamp, H. Follmann and R. K. Thauer, FEBS Lett., in press (1987).
20. F. K. Gleason and H. P. C. Hogenkamp, J. Biol. Chem., 245:4894-4899 (1970).
21. F. D. Hamilton, J. Biol. Chem., 249:4428-4434 (1974).
22. E. F. Carell and J. W. Seeger Jr., Biochem. J., 188:573-576 (1980).
23. R. Hofmann, W. Feller, M. Pries and H. Follmann, Biochim. Biophys. Acta, 832:98-112 (1985).
24. J. Harder and H. Follmann, results to be published (1987).

25. F. Stutzenberger, *J. Gen. Microbiol.*, 81:501-503 (1974).
26. M. Lowdon and E. Vitols, *Arch. Biochem. Biophys.*, 244:430-438 (1986).
27. M. Lammers and H. Follmann, *Eur. J. Biochem.*, 140:281-287 (1984).
28. M. Lammers and H. Follmann, *Arch. Biochem. Biophys.*, 244:430-438 (1986).
29. J. F. Whitfield and T. Youdale, *Pharmac. Ther.*, 29:407-419 (1985).
30. A. Gräslund, A. Ehrenberg and L. Thelander, *J. Biol. Chem.*, 257::5711-5715 (1982).
31. R. Kucera and H. Paulus, *Exp. Cell Res.*, 167:417-428 (1986).
32. N. M. Standart, S. J. Bray, E. L. George, T. Hunt and J. V. Ruderman, *J. Cell Biol.*, 100:1968-1976 (1985).
33. N. Standart, T. Hunt and J. V. Ruderman, *J. Cell Biol.*, 103:2129-2136 (1986).
34. M. Sahlin, A. Gräslund, A. Ehrenberg and B. -M. Sjöberg, *J. Biol. Chem.*, 257:366-369 (1982).
35. D. R. Averett, C. Lubbers, G. B. Elion and T. Spector, *J. Biol. Chem.*, 258:9831-9838 (1983).
36. B. E. Henry, R. Glaser, J. Hewetson and D. J. O'Callaghan, *Virology*, 89:262-271 (1978).
37. H. Lankinen, A. Gräslund and L. Thelander, *J. Virol.*, 41:893-900 (1982).
38. M. B. Slabaugh and C. K. Mathews, *J. Virol.*, 60:506 (1986).
39. E. A. Cohen, J. Charron, J. Perret and Y. Langelier, *J. Gen. Virol.*, 66:733-745 (1985).
40. B.-M. Sjöberg, H. Eklund, J. A. Fuchs, J. Carlson, N. M. Standart, J. V. Ruderman, S. J. Bray and T. Hunt, *FEBS Lett.*, 183:99-102 (1985).
41. M. A. Ator, J. Stubbe and T. Spector, *J. Biol. Chem.*, 261:3595-3599 (1986).
42. G. W. Ashley and J. Stubbe, *Pharmac. Ther.*, 30:301-329 (1985).
43. G. Harris, G. W. Ashley, M. J. Robins, R. L. Tolman and J. Stubbe, *Biochemistry*, 26:1895-1902 (1987).
44. I. W. Caras, B. B. Levinson, M. Fabry, S. R. Williams and D. W. Martin Jr., *J. Biol. Chem.*, 260:7015-7022 (1985).
45. S. Eriksson, B. -M. Sjöberg, H. Jörnvall and M. Carlquist, *J. Biol. Chem.*, 261:1878-1882 (1986).
46. B.-M. Sjöberg, S. Hahne, C. Z. Mathews, C. K. Mathews, K. N. Rand and M. J. Gait, *EMBO J.*, 5:2031-2036 (1986).
47. A. Larsson and B. -M. Sjöberg, *EMBO J.*, 5:2037-2036 (1986).
48. R. C. Scarrow, M. J. Maroney, S. P. Salowe and J. Stubbe, *J. Am. Chem. Soc.*, 108:6832-6834 (1986).
49. T. Barlow, R. Eliasson, A. Platz, P. Reichard and B. -M. Sjöberg, *Proc. Natl. Acad. Sci.*, USA, 80:1492-1495 (1983).
50. R. Eliasson, H. Jörnvall and P. Reichard, *Proc. Natl. Acad. Sci.*, 83:2373-2377 (1986).
51. R. L. Blakley, W. H. Orme-Johnson and J. M. Bozdech, *Biochemistry*, 18:2335-2339 (1979).
52. K. Wieghardt, V. Bossek, D. Ventur and J. Weiss, *Chem. Comm.*, 347-349 (1985).
53. C. L. Atkin, L. Thelander, P. Reichard and G. Lang, *J. Biol. Chem.*, 248:7464-7472 (1973).
54. A. Gräslund, M. Sahlin and B. -M. Sjöberg, *Environm. Health Perspect.*, 64:139-149 (1985).
55. R. Hofmann and H. Follmann, *Z. Naturforsch.*, 40c:919-921 (1985).
56. L. Thelander, A. Gräslund and M. Thelander, *Biochem. Biophys. Res. Comm.*, 110:859-865 (1983).
57. Y. Engström, S. Erikson, I. Jildevik, S. Skog, L. Thelander and B. Tribukait, *J. Biol. Chem.*, 260:9114-9116 (1985).
58. E. H. Rubin and J. G. Cory, *Cancer Res.*, 46:6165-6168 (1986).

CLONING, EXPRESSION AND NUCLEOTIDE SEQUENCE OF A GENE ENCODING

A SECOND THIOREDOXIN FROM CORYNEBACTERIUM NEPHRIDII

C.J. Lim[+], J.A. Fuchs[+], S.C. McFarlan*
and H.P.C. Hogenkamp*

[+]Department of Biochemistry, University of Minnesota
St. Paul, Minnesota, 55108

*Department of Biochemistry, Minneapolis
Minnesota 55455

INTRODUCTION

Thioredoxins are small, apparently ubiquitous proteins known to participate in thiol dependent redox reactions (Holmgren, 1985). One unifying feature of the thioredoxins is the presence of a disulfide bridge at the active site forming a fourteen-membered ring with the sequence -cys-gly-pro-cys-. The complete primary structures of thioredoxins from several sources have been determined and they exhibit a high degree of homology (Holmgren, 1968; Meng and Hogenkamp, 1981; Gleason, Whittaker, Holmgren and Jornvall, 1985; Tsugita, Maeda and Schormann, 1983). The T4 thioredoxin is an exception, its active site contains the tetrapeptide -cys-val-tyr-cys- (Sjoberg and Holmgren, 1972). The T4 protein shows little homology with the other thioredoxins and can function as a glutaredoxin as well as a thioredoxin. Thus far, single thioredoxins have been characterized in spinach chloroplasts: the f-type, which activates fructose-1,6-diphosphatase, and the m-type, which activates NADP-malate dehydrogenase (Schurmann, Maeda and Tsugita, 1981). Two thioredoxins have also been purified and characterized from Anabaena sp. 7119, a blue-green algae. The m-type resembles the E.coli protein while the f-type appears to be different from other thioredoxins (Gleason and Holmgren, 1981; Yee, De La Torre, Crowford, Lara, Carlson and Buchanan, 1981; Whittaker and Gleason, 1984).

A thioredoxin from C.nephridii was purified, characterized and sequenced in this laboratory (Meng and Hogenkamp, 1981). It consists of 105 amino acid residues, half of which are identical with those of the E.coli protein. This thioredoxin is able to serve as a reducing substrate for yeast methionine sulfoxide reductase, for E.coli thioredoxin reductase, and for L.leichmanii ribonucleotide reductase. However, it is unable to serve as a cofactor for ribonucleotide reductase from C.nephridii in the presence of either C.nephridii or E.coli thioredoxin reductase, suggesting that the thioredoxin is a general disulfide reductase.

In order to obtain larger quantities of C.nephridii thioredoxin to use in physical studies, a C.nephridii gene encoding thioredoxin was cloned in a high copy number E.coli vector, pUC13, and transformed in the E.coli trxA strains, JF510 and BH2012 (Holmgren, Ohlsson and Gronkvist, 1978; Lim,

Geraghty and Fuchs, 1985; Lim, Gleason and Fuchs, 1986). Strain BH2012 grows very slowly on minimal medium when methionine sulfoxide is the methionine source while strain JF510 produces significantly smaller colonies on enriched medium plates than trxA$^+$ strains. The transformants of these strains containing a thioredoxin gene were examined for their abilities to support the growth of F$^-$ specific T7 phage (Modrich and Richardson, 1975) and F$^+$ specific filamentous phages f1 and M13 (Russel and Model, 1985). A trxA gshA double mutant which requires glutathione for growth (Tuggle and Fuchs, 1985) was also used to characterize the E.coli trxA strains transformed by the C.nephridii thioredoxin gene. Preliminary results indicated that more than one type of thioredoxin gene was cloned. In this paper we present the cloning, characterization, and nucleotide sequence of a gene encoding a thioredoxin distinct from the C.nephridii thioredoxin previously reported as well as the purification and characterization of the new thioredoxin protein, termed C-2.

EXPERIMENTAL PROCEDURES

Bacterial Strains and Phages

Strains JF510 and BH2012 were used as recipients for transformation. Strain JF10 (Lim, Gleason and Fuchs 1986) is the ilvC::Tn5 trxA7004 derivative of 71-18 (Heidecker,Messing and Groneborn, 1980) and strain BH2012 (Lim, Gleason and Fuchs, 1986) is E.coli K12 F ara D139? galU galK hsr strA metE45 argH1 trxA7004 ilvC::Tn5. Strain BH5262, which is E.coli K12 F ara D139? galU galK hsr strA argH1 trxA7004 gshA srl::Tn10, was used for complementation testing (Lim, Gleason, and Fuchs, 1986). Plasmids pUC13, pUC18, and PUC19 were used as cloning vectors (Norrander, Kempe and Messing, 1983). Phages M13mp2 and wild-type f1 were used for infection testing (Tuggle and Fuchs, 1985) and Phages M13mp18 and M13mp19 were used for DNA sequencing (Norrander, Kempe and Messing, 1983). Strain JM101 was used as a host for M13mp18 and M13mp19. Minimal medium (Davis and Mingioli, 1950), appropriately supplemented, and the rich medium (Miller, 1972), supplemented with ampicillin (30µg/ml), were used for the growth of the E.coli K12 strains.

Cloning Procedures

The genomic DNA from C.nephridii was isolated by the procedure described for E.coli (Mandel and Higa, 1970) and was digested with endonuclease Sau 3A or Taq 1. The digested total genomic DNA was ligated to vector pUC13 that had been digested with BamH 1 or Acc 1 and treated with calf intestinal phosphatase. Ligations were performed at 14°C overnight. The ligation mixtures were used to transform JF510 and BH2012 by the CaCl$_2$ procedure (Mandel and Higa, 1970). The transformed cells of JF510 were plated on enriched plates with ampicillin, IPTG, and X-gal, whereas transformed BH2012 cells were plated on a minimal medium containing methionine sulfoxide and ampicillin.

DNA sequencing analysis

DNA restriction fragments were subcloned into M13mp18 (Norrander, Kempe and Messing, 1983) and were sequenced by the dideoxy chain termination method (Sanger, Micklen and Coulson, 1977) with a synthetic 17 base universal primer. The fragments sequenced in both M13mp18 and M13mp19 are indicated in Figure 2.

Assay for Thioredoxin Activity

Thioredoxin activity was determined in a coupled asay which depends on the reaction between yeast or E.coli methionine sulfoxide reductase and

C.nephridii thioredoxin as previously described (Meng and Hogenkamp, 1981). One unit of thioredoxin activity corresponds to the amount of thioredoxin that will deduce 1 nmol of L-[^{14}C]-methionine-(±)-sulfoxide in 15 min at 37°C.

The reduction of insulin by thioredoxin in the presence of DTT or C.nephridii thioredoxin reductase and NADPH was determined by the methods of Holmgren (Holmgren, 1979). The activation of the spinach enzymes by thioredoxin was also assayed by published procedures: fructose-1, 6-diphosphatase (Wolosiuk and Buchanan, 1977), NADP-malate dehydrogenase (Wolosiuk, Buchanan and Crowford, 1978), NADP-glyceraldehyde-3-phosphate dehydrogenase (Wolosiuk and Buchanan, 1978), and phosphoribulokinase (Wolosiuk, Crowford, Yee and Buchanan, 1979).

Purification of thioredoxin C-2

All operations were carried out at 0-4°C unless otherwise stated. Frozen cell paste was suspended in 5 volumes of 0.05 M potassium phosphate buffer, pH 7.0, containing 1 mM EDTA (buffer A), and the cells were disrupted by sonication. The cell debris was removed by centrifugation at 16,000 x g for 1 hour. To the cell free extract was added a 20% solution of streptomycin sulfate to give a final concentration of 1%. After the slow addition of the streptomycin sulfate the suspension was stirred an additional thirty minutes and the precipitate was then removed by centrifugation. The supernatant solution was heated at 65°C in a 75°C water bath for 3 min and rapidly cooled to 5°C in an ice salt bath. The precipitate was removed by centrifugation. The volume of the supernatant was reduced using an Amicon ultrafiltration apparatus and YM2 membrane and the concentrated solution was then applied to a Sephadex G-75 column (5 x 80cm) previously equilibrated with buffer A. The column was developed using the same buffer. The fractions containing thioredoxin activity were pooled and concentrated by ultrafiltration as described above and dialyzed against 0.05 M ammonium acetate, pH 7.8. Minor contaminants were removed by passing the preparation through a Sephadex G-50 column (2.5 x 80cm), previously equilibrated in 0.05 M ammonium acetate, pH 7.8.

RESULTS

Since the C.nephridii thioredoxin functions as a redox substrate in the E.coli thioredoxin reductase and methionine sulfoxide reductase systems (Meng and Hogenkamp, 1981), we anticipated that the C.nephridii thioredoxin gene cloned into a pUC vector would function in complementing an E.coli thioredoxin dificient mutant (trxA). We assumed that the C.nephridii thioredoxin gene could utilize the lac promoter of the vector if the C.nephridii promoters did not function in E.coli. The high copy number of the pUC vector would allow complementation even if the C. Nephridii thioredoxin functioned poorly in E.coli. Tow E.coli trxA phenotypes were used for initial selection. Transformants of JF510 that had an insert in pUC13 were white and white colonies with increased colony size on an enriched medium were selected. In addition, transformants of BH2012 that could utilize methionine sulfoxide as a methionine source on a minimal medium were selected. Eleven large white colonies were chosen among the JF510 transformants and 7 large colonies were found in a background of very small colonies of BH2012. All 18 transformants were analyzed for the presence of thioredoxin by the coupled assay using methionine sulfoxide reductase (Table 1). Four strains expressed unusally high levels of thioredoxin activity. Strains carrying plasmids p1C<n2 and pLCN4 appeared to differ from strains with plasmids pLCN1 and pLCN3 in the ability to support T7 phage replication and were chosen for further study. The plasmid p1CN2 was found to contain a 1.85 kb Sau 3A insert while plasmid pLCN4 was found to carry a 10 Kb Taq 1 insert. To verify that a thioredoxin gene had indeed been cloned, the biological properties of the

Table 1. Thioredoxin Activity of Heat Treated Extracts of Transformed E.coli Strains BH2012 and JF510.

Strains	Thioredoxin activity*
BH2012/pLCN1	77.2
BH2012/pLCN2	194.6
JF510/pLCN3	67.4
JF510/pLCN4	128.7
BH2012/pLCN5	16.1
BH2012/pLCN6	15.5
BH2012/pLCN7	15.1
BH2012/pLCN8	8.5
BH2012/pLCN9	11.6
JF510/pLCN10	11.3
JF510/pLCN11	15.5
JF510/pLCN12	13.3
JF510/pLCN13	11.3
JF510/pLCN14	11.4
JF510/pLCN15	14.7
JF510/pLCN16	17.5
JF510/pLCN17	15.6
JF510/pLCN18	18.0
JF510/pUC13	1.0

* Thioredoxin activity is expressed as nmoles L-methionine produced/min/ml extract.

strains carrying plasmid pLCN2 and pLCN4 were investigated. These plasmids were transformed into JF510, BH2012, and BH262. JF510 containing pLCN2 and JF510 containing pLCN4 produced colonies on an enriched medium that were indistinguishable from the TrxA$^+$ parent of JF510 and were significantly larger that JF510 colonies. BH2012 with pLCN2 and BH2012 with pLCN4 grew as well with methionine sulfoxide as with methionine on a minimal medium. BH5262 containing pLCN2 and BH5262 containing pLCN4 were able to grow on a supplemented minimal medium without glutathione. Unlike JF10, the JF510 strains transformed by pLCN2 and pLCN4 were able to support the growth of M13mp2 phage as well as f1 phage. However, f1 formed small plaques compared to a trxA$^+$ strain. Neither BH2012 strain containing pLCN2 or pLCN4 could support T7 phage replication but they were able to support the replication of strain T7tas1. This strain was isolated as a T7 mutant phage that is able to utilize Anabaena sp. 7119 thioredoxin (Lim, Gleason and Fuchs, 1986). These results indicate that a thioredoxin gene has been cloned.

A restriction map of plasmid pLCN2 is presented in Figure 1.

To locate the region of DNA encoding thioredoxin and to determine whether plasmids pLCN2 and pLCN4 contain a common region of DNA, the plasmids were digested with Hinf 1 and the ends were filled in by Klenow fragment of DNA polymerase 1 and ligated to pUC19 that had been digested with Hinc II. After transformation of JF510, large white colonies were assayed for thioredoxin activity. Plasmids from thioredoxin positive transformants derived from pLCN2 were analyzed and found to contain three different sized Hinf I inserts. These plasmids, pLCN21, pLCN22, and pLCN23, contained a region common to all three plasmids as indicated by restriction analysis and the different sized inserts were attributed to incomplete digestion. A 620 bp Hinf I insert in plasmid pLCN41 was obtained from plasmid pLCN4 and was used for further study. A restriction map of pLCN41 is presented in Figure 2.

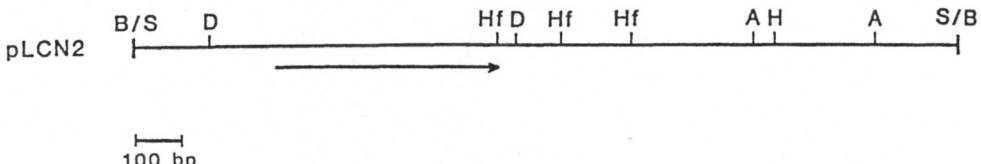

Fig. 1. The restriction map of the insert in plasmid pLCN2: B, BamHI; S, SAU3AI; Hf, HinfI; A, AvaII; d, DdeI; H, HaeII. The arrows indicate the region encoding thioredoxin.

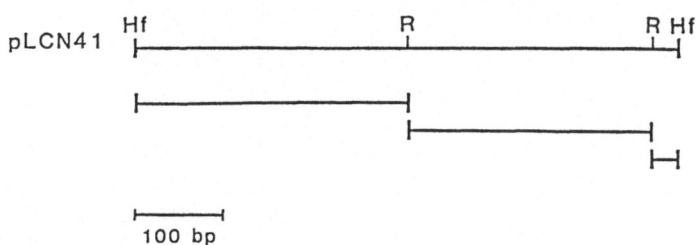

Fig. 2. The restriction map of the insert in plasmid pLCN41: Hf, HinfI; R, RsaI.

The internal Rsa fragment as well as the small Rsa I - Hinf I fragment is common to the three Hinf I subclones from pLCN2, indicating that pLCN2 and pLCN4 encode the same thioredoxin protein.

A cell free extract of strain JF 510/pLCN2 was heated at 65°C for 3 min and the soluble proteins were subjected to both native and SDS- poly- acrylamide gel electrophoresis. Figures 3A and 3B show a dense band of protein produced in the JF510 strain carrying pLCN2 that is not present in the control JF510 strain containing only the vector pUC13 and at a position distinct from that of either the E.coli or the first C.nephridii thio- redoxin.

The restriction fragments of pLCN41 shown in Figure 2 were ligated into pUC19 and the multiple restriction site regions were transferred into M13mp18 and M13mp19. These inserts as well as the 620 bp Hinf I insert were sequenced in both M13 mp18 and M13mp19. The nucleotide and the derived protein sequence of the Hinf I insert of pLCN41 is given in Figure 4. In contrast to all but one of the thioredoxin sequences previously reported, the active site in this thioredoxin has a -cys-ala-pro-cys- sequence instead of the sequence -cys-gly-pro-cys-.

The cloned C.nephridii thioredoxin is the first procaryote thioredoxin that contains a third -cystine residue. This protein was designated thioredoxin C-2 to distinguish it from thioredoxin C-1, the thioredoxin from C.nephridii previously sequenced (Meng and Hogenkamp, 1981).

Strain JF510 carrying pLCN2 was grown on enriched medium containing ampicillin (30 µg/ml) and the thioredoxin protein was purified. The results of a typical preparation starting with 25g of a frozen cell paste are shown in Table 2. The purified thioredoxin was homogeneous as judged by polyacrylamide gel electrophoresis in native and SDS gels and by HPLC.

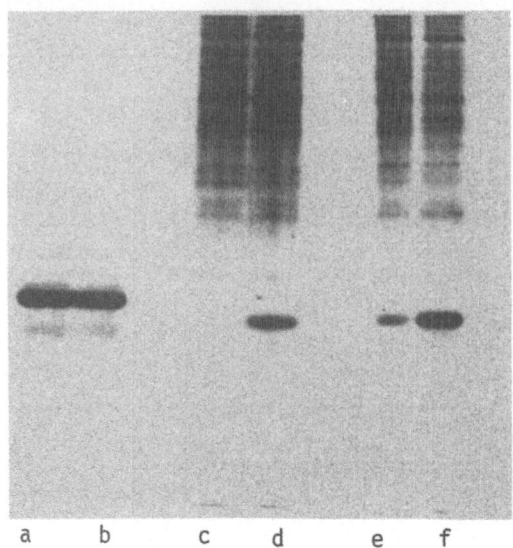

a b c d e f

Fig. 3A. 12.5% native polyacrylamide electrophoretic analysis of E.coli
thioredoxin, thioredoxin C-, and the heat-treated extracts of
JF510 containing pUC13, JF510 containing pLCN2, JF510 containing
pLCN40, and JF510 containing pLCN41. Lane a, E.coli thioredoxin;
lane b, C.nephridii thioredoxin C-1; lane c, JF510 containing
pUC13 heat-treated extract; lane d, JF510 containing pLCN2
heat-treated extract; lane e, JF510 containing pLCN40 grown in
the presence of IPTG; lane f, JF510 containing pLCN41 grown in
the presence of IPTG.

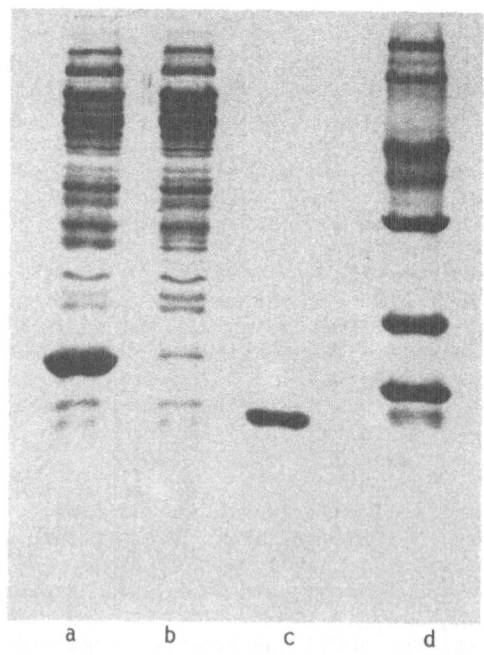

a b c d

Fig. 3B. SDS-polyacrylamide gel electrophoretic analysis of the
heat-treated extracts of JF510 containing pLCN2 and JF510
containing pUC13 and of thioredoxin C-1. (cont...)

32

Table 2. The Purification of Thioredoxin C-2.

Fraction	Volume	Protein	Thioredoxin activity*		Yield	Purification fold
	ml	mg	units	units/mg	%	
Crude extract	150.0	3937.5	612438	155.5	100	
Streptomycin sulfate	162.0	4212.0	646878	153.6	106	
Heat treatment	160.0	2900.8	592372	204.2	96.7	1.31
Sephadex G-75	25.0	302.5	572512	1892.6	93.6	12.17
Sephadex G-50	25.5	249.9	573670	2295.3	93.6	14.76

This molecular weight calculated from the DNA sequence is 15,803. The isoelectric point of thioredoxin C-2 determined by isoelectric focusing gel electrophoresis is 4.5 Thioredoxin C-2 functions as a substrate for the yeast and E.coli methionine sulfoxide reductases, the E.coli and C. nephridii thioredoxin reductases and the E.coli and L.leichmannii ribonucleotide reductases. When reduced with dithiothreitol, thioredoxin C-2

```
                                    CC TCT AGA GTC AGT CGA CCG GAT AAT

CGT CAC GGT GAA GTG GAT ACC ACC GCC AAG GCT GCA CCC ATT CAA CGC GGT GCA GAA GCG

AGC CGA ACT TTT CTG TCG ATG ATC AAC ATC GTG AGC ATC CAT TAG ACT TTT CAT TTT CAA

ACT TGA ATA ATC GGA ACA AAC CCT TAC ATT GAA TAA AGC GGC AAT GTA ATA TTC ATC GTG


Met Met Phe Lys Phe Ala Leu Tyr Phe Leu Asn Leu Glu Gln Pro Tyr Ser Ala Thr Ile
ATG ATG TTT AAA TTC GCA CTT TAT TTC TTA AAT TTG GAG CAA CCT TAT TCT GCG ACT ATT

Val Asn Thr Thr Asp Glu Asn Phe Gln Ala Asp Val Leu Asp Ala Glu Thr Pro Val Leu
GTT AAT ACT ACT GAT GAA AAC TTC CAA GCA GAT GTT TTA GAT GCT GAA ACG CCT GTA CTT

Val Asp Phe Trp Ala Gly Trp Cys Ala Pro Cys Lys Ala Ile Ala Pro Val Leu Glu Glu
GTA GAT TTC TGG GCT GGC TGG TGT GCA CCA TGT AAA GCA ATT GCC CCT GTC CTT GAA GAA

Leu Ser Asn Glu Tyr Ala Gly Lys Val Lys Ile Val Lys Val Asp Val Thr Ser Cys Glu
CTT TCA AAC GAA TAC GCG GGT AAA GTG AAA ATC GTT AAA GTT GAT GTG ACT TCA TGC GAA

Asp Thr Ala Val Lys Tyr Asn Ile Arg Asn Ile Pro Ala Leu Leu Met Phe Lys Asp Gly
GAC ACT GCG GTG AAA TAT AAT ATC CGT AAT ATT CCA GCA TTA TTG ATG TTT AAA GAT GGT

Glu Val Val Ala Gln Gln Val Gly Ala Ala Pro Arg Ser Lys Leu Ala Ala Phe Ile Asp
GAA GTT GTT GCT CAA CAG GTT GGC GCA GCA CCA CGT TCA AAA CTT GCA GCA TTT ATT GAC

Gln Asn Ile
CAA AAC ATC TAA
```

Fig. 4. The nucleotide sequence and derived protein sequence of the Hinf I insert of plasmid pLCN41.

Fig. 3B. (continued)
lane a, JF510 containing pLCN2 heat treated extract; lane b, JF510 containing pUC13 heat treated extract; lane c, C.nephridii thioredoxin C-1; lane d, marker proteins from top to bottom: phosphorylase B, M_r = 92,500; bovine serum albumin, M_r = 66,200; ovalbumin, M_r = 45,000; carbonic anhydrase, M_r = 31,000; soybean trypsin inhibitor, M_r = 21,500; and lysozyme, M_r = 14,400.

can activate spinach chloroplast NADP-malate dehydrogenase and phosphori-bulokinase but not fructose-1,6-diphosphatase nor NADP-glyceraldehyde-3-phosphate dehydrogenase. Thioredoxin C-2 reduces the disulfide bridges of insulin in the presence of E.coli or C.nephridii thioredoxin reductase and NADPH or in the presence of dithiothreitol.

DISCUSSION

The new thioredoxin identified in this paper represents the first example of a second thioredoxin found in a non-photosynthetic procaryote organism. In the earlier purification of thioredoxin from C.nephridii, a coupled enzyme assay with yeast methionine sulfoxide reductase was used and the only significant activity observed was in fractions containing the previously characterized thioredoxin (Meng and Hogenkamp, 1981). The C. nephridii culture used for purification was grown anaerobically under well defined conditions. It is possible that the thioredoxin described here is not synthesized under these conditions. The gene encoding the C.nephridii thioredoxin C-2 appears to be expressed at a relatively high level in E.coli grown aerobically. An insert with an orientation opposite to the lac promoter appeared to be expressed at one-half the level of the insert with the same orientation as the lac promoter, suggesting that the C. nephridii promoter is functioning in E.coli. The thioredoxin produced in vivo has sufficient structural similarity to E.coli thioredoxin that it can complement an E.coli mutant for all phenotypes except replication of T7 phage. Presumably, the structural difference is sufficient to prevent T7 DNA polymerase from utilizing thioredoxin C-2. However, as tas mutant of T7, isolated as a mutant that could use the Anabaena sp. 7119 thioredoxin, was also able to utilize C.nephridii thioredoxin C-2 as a subunit (Lim, Gleason and Fuchs, 1986). The tas I mutant requires a host strain with a thioredoxin but appears to have an altered specificity for the thioredoxin that can be used.

Thioredoxin C-2 is the first example of a thioredoxin with an active site that differs from the sequence -cys-gly-pro-cys- other than the bac-teriophage T4 protein, which serves as a glutaredoxin as well as a thiore-doxin. The substitution from glycine to alanine would not affect the overall conformation because both glycine and alanine are small neutral amino acids. The other unique feature of thioredoxin C-2 is the presence of a third half-cystine residue which has not been found in any other procaryote thioredoxins. All purified mammalian thioredoxins have, how-ever, two structural half-cystine residues in the C-terminal region of the molecule that may be involved in regulation of the thioredoxin system (Holmgren and Slaby, 1979).

The molecular weight of thioredoxin C-2 calculated from the nucleotide sequence is 15,803. However, the molecular weight as determined by SDS-polyacrylamide gel electrophoresis appears to be 15,000. Preliminary results indicate that part of the amino terminal sequence predicted from nucleotide sequence is not present in the isolated protein. Edman degra-dation sequence analysis of the protein suggests that the amino terminus is blocked.

Although the E.coli thioredoxin – thioredoxin reductase system was first purified as the hydrogen donor for ribonucleotide reduction, it is now apparent that many thioredoxins are efficient protein disulfide reductases that participate in other reduction reactions. Thioredoxin C-2 serves as the reducing substrate for yeast and E.coli methionine sulfoxide reductases, for E.coli and L. leichmanii ribonucleotide reductases. Surprisingly, neither thioredoxin from C.nephridii can reduce ribo-nucleotide reductase from the same organism. Likewise, rabbit bone marrow thioredoxin is unable to reduce its own ribonucleotide reductase. Thioredoxin C-2 is an excellent reducing agent of the disulfides of insulin in the presence of DTT or thioredoxin reductase and NADPH. This activity suggests that thioredoxin C-2 functions as a general protein disulfide reducing agent in vivo.

REFERENCES

Davis, B.D. and Mingioli, E.S., 1950, J. Bacteriol, 60; 17-28.
Gleason, F.K. and Holmgren, A. 1981 J.k Biol.Chem. 256; 8306-8309.
Gleason, F.K., Whittaker, M.M., Holmgen,A., and Jornvall, H. 1985 J. Biol. Chem. 260; 9567-9573.
Heidecker, G., Messing, J., and Gronenborn, B. 1980, Gene 10, 69-73.
Holmgren, A. 1968, Eur. J. Biochem. 54; 237-271.
Holmgren, A., Ohlsson, I., and Gronkvist, M.L., 1978, J. Biol. Chem. 253, 430-436.
Holmgren, A. 1979, J. Biol. Chem. 254; 9113-9119.
Holmgren, A. and Slaby, I. 1979 Biochemistry 18, 5592-5599.
Lim, C.-J., Geraghty, D., and Fuchs, J.A. 1985 J. Bacteriol. 163; 311-316.
Lim, C.-J., Gleason, F.K., and Fuchs, J.A. 1986; J. Bacteriol., 168; 1258-1264.
Lundell, D.J., and Howard, J.B. 1978; J.Biol. Chem. 253; 3422-3426.
Mandel, M., and Higa, A 1970; J. Mol. Biol. 53; 159-162.
Meng, M., and Hogenkamp, H.P.C. 1981 J. Biol. Chem. 256; 9174-9182.
Miller, J.H. 1972; Experiments in Molecular Genetics, pp. 431-435, Cold Spring Harbor Laboratory, Cold Spring Harbor, N.Y.
Modrich, P. and Richardson, C.C., 1975; J. Biol. Chem. 250; 5515-5522.
Norrander, J., Kempe, T., and Messing, J. 1983, Gene 26; 101-106.
Russel, M., and Model, P. 1984; J. Bacteriol. 157; 526-533.
Sanger, R., Micklen, S., and Coulson, A.R. 1977, Proc. Natl. Acad. Sci. USA. 74; 5463-5467.
Schurmann, P., Maeda, K., and Tsugita, A. 1981, Eur. J. Biochem. 116; 37-45.
Sjoberg, B.-M., and Holmgren, A. 1972, J.Biol. Chem. 247; 8063-8068.
Tsugita, A., Maeda, K. and Schurmann, P. 1983, Biochem. Biophys. Res. Commun. 115, 1-7.
Tuggle, C., and Fuchs, J.A.1985, J. Bacteriol. 162; 448-450.
Whittaker, M.M., and Gleason, F.K. 1984, J. Biol Chem. 259; 14088-14093.
Wolosiuk, R.A., and Buchanan, B.B. 1977, Nature 266; 565-567.
Wolosiuk, R.A., and Buchanan, B.B. 1978, Plant Physiol. 61; 669-671.
Wolosiuk, R.A., Buchanan, B.B., and Crowford, N.A. 1978, FEBS lett. 81; 253-258.
Wolosiuk, R.A., Crowford, N.A., Yee, B.C., and Buchanan, B.B. 1979, J. Biol. Chem. 254; 1627-1632.
Yee, B.C., De La Torre, A., and Crowford, N.A., Lara, C., Carlson, D.E. and Buchanan, B.B. 1981, Arch. Microbiol. 130; 14-18.

THIOREDOXIN FROM STREPTOMYCES AUREOFACIENS

M. Kollárová, E. Kašová*, K. Szuttorová*, H. Follmann**
and J. Zelinka*

Department of Biochemistry, Faculty of Science
Comenius University, 842 15 Bratislava, Czechoslovakia
*Institute of Molecular Biology, Slovak Academy of Sciences
842 51 Bratislava, Czechoslovakia
**Fachbereich Chemie/Biochemie, Philipps Universität
D-3550 Marburg, FRG

Deoxyribonucleotides are formed de novo by direct reduction of the corresponding ribonucleotides. The reduction of ribonucleotides is a stepwise process in which transfer of electrones from NADPH takes place first to thioredoxin and from it to ribonucleotides. The first step is catalysed by thioredoxin reductase and the second by ribonucleotide reductase.

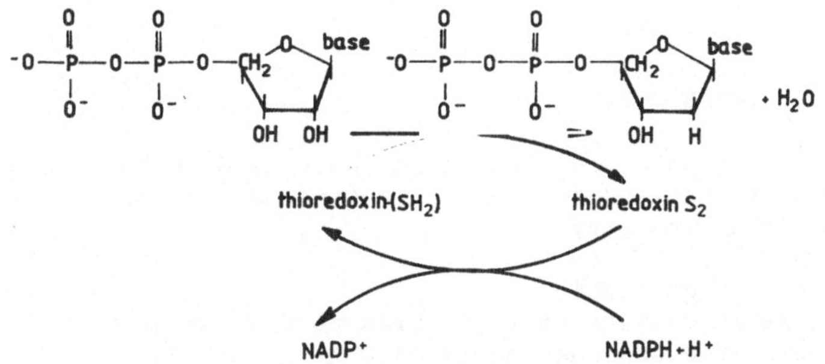

Fig. 1.   Ribonucleotides reduction (Kornberg, 1974).

Thioredoxin systems include NADPH, thioredoxin and thioredoxin reductase. Principal catalytical groups are disulfides with redox activity in thioredoxin and thioredoxin reductase. The role of thioredoxin is the reduction of disulfide to dithiol in the active site of ribonucleotide reductase. Thioredoxin serves as the hydrogen donor for the reduction of ribose.

From a chlortetracycline producing strain of Streptomyces aureofaciens we have isolated and characterized a coenzyme B_{12}-dependent ribonucleotide reductase catalysing the conversion of ribonucleoside diphosphates and ribonucleoside triphosphates to the corresponding deoxyribonucleotides Kollárová et al., 1980). Due to the complex study of ribonucleotides reduction in this strain of S.aureofaciens we have isolated and characterized its thioredoxin.

ASSAY OF THIOREDOXIN ACTIVITY

Reduced thioredoxin spontaneously reduces 5,5' –dithiobis– (2-nitrobenzoic acid) (DTNB), insulin or CDP via ribonucleotide reductase. The activity of thioredoxin can be measured also undirectly by measuring the activity of the enzymes which are stimulated by it (e.g. less active oxidated –S–S–enzymes of chloroplasts are converted into more active reduced –SH enzymes).

For the assay of thioredoxin activity we have used the reduction of DTNB to the yellow product of 5-thio-2-nitrobenzoic acid and the conversion of labeled (^{3}H) CTP into (^{3}H)dCTP.

Method 1:
The assay mixture contained 50 mmol.1^{-1} EDTA, 0.5 mmol.1^{-1} DTNB and thioredoxin in a final volume of 1 ml. The reaction was initiated by the addition of 30 µg S.aureofaciens thioredoxin reductase. The increase in absorbance was monitored at 412 nm during 30 minutes.

One unit of thioredoxin activity (1 U) was expressed as ΔA_{412} sample / ΔA_{412} control sample = 1.

Method 2:
The assay procedure used for ribonucleotide reductase is described in detail in the paper of Kollárová et al. (1980). The standard assay mixture contained 5–80 µg thioredoxin. Thioredoxin activity was observed as stimulation of ribonucleotide reductase activity.

PURIFICATION OF THIOREDOXIN

For the isolation of thioredoxin we have used cells of S.aureofaciens after 20 hours of cultivation. The isolation and purification procedure comprised the following steps:

- mycelium homogenization
- streptomycin sulfate precipitation to 0.8% saturation
- heating of the supernatant to 65°C
- salting out with ammonium sulfate from 50–90% saturation
- chromatography on 2',5'-ADP Sepharose
- chromatography on DEAE cellulose
- chromatography on Sephadex G-100

We have included 2',5'-ADP Sepharose chromatography into the purification procedure because we have found that thioredoxin reductase is bound to it and thus it was possible to use this step for separation of thioredoxin from thioredoxin reductase.

The effluent which did not contain any thioredoxin reductase activity was chromatographed on DEAE cellulose using a linear 0–03 mol.1^{-1} NaCl gradient. Figure 2 shows that the thioredoxin containing fractions were eluted from 0.10 to 0.14 mol. 1^{-1} NaCl.

The active fractions were pooled, concentrated, dialysed and chromatographed on Sephadex G-100 column (Figure 3).

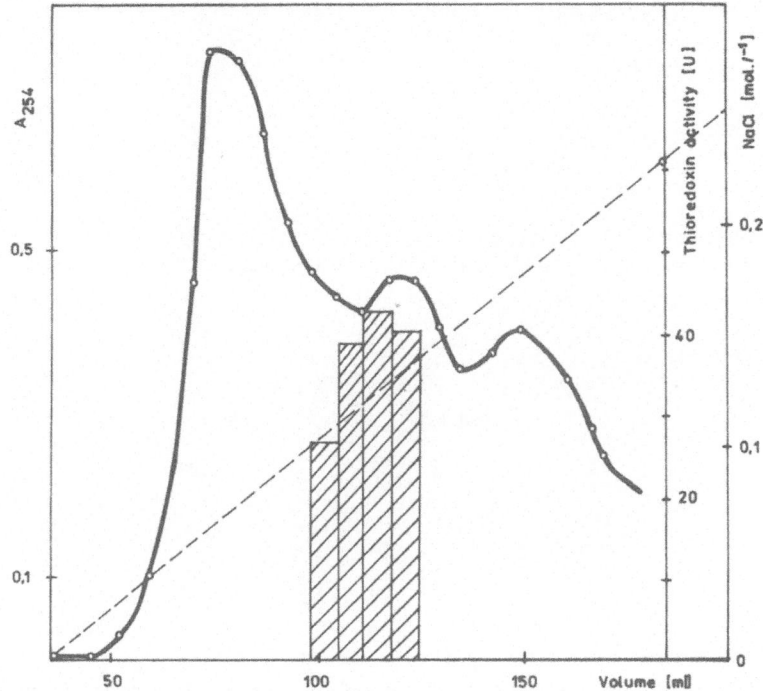

Fig. 2. Elution profile of thioredoxin from S.aureofaciens on DEAE cellulose. Column: 2 x 13 cm; elution: 160 ml NaCl gradient $(0-0.3 \text{ mol.l}^{-1})$ in 50 mmol.l^{-1} Tris-HCl, pH 8.0 with 2 mmol.l^{-1} EDTA; elution rate: 17 ml per hour.

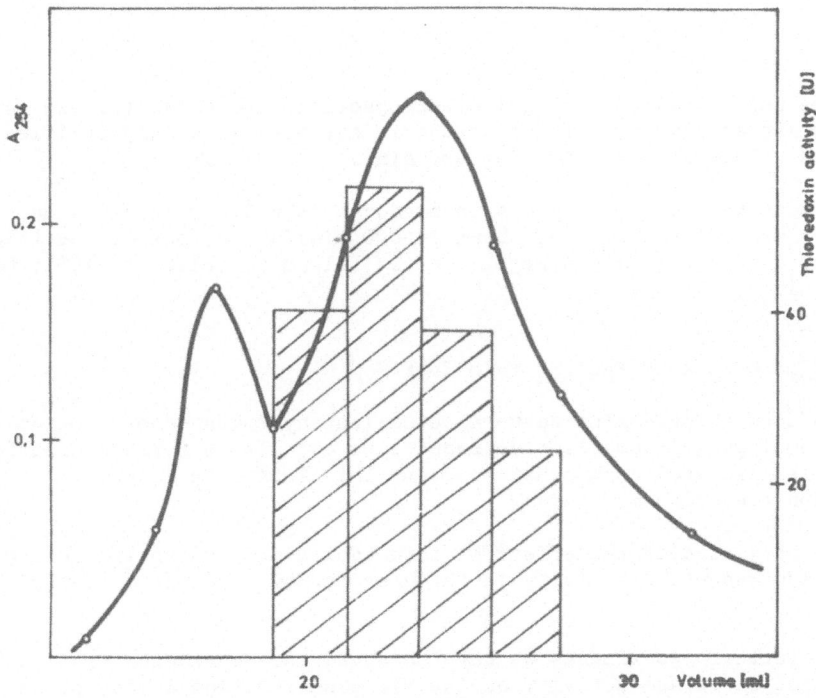

Fig. 3. Elution profile of thioredoxin from S.aureofaciens on Sephadex G-100. Column: 1.6 x 19 cm; elution: 50 mmol.l^{-1} Tris-HCl pH 8.0 with 2 mmol.l^{-1} EDTA; elution rate: 3.4 ml per hour.

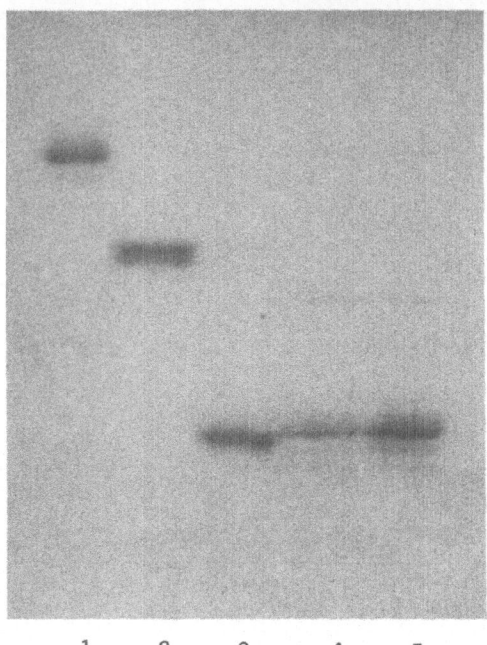

1 2 3 4 5

Fig. 4. SDS polyacrylamide gel electrophoresis of S.aureofaciens
thioredoxin. (1) albumin egg; (2) chymotrypsinogen; (3)
cytochrome C; (4) pooled fractions after DEAE cellulose column
chromatography; (5) pooled fractions after Sephadex G-100 column
chromatography.

The SDS polyacrylamide gel electrophoresis indicates (Figure 4) that
the protein preparation with thioredoxin activity is highly purified and
contains only four contaminating proteins.

As can be seen in Figure 4 thioredoxin from S.aureofaciens has a
relative molecular weight of about 12 000 similar to most of the thioredoxins isolated from prokaryotic cells (Gleason, Holmgren, 1981; Laurent
et al., 1964).

PROPERTIES OF S.AUREOFACIENS THIOREDOXIN

Reduced thioredoxins serve as excellent hydrogen donors not only for
their homologous ribonucleotide reductases but also for ribonucleotide
reductases isolated from other sources. This indicates identical thiol
oxidation reduction mechanism.

We have studied the effect of thioredoxin on S.aureofaciens ribonucleotide reductase activity in the presence of constant amount of DTE
(Figure 5).

In further experiments we have observed the dependence of ribonucleotide reductase activity on the DTE concentration in the presence of
different amount of thioredoxin (Figure 6).

In both sets of experiments stimulation of ribonucleotide reductase
activity was observed.

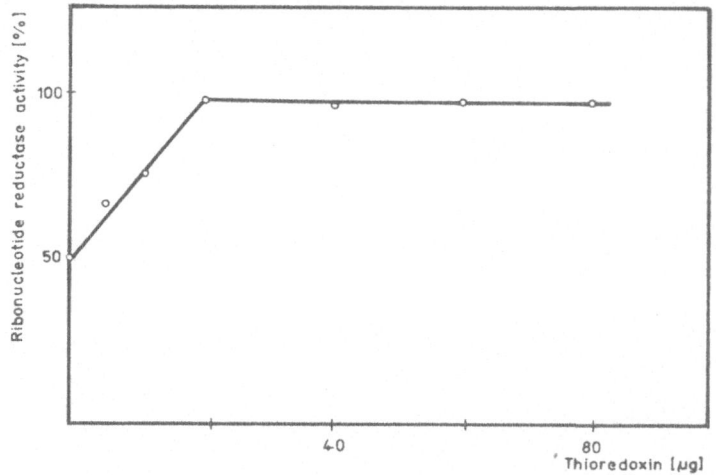

Fig. 5. Effect of thioredoxin concentration on activity of S.aureofaciens ribonucleotide reductase.

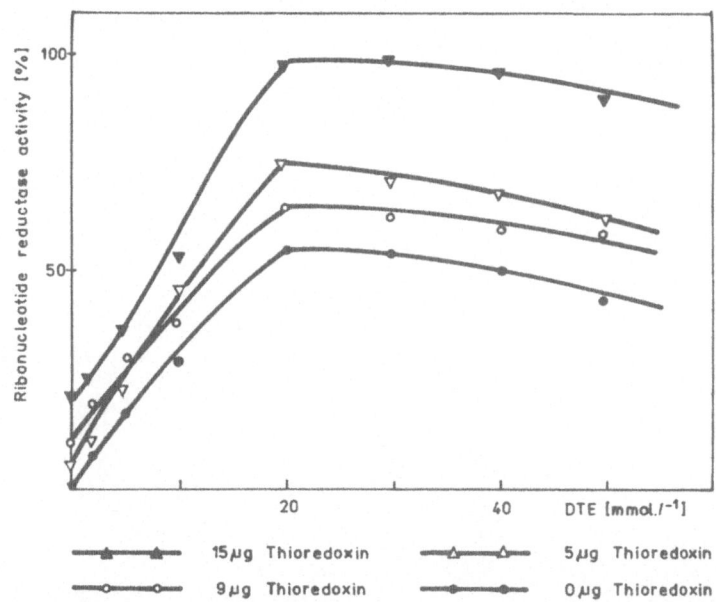

Fig. 6. Effect of dithioerythritol concentration on activity of S.aureofaciens ribonucleotide reductase in the presence of thioredoxin.

Reduced thioredoxin from S.aureofaciens is not only the hydrogen donor for homologous coenzyme B_{12}-dependent ribonucleotide reductase but also stimulates the activity of E.coli ribonucleotide reductase.

REFERENCES

Gleason, F. K., Holmgren, A. J., 1981, Biol. Chem., 256, 8306–8309
Kollárová, M., Perečko, D., Zelinka, J.,1980, Biológia, 35, 907–910.
Kornberg, A. 1974, in: "DNA synthesis," W.H. and Company eds., San Francisco, USA), 29–55.
Laurent, T.C., Moore, E.C., Reichard, P. 1964, J.Biol Chem. 239, 3436–3444.

TWO–DIMENSIONAL ^{1}H–NMR INVESTIGATION OF RIBONUCLEASE T$_1$ AND THE COMPLEXES OF RNASE T$_1$ WITH 2'– AND 3'–GUANOSINE MONOPHOSPHATE

Heinz Rüterjans, Eberhard Hoffmann, Jurgen Schmidt and Jeannette Simon

Institute of Biophysical Chemistry, Johann Wolfgang Goethe University, D-6000 Frankfurt a.M. 70, Germany

Ribonuclease T$_1$ (RNase T$_1$) cleaves the phosphordiester bond of RNA, specifically at the 3' end of guanosine. monophosphates (2'- or 3' -GMP) as well as other purine nucleotides act as inhibitors for this reaction (Takahashi and Moore, 1982). The tertiary structure of the RNase-T$_1$ - 2' -GMP complex was determined by X-ray crystallography (Heinemann and Saenger, 1982; Sugio et al., 1985; Arni et al., in press). The present study revealed not only the folding manner of the enzyme but also some interesting features of protein - nucleotide interactions. We report here on the investigation of the solution structure of RNase T$_1$ and of complexes with 2'- GMP and 3' -GMP, using 2D-NMR spectroscopy. Using the distance parameters obtained from this study, a preliminary molecular dynamics calculation was carried out to arrive at a tertiary structure of the complexes comparable to that obtained from crystals.

The immense complexity of ^{1}H–NMR spectra of RNase T$_1$ has hitherto restricted ^{1}H–NMR studies to the observation of the aromatic resonances or isolated methyl resonances of the protein spectrum (Ruterjans and Pongs, 1971; Arata et al., 1976, 1979; Fulling and Ruterjans, 1978; Kimura et al., 1979; Inagaki et al., 1981, 1985; Nagai et al., 1985). Interactions between histidines and tyrosines of the active site and of the base and phosphate moieties of nucleotides were already derived from observations of the pH dependence of some protein proton resonances and of ^{15}N, ^{31}P, and ^{1}H resonances of the nucleotide (Maurer, 1972; Kyogoku et al., 1982; Menke, 1984). The high magnetic field strengths currently used as well as the application of 2D-NMR techniques make it possible to gain useful additional information about the structure of the free enzyme and about the changes in structure induced by nucleotide binding.

On the basis of the 2D procedures developed by Wüthrich and others (Aue et al., 1976; Wüthrich, 1986) we were able to identify most of the amino acid spin systems and to assign about 90% of the side chain and backbone proton resonances to specific amino acid residues in this protein. We used the following 2D experiments for the sequential assignment of the ^{1}H resonances: 1. COrrelated SpectroscopY (COSY; Nagayama et al.,1980; Bax and Freeman, 1981; Marion and Wuthrich, 1983); 2. Nuclear Overhauser Enhancement and exchange SpectroscopY (NOESY; Kumar et al., 1981; Macura et al., 1982); 3. Relayed Coherence Transfer spectroscopy (RCT; Eich et al., 1982; Wagner, 1983; Bax and Drobny, 1985). All spectra were acquired on a Bruker AM-500spectrometer. The pulse sequence $(T-90^\circ-t_1-90^\circ-t_2)_n$ was used

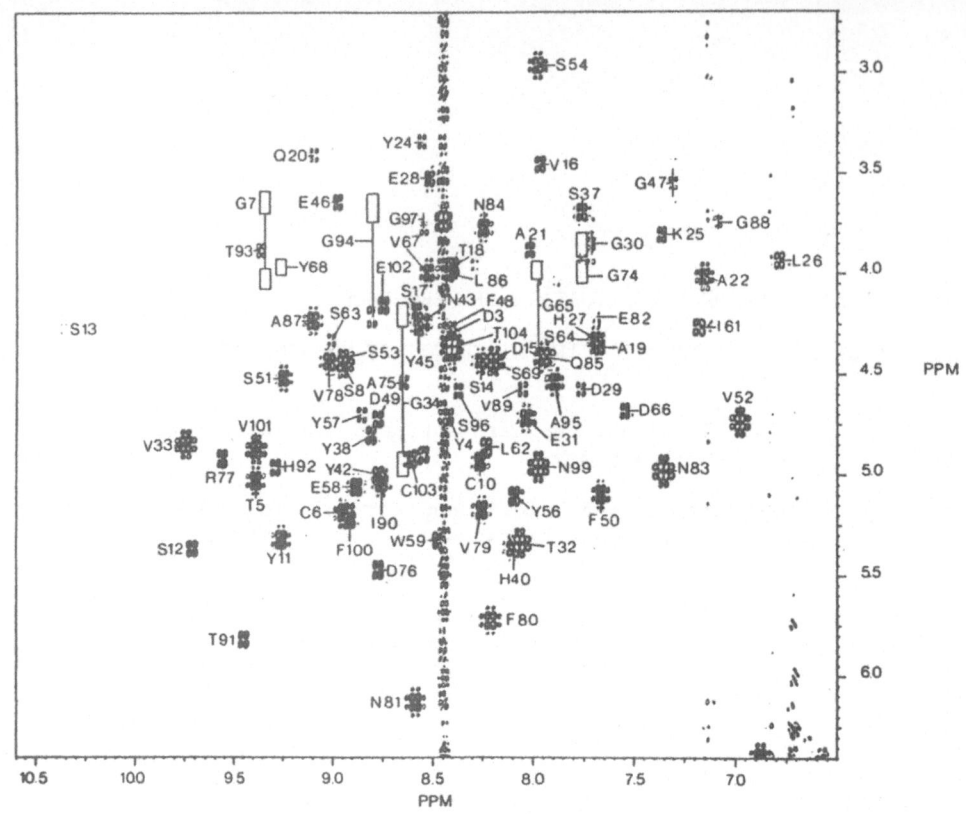

Fig. 1A. Phase-sensitive COSY spectrum of RHase T_1. The region depicted
contains cross peaks originating from the scalar coupling between
$C_\alpha H$ and amide protons (fingerprint region). The assignments are
indicated. The cross peaks of G 23 are outside the plotted
region, approximately at 7.3 ppm and 2.0 ppm.

for the acquisition of the COSY spectra. A typical Cosy spectrum was
recorded within about 36 h, with 144 scans accumulated per free induction
decay (FID). The NOESY spectra were recorded with the pulse sequence
$(T-90°-t_1-90°-t_m-90°-t_2)_n$. The mixing time t_m was set to 150 ms. In order
to eliminate artifacts resulting from coherence transfer, t_m was varied
randomly by 10%. 256 scans were accumulated per FID. A typical NOESY
spectrum was obtained within about 68 h. RCT experiments were performed
with the pulse train $(T-90°-t_1-90°-_m/2-180°-_m/2-90°-t_2)_n$. Undesired
artifacts were eliminated by phase cycling as described by Bodenhausen et
al. (1984) and Bain (1984). As an important advantage an RCT spectrum
contains information about connectivities between two spins that exhibit
negligible direct coupling, but do both exhibit significant coupling to a
common third spin.

For experiments in H_2O, the solvent resonance was saturated by continuous selective irradiation at all times except t_1 and t_2.

The procedure for sequence-specific assignments of proton resonances
of proteins has been developed by Wüthrich (1986). The basic concept is to
utilize scalar coupling phenomena to identify complete spin systems. The
type of amino acid is then assigned via the characteristic chemical shifts
of the coupled protons. After this identification of the spin system the

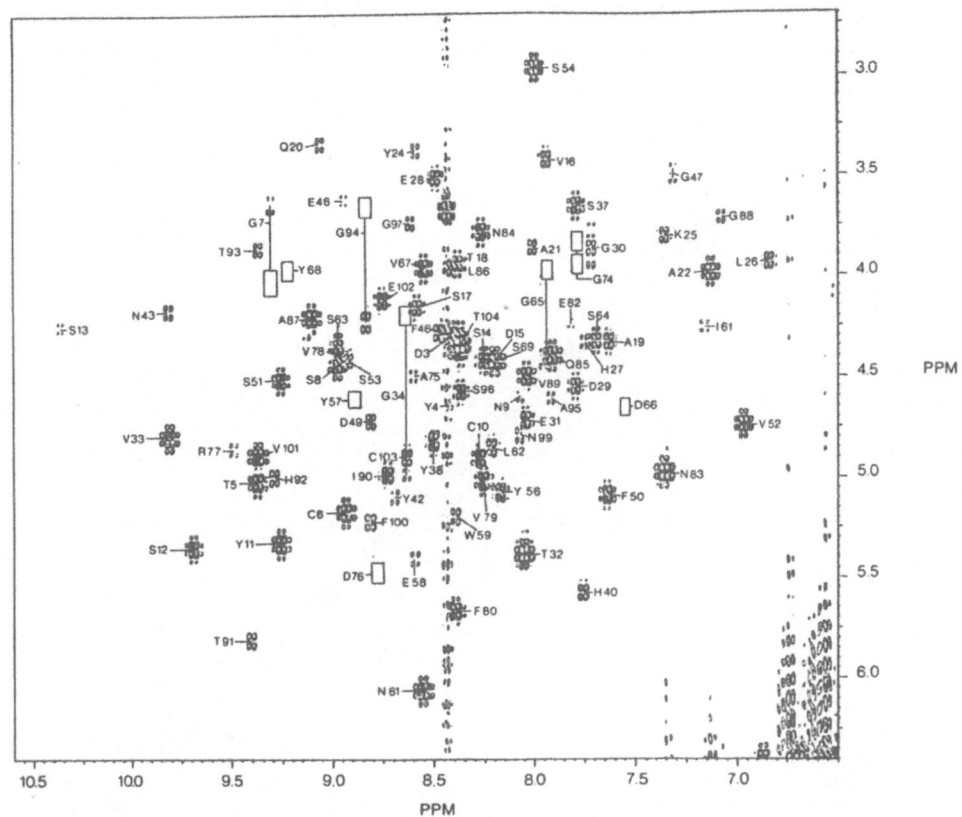

Fig. 1B. COSY spectrum of the RNase-T$_1$ - 2'-GMP complex (fingerprint region).

C$_\alpha$H and backbone amide protons of the amino acid in question have to assigned. Through-space connectivities between C$_\alpha$H and backbone amide protons are monitored via NOESY spectroscopy. In this manner successive residues can be identified step by step. With the help of the primary structure of the molecule, the sequence-specific assignment can be carried out.

RESULTS AND DISCUSSION

In Figs. 1A and B the so-called fingerprint peak region, i.e. the connectivities between the NH and the C$_\alpha$H resonances of phase-sensitive-Cosy spectra, is shown for free RNase T$_1$ and for the RNase-T$_1$-2'-GMP complex. About 100 cross peaks of this area could be assigned to specific amino acid residues.

From corresponding NOESY spectra, NOE values of adjacent protons in the molecule were derived. In Figure 2 these NOE data are presented in a diagonal plot. Since the NOE values are related to distances between the corresponding protons, such a plot is particularly helpful for the determination of the secondary structure. The filled squares indicate NOEs between backbone protons (NH or C$_\alpha$H). The sequential NOEs appear as a nearly continuous line of squares immediately adjacent to the diagonal. The helical regions are characterized by lines of squares running parallel to the diagonal at a distance of 3 sequence positions, with some additional

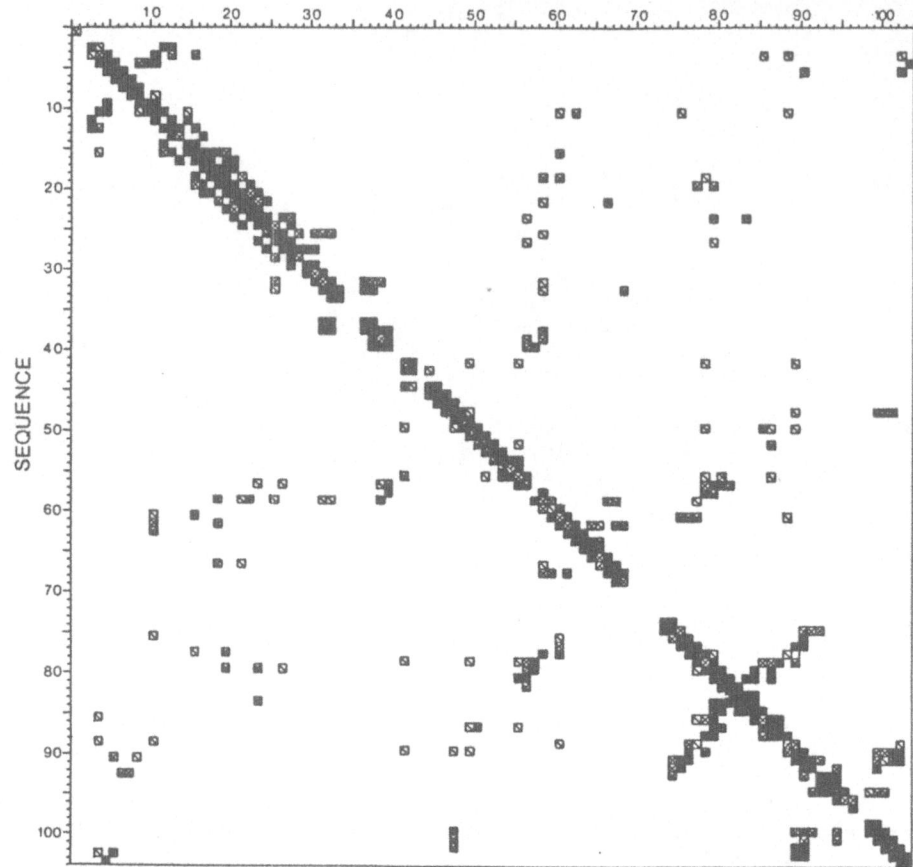

Fig. 2. Diagonal plot of the NOEs observed in RNase T_1. The two axes represent the amino acid sequence. (■) NOEs between backbone protons (NH or $C_\alpha H$); (⊠) NOEs between NH or $C_\alpha H$ and any side chain proton; (⊠) NOEs between the side chain protons.

NOEs at 2 or 4 positions from the diagonal. The regular antiparallel β sheet formed between four strands in manifested by a continuous line of squares perpendicular to the diagonal. A parallel β sheet would produce a line of squares parallel to the diagonal at a distance determined by the relative sequence locations of the two neighbouring strands. Hence, a diagonal plot presentation of the NOE data can provide a clear illustration of the characteristic patterns of short ^1H–^1H distances for the common polypeptide secondary structures. In the diagonal plot, NOEs between NH or $C_\alpha H$ and a side chain proton are also indicated, as well as NOEs between side chain protons of two different residues.

From a comparison of COSY spectra of the free RNase T_1 and its 2'-GMP and 3'-GMP complexes (Figures 1A and B; Figures 3A and B) two essential features for conformational changes induced by complex formation were derived:

1. Interaction of the inhibitor and the substrate with the active site leads to considerable changes of the side-chain and backbone positions. Correspondingly larger changes are observed in the proton resonance positions of amino acid residues directly involved in binding.

46

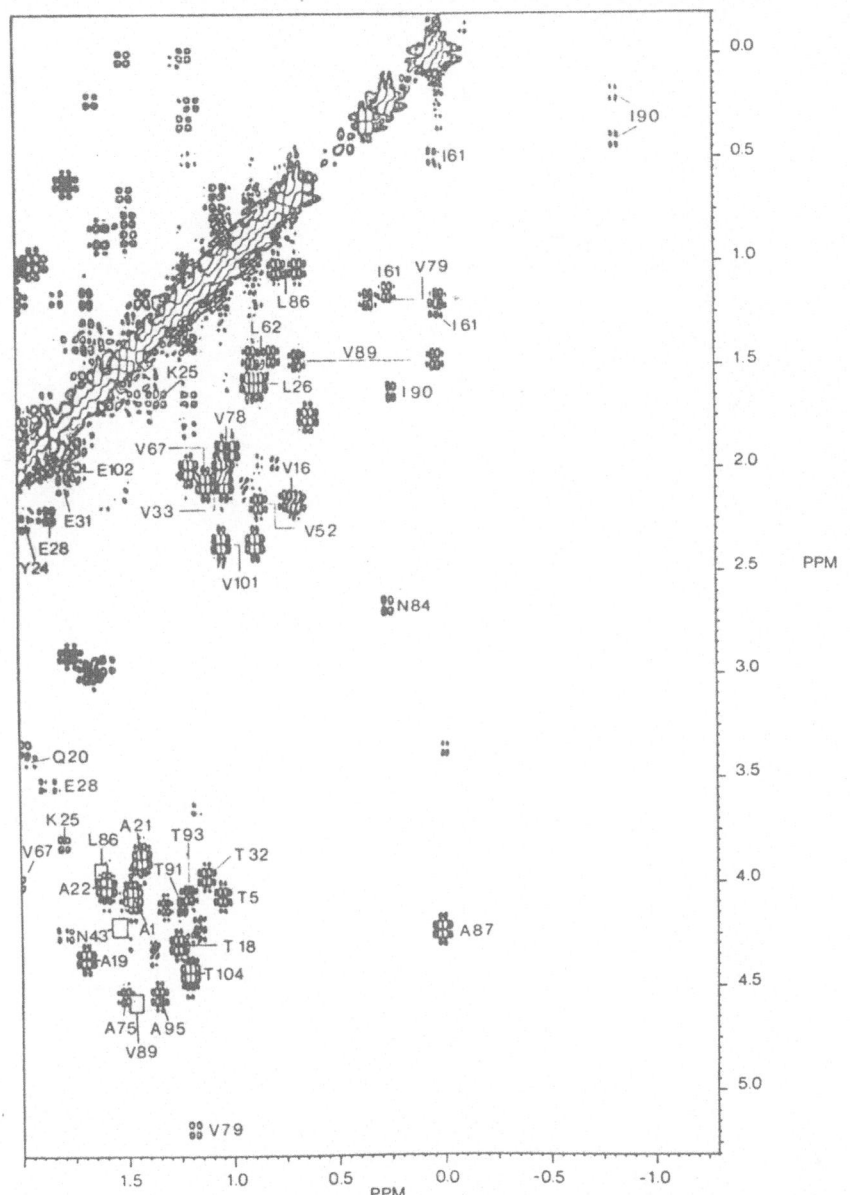

Fig. 3A. Phase-sensitive COSY spectrum of RNase T$_1$. The cross peaks in this area originate from the scalar coupling between the methyl protons of alanine, threonine, valine, isoleucine, or leucine, and the C$_\alpha$, C$_\beta$, or C$_\gamma$ protons.

2. Smaller changes occur also inside the protein domain, in particular in the contact area between the α helix and the β-sheet structure. Proton resonances of amino acids located in this area shift their respective positions only slightly.

T 38, H 40, Y 42, N 43, Y 45, Y 56, E 58, W 59, R 77, H 92, N 99, and F 100 proton resonances shift considerably (Figs. 1A and B). The side-chain resonances of V 79 and I 90 (Figs. 3A and B), which are located

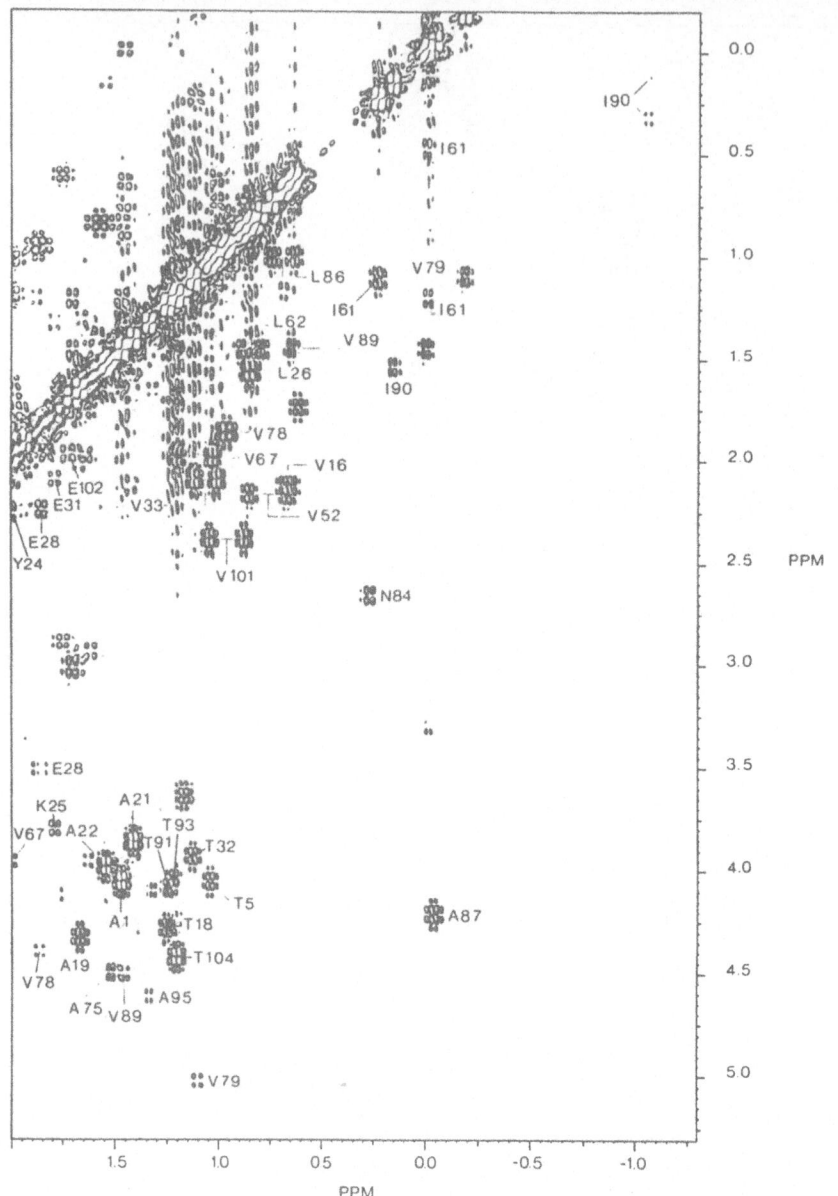

Fig. 3B. Phase-sensitive COSY spectrum of the RHase-T_1 – 2'-GMP complex. Ala, Thr, Val, Ile, and Leu methyl cross peaks.

between the central β sheet and the active center, are also shifted. These residues are more or less directly involved in nucleotide binding or are near the active site.

Smaller high- or low-field shifts can be observed for the ^1H spin systems of A 19, Q 20, G 23, Y 24, L 26, H 27, T 32, V 33, P 39, Y 57, F 80, and N 81. These resonances are located in the contact area between the α helix and the β-sheet structure.

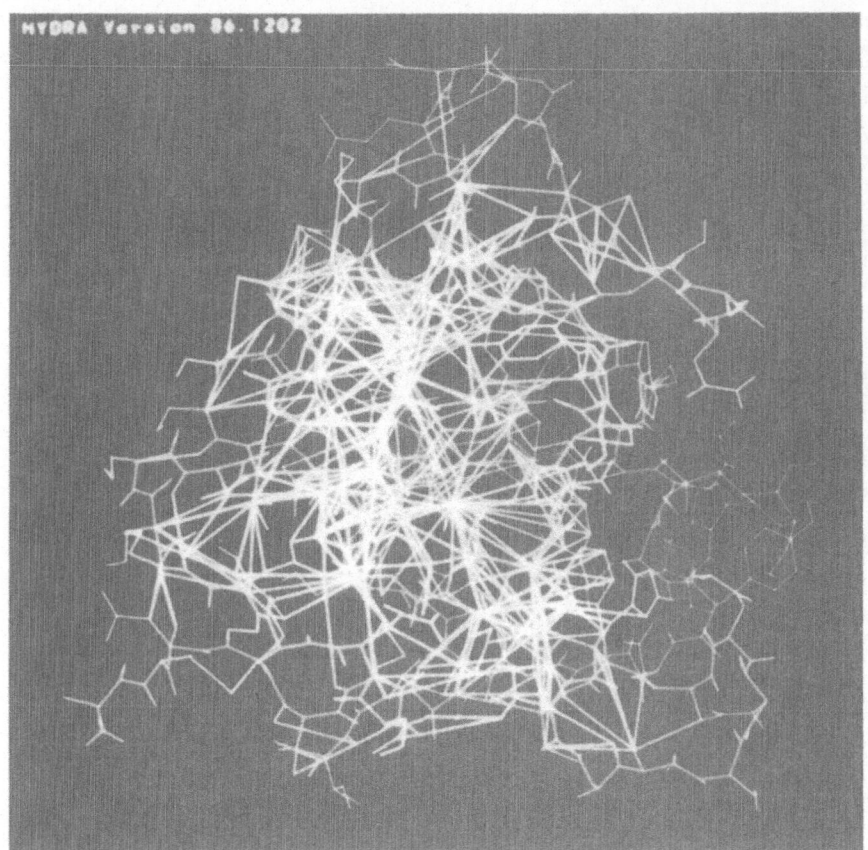

Fig. 4. Structure of the RNase-T_1 - 2'-GMP complex obtained from a
molecular dynamics calculation, using about 300 long-range NOEs as
distance constraints. The distance constraints are indicated by
white bars.

Distance changes between protons attached to aromatic side chains and
protons of adjacent groups also lead to considerable shift changes.

From a comparison of the COSY spectra it is evident that the conformation of the RNase-T_1 -3'-GMP complex is more similar to the conformation
of the free enzyme than to that of the RNase-T_1-2'-GMP complex (spectra of
the RNase-T_1 -3'-GMP complex not shown).

The so-called long-range NOEs contain valuable information about
distance constraints in the macromolecule. They constitute the major NMR
information on spatial protein structures and connect hydrogen atoms
located anywhere in the sequence. Therefore a distance geometry program or
a molecular dynamics calculation for structure determination from such data
should lead to the tertiary structure of a protein. For our study of RNase
T_1 we used the Gromos molecular dynamics program developed by van Gunsteren
(Van Gunsteren et al., 1984). In this program the bond lengths are kept
fixed, whereas all bond and torsional angles are accessible to changes at
any time. Violations of upper distance limits obtained by ^{1}H NMR or of
lower limits imposed by the atomic van der Waals' volumes are taken into
account in the protein structure. Figure 4 shows the refined structure of
the RNase-T_1-2'-GMP complex calculated with about 300 NOE long-range distance constraints. The distance-constrained atom pairs are indicated by
white lines. Owing to NMR-inherent reasons, intermolecular NOEs between

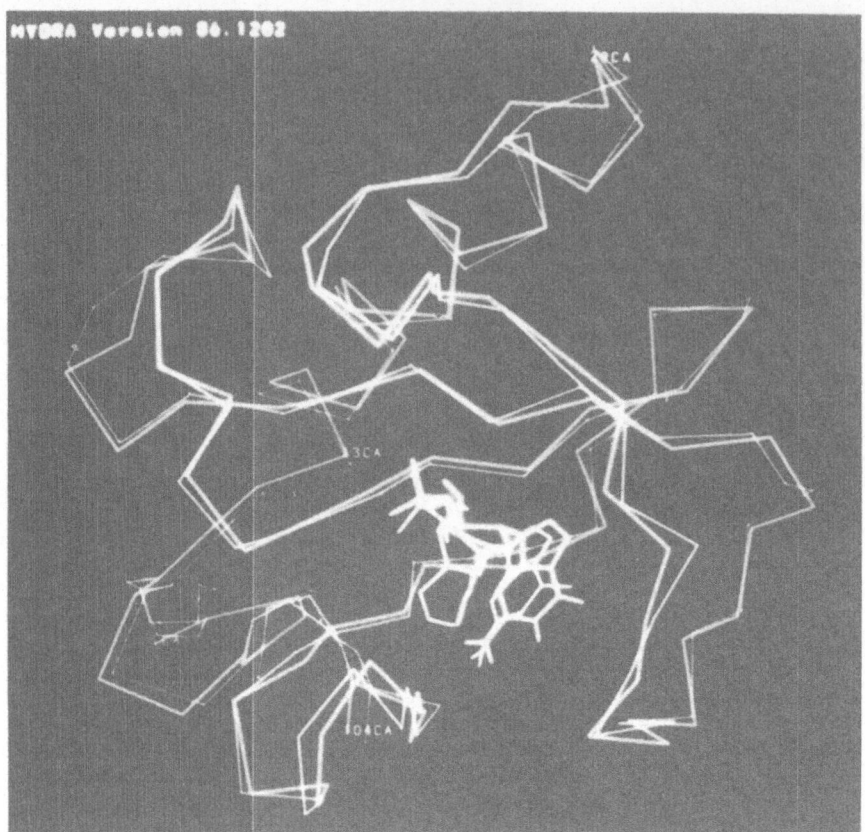

Fig. 5.   Comparison of structures of the RNase-T$_1$ – 2'-GMP complex obtained
          by X-ray crystallography and by a molecular dynamics calculation
          using NOE distance contraints.

the enzyme and the substrate are not observable, and intramolecular NOEs in
the loop regions of the active site are rare.

    The comparison of the refined structure and the X-ray structure
(Figure 5) shows a good agreement with respect to the backbone folding.
For reasons of clarity, the side-chain atoms exhibiting the largest r.m.s.
fluctuations are not displayed.

    As found in the crystal structure, the nucleotides 2'-GMP and 3'-GMP
take the syn conformation for the glycosidic bond (C_8-N_9-C'_1-O_4'). The syn
conformation of the nucleotide and the arrangement of the base moiety of
2'-GMP in the active site is depicted in Figure 6. According to our
results, the peptide CO group of Asn 98 is located in H-bond distance to
the amino group of the guanine base. The carboxyl group of Glu 46 is
apparently bound to the Guo-N_1 position, whereas the Guo-O_6 position forms
a hydrogen bond with the peptide NH of Asn 44. Contrary to the X-ray
structure, the N_7 position is not hydrogen-bonded to the amide group and
the peptide NH of Asn 43. Indeed the formation of this H-bond seems
unlikely also according to ^{15}N-NMR studies (Menke, 1984). The arrangement
of the phosphate group in the active site as derived from our combined
2D-NMR – MD study is shown in Figure 7. In agreement with previous 1D-NMR
studies, the phosphate group of 2'-GMP interacts with the imidazole ring of
His 40. The guanidinium group of Arg 77 is located near the phosphate
group, but too far away to form a hydrogen bond. Neither can His 92 reach

50

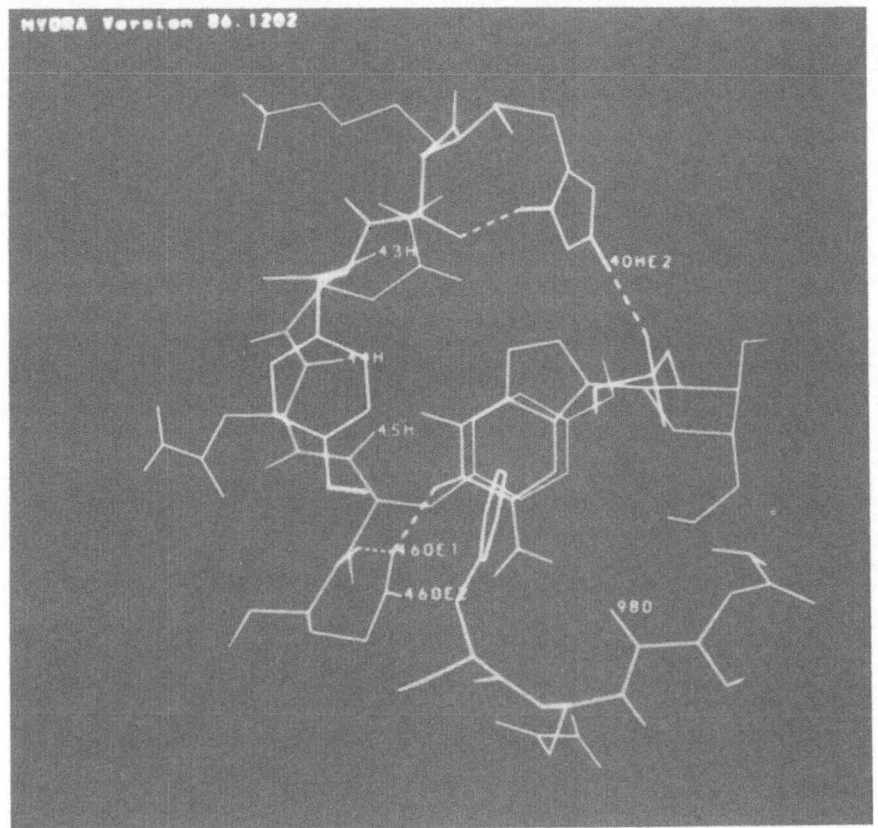

Fig. 6. Binding site of the base moiety of 2'-GMP in the RNase-T_1 - 2'-GMP
complex, obtained with an MD calculation using 2D-NMR data.

the phosphate group. Like in the crystal, the phenol ring of Tyr 45 is
located on top of the guanine ring. Unfortunately, it was not possible to
detect any NOE values for the interaction through space of the ring protons
of Tyr 45 with any of the surrounding protons in the RNase-T_1 - 2'-GMP
complex. Not even the COSY or NOESY cross peaks of the backbone NH and $C_\alpha H$
and of the side chain proton resonances were observed in the complex.
Apparently owing to exchange phenomena, the resonances are broadened.

For the initial transesterification step of RNase-T_1-catalysed
hydrolysis, a possible mechanism is proposed in Figure 8. According to
this mechanism, the carboxylate group of Glu 58 withdraws a proton from the
2'-hydroxyl group, which is then ready for a nucleophilic attack of the
phosphate group, possibly activated by an increase in electrophilicity of
the phosphate moiety via an interaction with His 92 or Arg 77. The O_5'
leaving group takes over a proton from the protonated imidazole group of
His 40. It appears that the side chains of Glu 58 and His 40 are connected
via a hydrogen bond in the free enzyme. The hydrolysis of the 2'-3' cyclic
phosphate should occur in a similar way. Glu 58 transfers its proton to
form the 2'-hydroxyl group, whereas His 40 activates the water molecule.
Possibly, contrary to this reaction scheme, His 92 takes the role of His
40, since it is also located near the phosphate group. Further studies
with mutants of RNase T_1, presently in progress, may clarify this point.

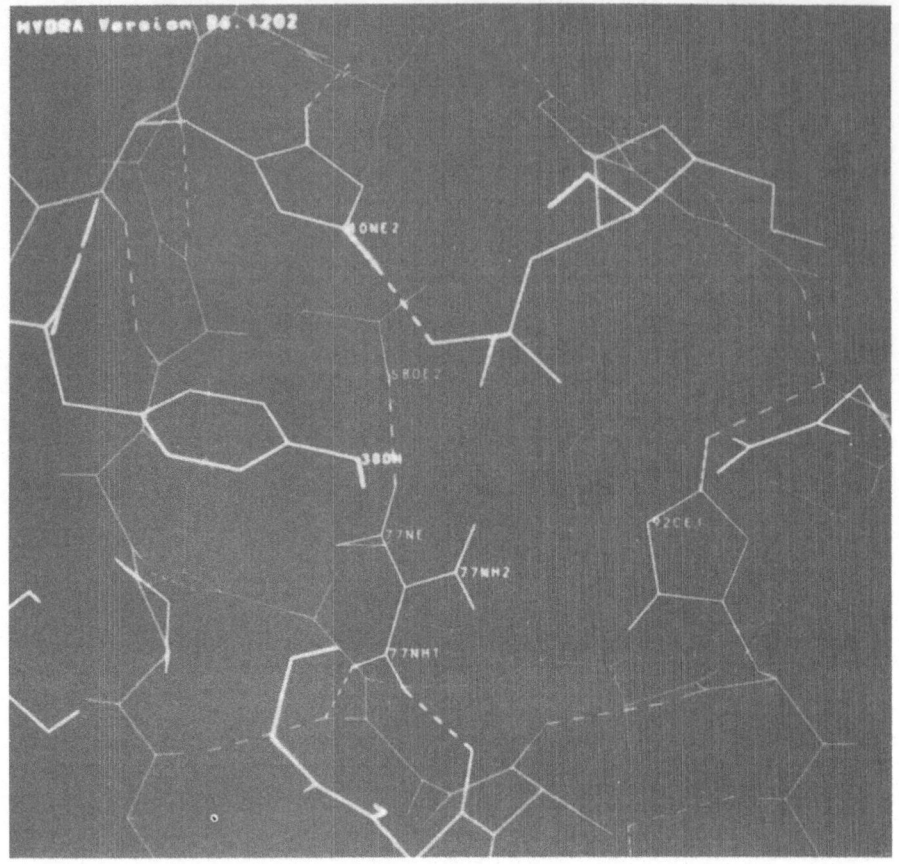

Fig. 7.   Binding site of the phosphate moiety of 2'-GMP of the RNase-T$_1$ – 2'-GMP complex, obtained with an MD calculation using 2D-NMR data.

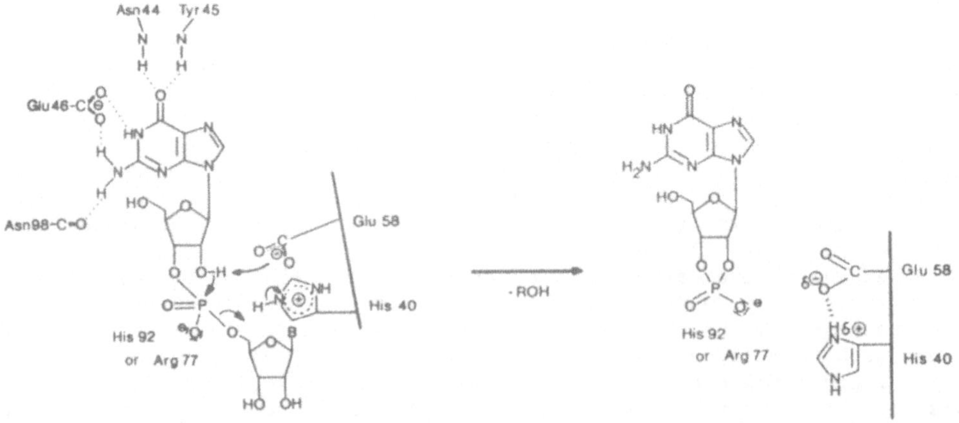

Fig. 8.   Possible arrangement of hydrogen bonds for the binding of the base of 2'-GMP in the active site of RNase T$_1$, and presumable mechanism of RNase-T$_1$ action.

Acknowledgements

We thank the Deutsche Forschungsgemeinschaft for a grant (Ru 145/8-2). Thanks are due to Miss H. Tietz for revising the English text and typing the manuscript.

REFERENCES

Arata, Y., Kimura, S., Matsuo, H., and Narita, K. 1976, Biochem.Biophys. Res. Commun. 73; 133-140.

Arata, Y., Kimura, S., Matsuo, H., and Narita, K. 1979, Biochemistry, 18; 18-24.

Arni, R., Heinemann, U., Maslowska, M., Tokuoka, R., and Saenger, W.(in press).

Aue, W.P., Bartholdi, E., and Ernst, R.R. 1976, J. Chem. Phys, 64; 2229-2246.

Bain, A.D. 1984, J Magn. Reson. 56; 418-427.

Bax, A. and Freeman, R. 1981, J. Magn. Reson. 44; 542-561.

Bax, A and Drobny, G. 1985, J. Magn.Reson. 61; 306-320.

Bodenhausen, G., Kogler, H., and Ernst, R.R. 1984, J. Magn. Reson. 58; 370-388.

Eich, G., Bodenhausen, G., and Ernst, R.R. 1982, J. Am. Chem Soc. 104, 3731-3732.

Fulling, R. and Ruterjans, H. 1978, FEBS Lett. 88; 279-282.

Heinemann,U. and Saenger, W. 1982, Nature, 299; 27-31.

Inagaki, F., Kawano, Y., Shimada, J., Takahashi, K., and Miyazawa, T. 1981, J. Biochem. (Tokyo) 89; 1185-1195.

Inagaki, F., Shimada, I., and Miyazawa, T. 1985, Biochemistry 24; 1013-1020.

Kimura, S., Matsuo, H., and Narita, K. 1979, J. Biochem. (Tokyo) 85; 301-310.

Kumar, A., Wagner, G., Ernst, R.R., and Wuthrich, K. 1981, J. Am. Chem. Soc. 103; 3654-3658.

Kyogoku, Y., Watanabe, M., Kainosho, M., and Oshima, T. 1982, J. Biochem. (Tokyo) 91; 675-679.

Macura, S., Wuthrich, K., and Ernst, R.R. 1982, J. Magn. Reson. 47; 351-357.

Marion, D. and Wuthrich, K. 1983 Biochem. Biophys. Res. Commun. 113; 967-974.

Maurer, W. 1972, Dissertation, Munster.

Menke, G. 1984, Dissertation Frankfurt.

Nagai, H., Kawata, Y., Hayashi, F., Sakiyama, F., and Kyogoku, Y. 1985, FEBS Lett. 189; 167-170.

Nagayama, K., Kumar A., Wuthrich, K., and Ernst, R.R. 1980, J. Magn. Reson. 40. 321-334.

Rüterjans, H. and Pongs, O. 1971, Eur. J. Biochem. 18; 313-318.

Sugio, S., Amisaki, T., Ohishi, H., Tomita, D., Heinemann, U., and Saenger, W. 1985, FEBS Lett. 181; 129-132.

Takahashi, K. and Moore, S. 1982, pp.435-468 in "The Enzymes" Vol. XV: Nucleic Acids, Part B (3rd ed.; P.D. Boyer, ed.). Academic Press, New York.

Van Gunsteren, W.F., Kaptein, R., and Zuiderweg, E.R.P. 1984, Proc, NATO/-CEGAM Workshop on Nucleic Acid Conformation and Dynamics, Orsay (W.K. Olson, ed.), 79-92.

Wagner G. 1983, J. Magn. Reson. 55; 151-156.

Wuthrich, K. 1986, "NMR of Proteins and Nucleic Acids". John Wiley and Sons, New York.

NUCLEOSIDE PHOSPHOTRANSFERASE AND NUCLEASE S_1, TWO ENZYMES

WITH ACYLPHOSPHATE INTERMEDIATES, BUT DIFFERENT MECHANISMS

H. Witzel, A. Billich, W. Berg, O. Creutzenberg
and A. Karreh

Institute of Biochemistry, Univ. of Münster
D-4400 Münster, FRG

This is a report on two enzymes hydrolyzing the phosphomonoester bond of nucleotides via acylphosphate intermediates [1,2]: nucleoside phosphotransferase with retention [3] and nuclease S_1 with inversion of absolute configuration [4].

Phosphotransferase isolated from malt sprouts, M_r 51 000, two identical subunits, pH-optimum 5-7, containing 1 Mg^{2+}, hydrolyses 3'- and 5'-nucleotides as well as activated phosphomonoesters like p-nitrophenyl- or naphthylphosphates. Besides of hydrolyses a transfer occurs, however, only in the 5'- position of added nucleosides. Retention of absolute configuration in the transfer reaction [3] and incorporation of phosphate in the denatured protein [5] has been observed by Ives group in Columbus/Ohio.

In order to clarify the nature of the incorporated phosphate group we examined the liberation of inorganic phosphate after incubation at different pH-values. The relative stability in the range of pH 4-6 and the sensitivity below and above these pH values is typical for an acylphosphate intermediate.

Acylphosphate intermediates should be trapped by addition of hydroxylamine during reaction in presence of substrates. Two ways are possible: The enzyme produces inorganic phosphate but is inactivated by formation of hydroxamic acid or the enzyme remains active and produces N-phosphoryl-hydro-xylamine. We found the enzyme remained active.

Analysis of the reaction products by ion exchange chromatography yielded N-phosphorylhydroxylamine in dependence of the hydroxylamine concentration.

The second experiment we did to trap the acylphosphate intermediate was the reduction of the protein after denaturation during the reaction in presence of substrates. With sodiumcyanoborotritiid a proposed mixed anhydride of glutamic acid should be reduced to 2-amino-5-hydroxy-valeric acid, while that of aspartic acid should give homoserine after hydrolysis of the protein with 6N HCl. We found in acid solution the lactone of 2-amino-5-hydroxy-valeric acid which opens after alkaline treatment.

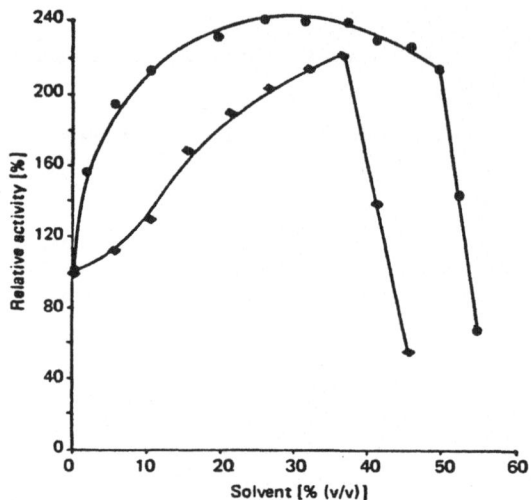

Fig. 1. Activity of nucleoside phosphotransferase in dependence of concentration of acetone (●) or methanol (◆) in the standard phosphatase assay [1].

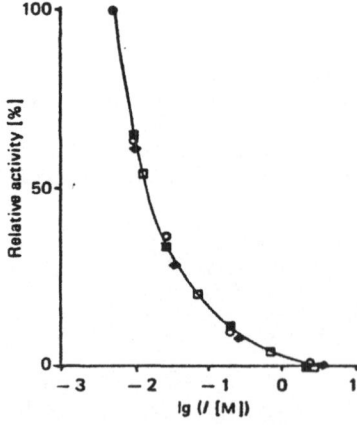

Fig. 2. Relative activity of the nucleoside phosphotransferase in 5mM NaAc pH 5 (●) or buffer containing different concentrations of NaCl (o), KCl (□), $MgCl_2$ (◆) or Na_2SO_4 (■). The indicated ionic strength (I) is the sum of buffer and salt contributions. The activity was tested with 10^{-4}M 4-nitrophenyl phosphate using 0.1 µM enzyme.

Table 1. Activity of nucleoside phosphotransferase in presence of organic solvents in the standard assay mixture [1].

Solvent	Concentration for maximal increase of activity %	Activity %
No addition	–	100
Acetone	25	245
Acetonitrile	8	240
Dimethylformamide	5	180
Methanol	30	230
Dimethyl sulfoxide	20	180
Glycerol	60	170

Table 2. Kinetic parameters of hydrolysis of 4-nitrophenyl phosphate.

Buffer	k_{cat} (s^{-1})	$10^5 \times K_m$ (M)	$10^{-5} \times k_{cat}/K_m$ $(M^{-1} \times s^{-1})$
0.01M NaAc pH 5, 10% acetone	77	3.8	20.1
0.01M NaAc pH 5,	70	8.1	8.6
0.01M NaAc pH 5, 0.1M NaCl	76	34	2.2
0.01M NaAc pH 5, 0.1M NaCl	73	100	0.73

Table 3. Reactivation of the apoenzyme by addition of 2mM of divalent cations.

Enzyme	Specific activity %
Native enzyme	100
Apoenzyme	0
Apoenzyme + $Mg^{2\oplus}$	100
Apoenzyme + $Co^{2\oplus}$	95
Apoenzyme + $Mn^{2\oplus}$	40
Apoenzyme + $Zn^{2\oplus}$	35
Apoenzyme + $Cu^{2\oplus}$	30
Apoenzyme + $Ni^{2\oplus}$	25
Apoenzyme + $Fe^{2\oplus}$	20
Apoenzyme + $Ca^{2\oplus}$	2
Apoenzyme + $Sr^{2\oplus}$, $Ba^{2\oplus}$, $Cd^{2\oplus}$, $Hg^{2\oplus}$, $Pb^{2\oplus}$	0

S_P-cAMP [S]   cAMP R_P-cAMP [S]
$I_{50} = 2 \times 10^{-7}$M $I_{50} = 1.4 \times 10^{-6}$M $I_{50} = 1.3 \times 10^{-5}$M

From these experiments we can conclude that a carboxylate group of glutaminic acid attacks the phosphate group, if it is a highly electrophic phosphomonester. If it is less electrophilic as in the nucleotides, it needs additional binding and pre-orientation by the base obviously by hydro-phobic interactions [1].

I

Uridine 3'-phosphate
$K_m = 9.2 \times 10^{-5}$M
$k_{cat} = 12$ s^{-1}
$k_{cat}/K_m = 1.3 \times 10^5$M^{-1}s^{-1}

II

3'-Deoxyuridine
3'-methanephosphonate
$K_i = 3.9 \times 10^{-2}$M

III

Adenosine 5'-phosphothioate
$K_m = 2.2 \times 10^{-4}$M, $k_{cat} = 0.4$ s^{-1},
$k_{cat}/K_m = 2 \times 10^3$M^{-1}s^{-1}

Since retention is found we can furthermore conclude that the carboxylate group attacks the phosphate opposite to the leaving group and that the water or hydroxylamine or the 5'- position of a second nucleoside enters opposite to the carboxylate group.

Modification studies on the enzyme [1] revealed that besides the carboxylate group only a histidine is essential for the reaction. Activity is lost after photooxidation with a relatively high pK value of 7.6 indication an interaction of an imidazolium cation with a carboxylate group in form of a salt bridge. This is supported by kinetic data in presence of organic solvents increasing the rates by stabilizing the salt bridge and in presence of salts decreasing the rate by destabilizing the salt bridge.

Mg^{2+} is essential. Removal by EDTA is time dependent and leads to an inactive apoenzyme which can be reactivated by different two-valent metal cations. Formation of the inactive apoenzyme is very slow in presence of inhibitors, however, inhibitors do not interfer with the reactivation. This indicates the Mg^{2+} binds to the phosphate group of inhibitors and substrates. Evidence for the stereochemistry of the interaction has been obtained by using nucleotide phosphothioates.

From all these results and kinetic data (here not shown) which are interpreted on the basis that the k_{cat} - value corresponds to the decay of the anhydride while the k_{cat}/K_m value reflects the formation step we postulate a mechanism, shown in the following scheme: The substrate is bound electrostatically to Mg^{2+} and by hydrophobic interaction of the base.

The carboxylate group attacks the phosphorus, simultaneously the imidazolium cation protonates the phosphoryl oxygen to lower activation energy for passing the pentacoordinate transition state. The ester oxygen of the nucleotide is released from an apical position, back-reaction might occur or transfer of the phosphate group to another nucleoside which has

Tentative reaction mechanism of the nucleoside phosphotransferase (B_1, B_2: base; N_1, N_2: nucleoside; *: base binding site).

replaced the first nucleoside. In absence of nucleosides hydrolysis occurs. This mechanism requires retention of the absolute configuration.

A second phosphotransferase from carrots, intensively studied by Chargaff's group [6], although different in M_r (=45 000) and amino acid composition (two nonidentical sub-units) appears to follow the same reaction path. Phosphorylhydroxylamine is formed. Instead of glutamic acid, however, aspartic acid is involved demonstrated by the formation of homoserine after reduction with sodiumcyanoborotritiid [7].

Nuclease S_1 is a monomeric Zn^{2+} enzyme (M_r 36 000) with an optimum at pH 4.5-5.5. It splits the 3'-0-P bond of nucleotides whether X is H, CH_3, OH or OR.

Again we found incorporation of ^{32}P when the protein is denatured in presence of substrate.

Using the same procedure as before we found no phosphorylhydroxylamine, but 2-amino-5-hydroxy-valeric acid indicating formation of an anhydride with glutamic acid.

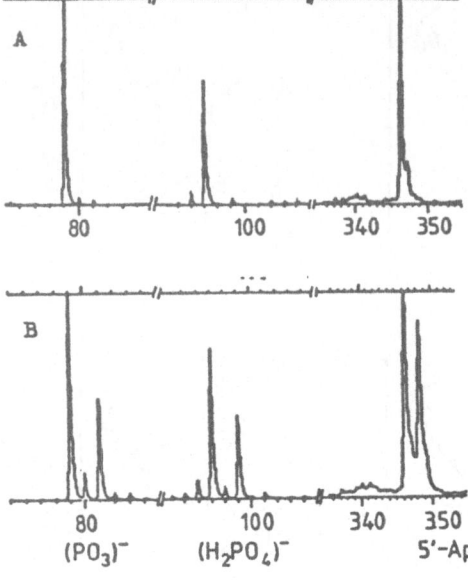

SIMS-spectra of the fragments PO_3^-, $H_2PO_4^-$ and 5'-Ap. A: $^{16}OH_2$ pretreated; B: $^{18}OH_2$ pretreated enzyme.

Since reaction occurs with inversion of absolute configuration (4) the only reasonable interpretation is an attack of the water molecule at the carboxylate group. This could be proved by reaction in $H_2^{18}O$, where ^{18}O should be incorporated in the carboxylate group. After gelfiltration to remove substrates, products and $H_2^{18}O$ we mixed the enzyme, now in $H_2^{16}O$ with 1.5 equivalent of substrate (UpA) in a stopped flow apparatus and analyzed the products by SIMS- spectroscopy. ^{18}O is incorporated in the phosphate groups confirming that water must have attacked the carboxylate group [2].

Furthermore we could demonstrate that a lysine is involved in an intramolecular peptide bond which is formed with a carboxylate group in presence of carbodiimide.

Kinetic data (not shown here) indicate that the rates are lowered the higher the electrophilicity of the phosphate group is (e.g. naphtyl esters). They are higher the more nucleophilic the phosphate oxygens are (e.g. in phosphoric acid esters or phosphonates). We therefore conclude that the substrates bind by the 3'-base and electrostatically with a lysine NH_3^+ while the Zn^{2+} hydrolyses specifically the anhydride at the carboxylate group.

60

REFERENCES

1. A. Billich and H. Witzel, Biol. chem. Hoppe-Seyler, 367; 267-300. 1986.
2. H. Witzel, W. Berg, O. Creutzenberg, A. Karreh in Zink Enzymes (I. Bertini, C. Lucinat, W. Maret, M. Zeppezauer eds.) Birckhauser-Verlag, Basel, 295-306, (1986).
3. J.P. Richard, D.C. Prasher, D.H. Ives and P.A. Frey J. Biol. Chem. 254; 4339-4341, (1979).
4. B.V.L. Potter, P.J. Romaniuk and F. Eckstein J. Biol. Chem. 258; 1758-1760 (1983).
5. D.C. Prasher, M.C. Carr, H.H. Ives, T.-C Tsai and P.A. Frey J. Biol. Chem. 257; 4931-4939 (1982).
6. G. Brawermann and E. Chargaff J. Biol. Chem. 210; 445-454 (1954). E.F. Brunngraber and E. Chargaff J. Biol. Chem. 242; 4834-4840, (1967). R. Rodgers and E. Chargaff J. Biol. Chem 247; 5448-5455 (1972).
7. B. Stelte and H. Witzel Eur. J. Biochem. 155; 121-124 (1986).

PROBING THE MECHANISM OF ENZYME ACTION

THROUGH SPECIFIC AMINO ACID SUBSTITUTIONS

A. J. Wilkinson

Department of Chemistry
University of York
York YO1 5DD, UK

ABSTRACT

The technique of site-directed mutagenesis mediated by specific oligonucleotides is a rapid, simple and powerful tool with which to study the
roles played by particular amino acid side chains in proteins. The information yielded from such studies is in proportion to how much is already
known structurally and functionally about the molecule under investigation.
For DNA gyrase of Escherichia coli the essentiality of the "active site"
tyrosine 122 has been tested by making phenylalanine and serine replacements. For the tyrosyl-tRNA synthetase of Bacillus stearothermophilus, an
enzyme whose three dimensional structure is known at high resolution, the
subtle contributions to binding and catalysis made by residues well removed
from the actual reaction center have been determined. In addition, the
importance for specificity of different types of hydrogen bonds has been
investigated. Studies of this enzyme have shown it is possible to engineer
a mutant with enhanced substrate affinity and these and other studies
indicate that there are good prospects for designing enzymes with greater
stabilities and with altered substrate specificities.

SITE-SPECIFIC RESTRICTION ENDONUCLEASE Sau BMK I FROM

STREPTOMYCES AUREOFACIENS

J. Timko, J. Turňa* and J. Zelinka

Institute of Molecular Biology, Slovak Academy of Sciences
Dúbravská, cesta 21, 842 51 Bratislava, Czechoslovakia

*Comenius University, Department of Biochemistry
Mlynská dolina, Ch I,842 15 Bratislava, Czechoslovakia

Type II restriction endonucleases have been isolated from many bacteria including Streptomycetes and their specifities have been characterized (Roberts, 1985).

From Streptomyces aureofaciens restriction endonucleases Sau I (Timko et al., 1981) and Sau 3239 I (Gašperík et al., 1983; Šimbochová et al., 1986) have been isolated and characterized. In this paper we describe a new restriction endonuclease, Sau BMK I from Streptomyces aureofaciens, strain BM-K, producing about 2500 µg of CTC per ml.

ISOLATION OF Sau BMK I

Identical cultivation conditions and media were used as described by Zelinková and Zelinka (1969). Sau BMK I activity was detected in the supernatant after PEI precipitation and in the sediment after ammonium sulfate precipitation (Figure 1). For the determination of the enzymatic activity the following reaction mixture (50 µl) was used: 10 mmol.l^{-1} Tris-HCl, pH 8.0; 10 mmol.l^{-1} MgCl$_2$; 50 mmol.l^{-1} NaCl; 7 mmol.l^{-1} 2-mercaptoethanol; 0.5 - 1 µg DNA pAT 153 and 1 - 2 µl of the sample. After incubation for 1 - 2 h at 37°C the reaction was stopped by the addition of 5 µl of 0.1 mol.l^{-1} EDTA, 0.05% bromphenol blue and 50% glycerol.

For the isolation 100 g wet cells were used; they were suspended in buffer A (10 mmol.l^{-1} Tris-HCl, pH 7.6; 20 mmol. l^{-1} ammonium sulfate; 5 mmol.l$^-$ 2-mercaptoethanol; 1 mmol.l^{-1} EDTA and 0.02 mmol. l^{-1} PMSF). Nucleic acids were precipitated by PEI to a final concentration of 0.5% and proteins were precipitated with ammonium sulfate to 80% saturation. The proteins were dissolved in buffer A, desalted on a column of Sephadex G-25 (2.5 x 37 cm) and chromatographed on a column of DE-52 cellulose (2.5 x 25 cm) using 500 ml of a linear gradient 20 - 500 mmol.l^{-1} ammonium sulfate in buffer A; fractions of 3.3 ml/5 min. Active fractions were eluted from 60 to 90 mmol.l^{-1} ammonium sulfate, dialysed against buffer A and chromatographed on a column of Blue-Sepharose (1.5 x 18cm) using 100 ml of a linear gradient of 0 - 1.5 mol.l^{-1} KCl in buffer PC (10 mmol.l^{-1} potassium phosphate, pH 7.6; 10 mmol.l^{-1} 2-mercaptoethanol; 1 mmol.l^{-1} EDTA; 0.02 mmol.l^{-1} PMSF and 10% glycerol) to which 100 µg/ml of BSA were added; fractions of 4 ml/20 min. The active fractions were eluted at 1.25 -

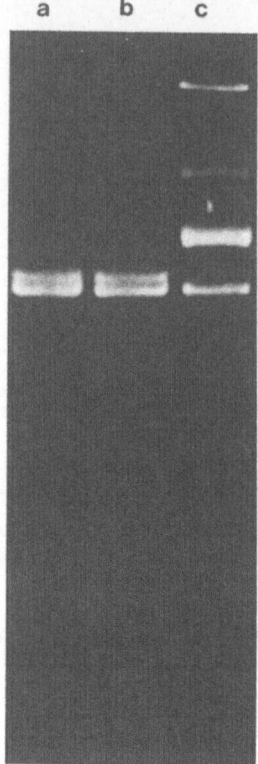

Fig. 1. Electrophoresis of pAT 153 cleaved by the enzyme fraction of Sau
BMK I in 1% agarose gel. a: supernatant after PEI precipitation;
b: sediment after ammonium sulfate precipitation; c: sediment
after ammonium sulfate precipitation without Mg^{2+}.

1.4 mol.1^{-1} KCl, dialysed against buffer PC and chromatographed on a
phosphocellulose P-11 column.

For the elution 40 ml of a linear gradient 0 - 1.0 mol.1^{-1} KCl in PC
buffer was used; fractions 2 ml/10 min. The active fractions were eluted
from 0.25 - 0.4 mol.1^{-1} KCl (Figure 2), pooled and dialysed against 50%
glycerol in PC buffer. The concentrated enzyme was stored at -20°C and was
found to be stable for 6 months. On the average a yield of 10% was obtained
with 300-fold purification (Table 1). Usually 150 - 200 U were isolated
from 1 g of wet cells. After 16 h of incubation of 20 U of the enzyme with
1 μg DNA pAT 153 no presence of contaminating non-specific endonucleases
was observed.

CLEAVAGE OF DNA pAT 153 BY Sau BMK I

For the determination of the cleavage sites on pAT 153 by endonuclease Sau BMK I endonuclease Hae III was used of which the cleavage
sites of plasmid pBR 322 are known (Sutcliffe, 1978). The plasmid pAT 153
is a derivate of plasmid pBR 322 with a deletion of 705 nucleotides (from
1647 to 2352 bp) corresponding to Hae II-B and Hae II-H fragments.

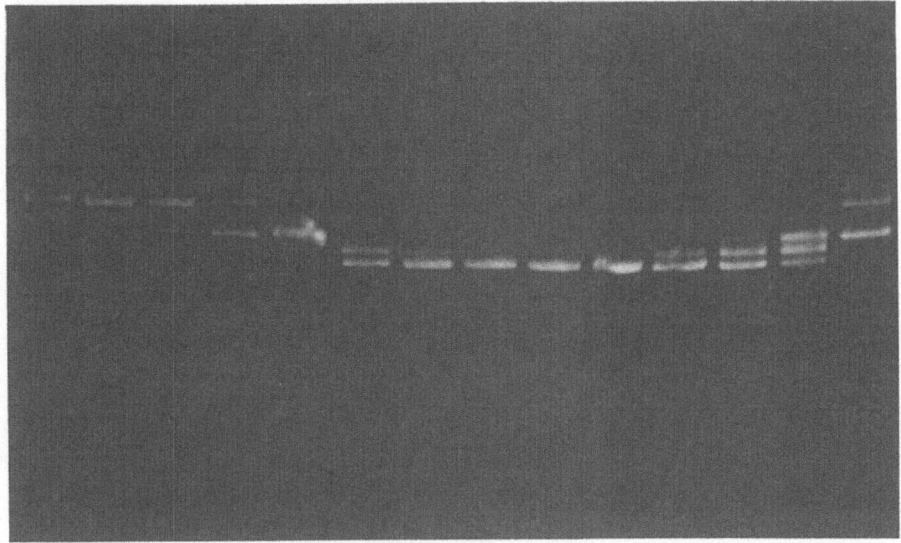

Fig. 2. Electrophoretical pattern of the activity of endonuclease Sau
BMK I in individual fractions after elution from the P-11
phosphocellulose column.

Table 1. Purification of Endonuclease Sau BMK I.

Isolation step	Volume (ml)	Proteins (mg)	Total activity (x10^3, U)	Specific activity (U/µg)	Yield (%)	Purification (fold)
(NH$_4$)$_2$SO$_4$ precipitation	190	1960	142.5	72.7	100	1
Chromatography on DE-52 cellulose	30	63.3	90.0	1421.8	63.1	19.5
Chromatography on Blue – Sepharose	9.6	7.72	33.25	4306.9	23.3	184.8
Chromatography on P – 11 cellulose	8	0.7	16.0	22857.14	11.2	314.4

From the cleavage of the pAT 153 plasmid by Hae III, the double cleavage by Hae III + Sau BMK I and from the known sequence of pBR 322 nucleotides we have established the size of the fragments (Figure 3, Table 2).

As DNA fragments smaller than 0.5 kb cannot be satisfactorily separated on agarose gels, 5% acrylamide gel was used. On Figure 3 the separation of fragments formed by the cleavage of pAT 153 by Hae III and Sau
BMK I can be clearly seen up to fragments Hae III - J + K which is a double
fragment as the difference between them is only one nucleotide (Table 2).

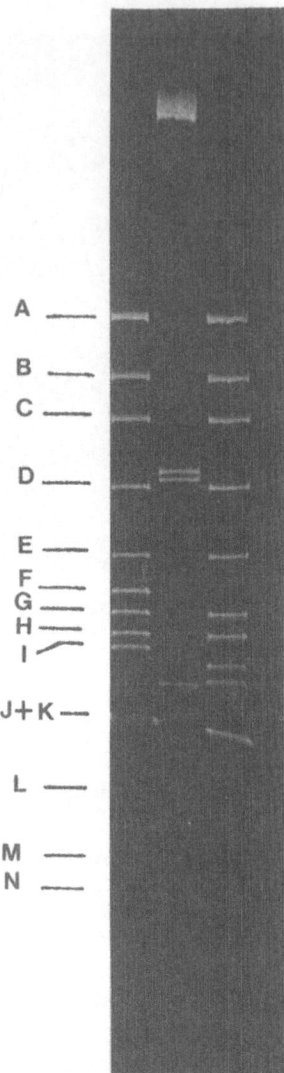

Fig. 3. Restriction analysis of pAT 153 plasmid. Electrophoretic pattern
of part of 5% polyacrylamide gel (40 x 20 x 0.15 cm; 160 V; 16
hours). 1: pAT 153 + Hae III; 2: pAT 153 + Sau BMK I; 3: pAT
153 + Hae III + Sau BMK I.

Further fragments smaller than J and K are not well discernible due to
their small relative molecular weight. By comparing lines 1 and 3 on Figure
3 it can be clearly seen that endonuclease Sau BMK I cleaves fragment Hae
III-F (234 bp) into two fragments, F_2 (173 bp) and F_2 (61 bp). Fragment
Hae III-I (184 bp) is also cleaved into two fragments, I_1 (162 bp) and I_2
(22 bp). Fragments F_1 and I_1 can be clearly seen on Figure 3, line 3 and
are situated closely over the last Sau BMK I-D fragment (160 bp) in line 2.

Based on results presented in Figure 3, Table 2 and known cleavage
sites of pBR 322 we were able to construct the physical map of pAT 153 with

Table 2. Size of fragments of pAT 153 plasmid after cleavage with Hae III and Sau BMK I.

| pAT 153 + Hae III | | pAT 153 + SAU BMK I | |
fragment	b.p.	fragment	b.p.
A	587	A	2 776
B	458	B	395
C	434	C	354
D	339	D	160
E	267		
F	234		
G	213		
H	192		
I	184		
J	124		
K	123		
L	104		
M	89		
N	80		
O	57		
P	54		
R	51		
S	21		
T	18		
U	11		
V	8		

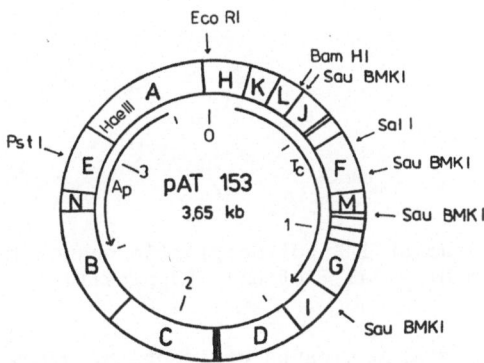

Fig. 4. Physical map of pAT 153 plasmid.

Hae III. We could also determine the cleavage sites of Sau BMK I on plasmid pAT 153 (Figure 4).

It is evident that endonuclease Sau BMK I cleaves plasmid pAT 153 at 4 sites at 401, 769, 929 and 1283 bp forming four fragments A = 2776, B = 395, C = 354 and D = 160 bp.

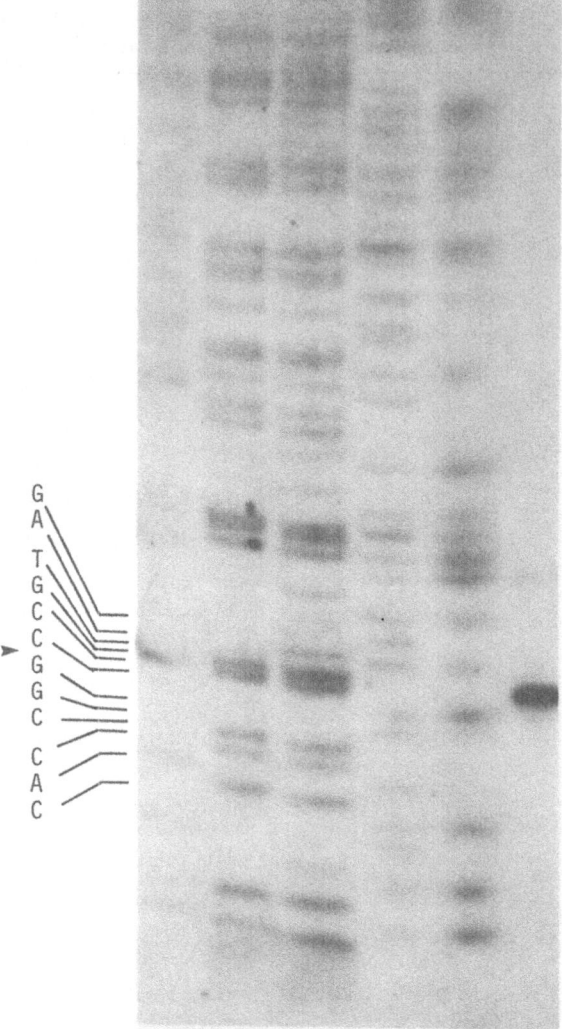

Fig. 5. Autoradiogram of 12% polyacrylamide sequencing gel. Fragment
3'-^{32}P Bam HI - Sal I of pAT 153 plasmid.

Plasmid pAT 153 is also cleaved at identical sites by endonuclease Nae
I from Nocardia aerocolonigenes ATCC 23 870 recognizing the 5'- GCC/GGC -
3' sequence (Roberts, 1985) indicating that Sau BMK I could be an isoschisomer of Nae I. This presumption could be verified only by DNA sequence
analysis.

Sau BMK I SPECIFICITY

Sau BMK I cleaves plasmid pAT 153 at four sites and thus it was easy
to select a suitable fragment of this plasmid containing a cleavage site
for Sau BMK I. According to the physical map of pAT 153 it is advantageous
to use the fragment Bam HI - Sal I of 276 bp for DNA sequencing; this
fragment contains a cleavage site for Sau BMK I close to the Bam HI end.

70

Sau BMK I

$\downarrow$

5´GATCCTCTACGCCGGACGCATCGTGGCC GGCATCAC...AGCG3´

3´GAGATGCGGCCTGCGTAGCACGG CCGTAGTG...TCGCACGT5´

Fig. 6. Part of nucleotide sequence of the fragment Bam HI – Sal I of pAT
153 plasmid (276 bp).

We have cleaved pAT 153 by Bam HI and labeled it on the 3' ends by DNA
polymerase I (Klenow fragment) and ^{32}P dCTP and then we have cleaved the
linearized plasmid by endonuclease Sal I. We have thus obtained two fragments labeled only on one end. These fragments were separated on 5% polyacrylamide gel and the smaller fragment [α– ^{32}P] Bam HI – Sal I was isolated and sequenced by the method of Maxam – Gilbert (1980) through the
Sau BMK I site. The labeled fragment was simultaneously cleaved by Sau BMK
I and loaded on the sequencing gel side by side with individual reaction
products (Figure 5). By establishing the sequence of the 3' end of the
fragment ^{32}P Bam HI – Sal I and by comparing it with the fragment formed by
the cleavage of ^{32}P Bam HI – Sal I by endonuclease Sau BMK I it became
evident that cleavage takes place between G and C under the formation of
blunt ends.

Taking into account the cleavage sites of Sau BMK I on pAT 153 plasmid
(Figure 4) and the fact that two-fold rotational symetry is present only in
three nucleotides on the right and left from the cleavage site (Figure 6),
it is possible to determine the exact recognition sequence and the cleavage
site of endonuclease Sau BMK I. On the double stranded DNA this endonuclease recognizes the sequence

$$5' - G\ C\ C\ /\ G\ G\ C - 3'$$
$$3' - C\ G\ G\ /\ C\ C\ G - 5'$$

and cleavage takes place as indicated and blunt end DNA fragments are
formed. Thus endonuclease Sau BMK I is an isoschisomer of endonuclease Nae
I isolated from Nocardia aeorocolonigenes (Comb and Wilson, unpublished
observation).

REFERENCES

Gašperík, J., Godány, A., Hostínová, E., Zelinka, J. (1983) Biológia
    (Bratislava) 38, 315–319.
Maxam, A.M., Gilbert, W. (1980) Methods in Enzymology, Vol. 65, L. Grossman
    and K. Moldave (eds.), Academic Press, (New York) 499–560.
Roberts, R.J. (1985) Nucl. Acids Res. 13, r165–r200.
Sutcliffe, J. G. (1978) Cold Spring Harbor Symp. Quan. Biol. 43, 77–90.
Šimbochová, G., Timko, J., Zelinková, E. Zelinka, J. (1986) Biológia
    Bratislava) 41, 357–365.
Timko, J., Horwitz, A.H., Zelinka, J., Wilcox, G. (1981) J. Bacteriol. 145,
    873–877.
Zelinková, E., Zelinka, J. (1969) Biológia (Bratislava) 24, 456–461.

THE EFFECTS OF N4- AND C5-METHYLATION OF CYTOSINE ON DNA

PHYSICAL PROPERTIES AND RESTRICTION ENDONUCLEASE ACTION

V. Butkus, S. Klimašauskas, L. Petrauskienė, Z. Manelienė,
L. E. Minchenkova,* A. K. Schyolkina and A. Janulaitis

Institute of Applied Enzymology, Fermentu 8, Vilnius 232028
Lithuanian SSR, USSR

*Institute of Molecular Biology, USSR Academy of Sciences
Vavilova 32, Moscow 117984, USSR

INTRODUCTION

Not long ago a new type of methylase, generation N4-methylcytosine
(4mC) in DNA was isolated in our laboratory from Bacillus centrosporus
(Janulaitis et al.,1983). At present a number of the 4mC type methylases
are isolated (Janulaitis et al., 1984; Butkus et al., 1985; 1987) and a
wide spread of N4-methlcytosine in DNA of thermophilic and mesophilic
bacteria is well documented (Ehrlich et al., 1985,1987). Since the two
methylated cytosine bases occure in native DNA, a comparison of the effects
of N4- and C5-methylation on DNA physico-chemical properties and enzyme
action seems to be of great interest.

RESULTS AND DISCUSSION

We used in our experiments synthetic oligodeoxyribonucleotides which
were synthesized by the modified phosphotriester method (Stawinski et al.,
1977). N4-Methylcytosine containing oligonucleotides were prepared following the published method (Petrauskienė et al., 1986).

Helix-coil transition. The melting curves (Figure 1) of the d-
(GGACCCGGGTCC), d(GGA5mCCCGGGTCC) and d(GGA4mCCCGGGTCC) dodecanucleotides
disclosed a differential effect of C5- and N4-methylation of cytosine on
the double helix stability. In agreement with the previously reported data
(Kemp et al., 1976) 5-methylcytosine increased the stability of the oligonucleotide duplexes. However, N4-methylcytosine caused a profound dest-
abilization of the double helix (Figure 1).

The stereochemistry of N4-methyl group of cytosine in the heterocyclic
moiety is quite similar to that of N6-methyladenine, as both of them replace one of the two hydrogen atoms of the exocyclic amino group which are
involved in the intrastrand Watson-Crick hydrogen bond linkage between
complementary nucleotides. The oligonucleotide duplexes containing N6-
methyladenine are characterized by a remarkably lower Tm value than unmethylated ones (Engel et al.,1978). Therefore it may be assumed that poorer
statistical capabilities of methylamino group to adopt a required trans-

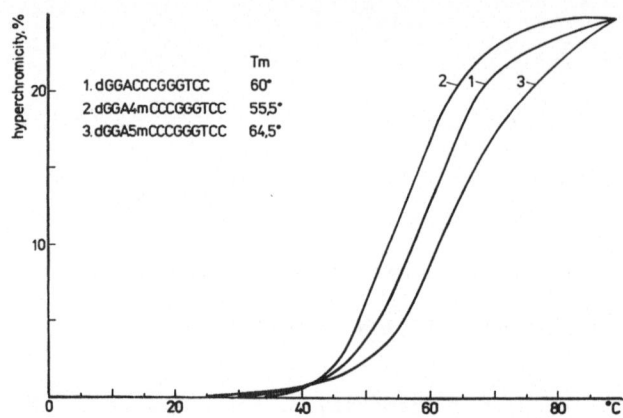

Fig. 1.   The melting curves of the dodecanucleotides d(GGACCCGGGTCC) (1),
          d(GGA4mCCCGGGTCC) (2) and d(GGA5mCCCGGGTCC) (3). Conditions: 0.1
          M NaCl, 0.2 mM EDTA, the heating rate 0.7 deg/min. Concentrations
          of the duplexes were 0.23-0.25 mM in nucleotides.

conformation, would tend to force the helix<-->coil equilibrium towards the
single stranded species.

     The B-->A transition. The same dodecanucleotides were also compared
in respect of their ability to form the A-helix of DNA. The diagram of the
B<-->A equilibrium as a function of trifluoroethanol (TEE) concentration is
presented in Figure 2. The midpoints of the transition in the case of
d(GGACCCGGGTCC) was at 61.4% and in the case of d(GGA5mCCCGGGTCC) and
d(GGA4mCCCGGGTCC) at 65.3% of TFE, respectively. Thus, 5mC and 4mC altered
the stability of A-helix in the oligonucleotide duplexes negatively.

     The B-->Z transition. The circular dichroism spectra at low salt
concentrations of the three investigated octanucleotides with the alternating CG sequence - d(CGCGCGCG), d(CG5mCGCGCG) and d(CG4mCGCGCG) were
characteristic to that of DNA in B-form (Figure 3). At higher salt concentrations the CD spectra, characterized by positive maxima at 265-270 nm and
important negative maxima at about 290 nm, were quite different, indicating
the predominance of Z-form (Figure 3).

     The plots of the Z-conformer fraction against the molarity of NaCl
revealed that 4mC facilitate the B-->Z transition (Figure 4). The half-
transition in our experiments was at 2.3 M for the d(CGCGCGCG) and at 1.9
and 2.1 M of NaCl for 5mC and 4mC containing species, respectively. The
Z-stabilizing effect of 5mC is well documented on a polymer (Klysik et al
1983) as well as oligomer (TRAN-DINH et al 1983) levels. The capability
of N4-methylcytosine, a new biomethylated base, to stimulate the B-->Z
transition is reported here for the first time.

     Chemical stability. The residues of 5mC in DNA are known to easily
undergo spontaneous deamination (5mC-->T) resulting in the substitution of
the AT pair for the GC pair, which causes point mutations of DNA. A comparative stability of 4mC in this respect was recently found being
4mC>C>5mC (EHRLICH et al., 1986). In addition, resulting 4mC-->U deamin-

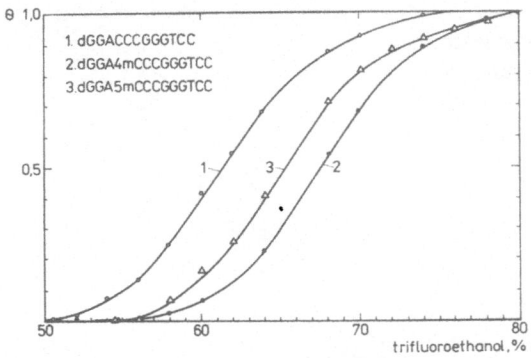

Fig. 2. The B-->A transition curves of the dodecanucleotides
d(GGACCCGGGTCC) (1), d(GGA4mCCCGGGTCC) (2) and d(GGA5mCCCGGGTCC)
(3). Θ is for the fraction of A form which was estimated from the
CD spectra. Conditions: -20°C, 1 mM NaCl and 0.1 mM EDTA.
Concentrations of the duplexes 0.23-0.25 mM in nucleotides.

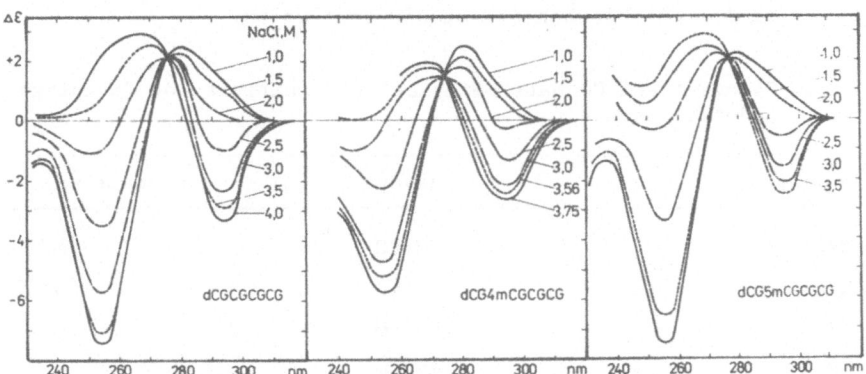

Fig. 3. The circular dichroism spectra of the octanucleotides d(CGCGCGCG)
(1), d(CG4mCGCGCG) (2) and d(CG5mCGCGCG) (3). At different
concentrations of NaCl (the curves' labels correspond to NaCl
molarities). Conditions: 20°C, 0.4 mM EDTA; concentration of the
duplexes were 0.2-0.3 mM in nucleotides.

ation did not lead to a DNA mutation since uridine could be removed by
reparation enzymes.

Interaction of 4mC and 5mC containing substrates with restriction
endonucleases. Recognition and cleavage specificities of the investigated
endonucleases and their cognate methylated sites are presented in Table 1.

As substrates we used the synthetic parent dodecanucleotide
d(GGACCCGGGTCC) and a set of dodecanucleotides containing all possible
substitutions of either 4mC or 5mC for C in the CCCGGG sequence. All the
endonucleases gave expected cleavage products which were analyzed by homo-

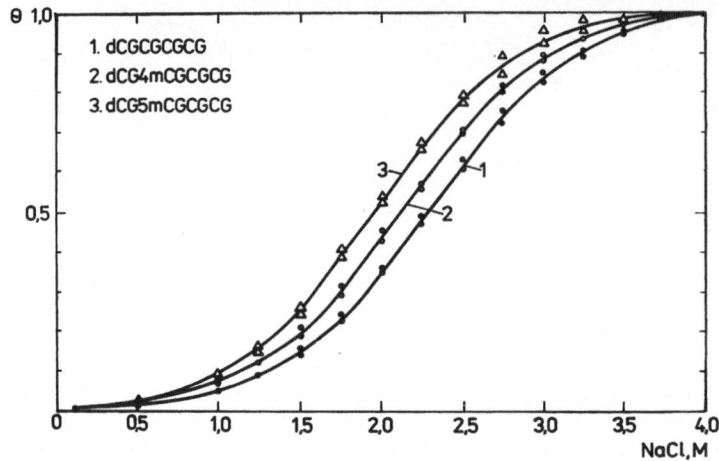

Fig. 4. The B to Z transition curves of the octanucleotides d(CGCGCGCG) (1), d(CG4mCGCGCG) (2) and d(CG5mCGCGCG) (3). Θ is for the fraction of the Z-conformer, calculated from UV spectra after ratioing of absorbancies at 260 and 295 nm.

Table 1. Cleavage and Cognate Methylation Specificity of Investigated Endonucleases.

Endonuclease	Cleavage	Cognate methylation
MspI	CC↓GG	5mCCGG
HpaII	CC↓GG	C5mCGG
SmaI	CCC↓GGG	?
XmaI	C↓CCGGG	?
Cfr9I	C↓CCGGG	C4mCCGGG

chromatography (Figure 5). The methylated oligonucleotide duplexes were tested under the conditions found optimal for unmethylated substrates.

The data presented in Table 2 indicate clearly that non-cognate methylations, occuring in a recognition sequence, display a wide range of cleavage inhibition. For example, the introduction of N4-methylcytosine blocked the action of most endonucleases under investigation (HpaII, SmaI, XmaI and Cfr9I). The inhibitory effect of 5-methylcytosine was significantly lower. Although both the methyl groups were exposed in the major groove of B-DNA, the endonucleases strictly differentiated these methylations, displaying different sensitivity to the substrates. Our assumption would be that N4-methylation of the cytosine base not only hinders stericaly the interaction with enzymes but also destroys the essential C:G base pair recognition contact. In this respect methylation of cytosine at the N4-position is essentially different from methylation at the C5-position.

CONCLUDING REMARKS

Methylated base residues in DNA may serve as specific signals for DNA-protein interactions. However, they must meet certain requirements to

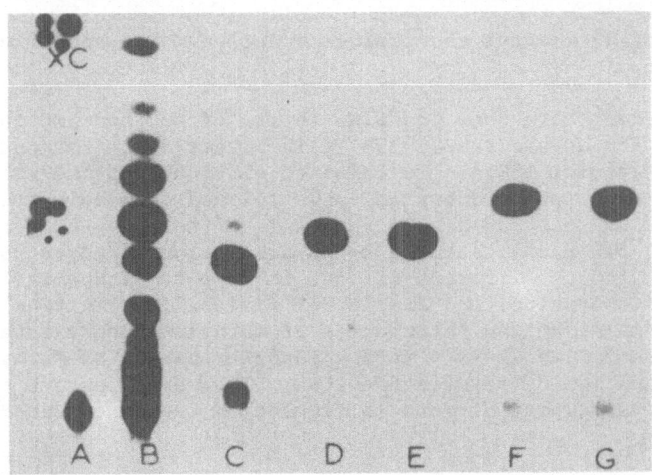

Fig. 5. Cleavage of the 5'-^{32}P-labeled dodecanucleotide substrate d(GGACCCGGGTCC) with restriction endonucleases under investigation. Line (A) — initial dodecanucleotide; (B) — VPDE hydrolyzate; (C) — SmaI; (D) — HpaII; (E) — MspI; (F) — XmaI; (G) — Cfr9I. The reaction products were analyzed by TLC on DEAE—cellulose in homomix VI.

Table 2. Cleavage of methylated substrates with restriction endonucleases. The cleavage efficiency is expressed in per cents.

| Site | Restriction endonuclease | | | | |
	MspI	HpaII	SmaI	XmaI	Cfr9I
CCCGGG	97	94	92	91	51
5mCCCGGG	96	92	5	1	5
C5mCCGGG	0	20	70	10	8
CC5mCGGG	50	0	0	23	46
4mCCCGGG	89	80	0	0	0
C4mCCGGG	31	0	0	0	0
CC4mCGGG	16	0	0	0	0

perform the regulatory function. So, a methylated moiety should be accessible to proteins, i.e. it should be exposed in the major or at least in the minor groove of DNA double helix; alternatively, methylation can induce a conformational shift of DNA, causing a differentiated interaction with a protein molecule. Further, methylation should not impede the main function of DNA — storage and transfer of the genetic message; for this methylation must not lead to a significant drop in the chemical stability of a polynucleotide chain as well as of a modified nucleotide itself.

All the biomethylated DNA bases, i.e. 4mC, 5mC and also N6—methyladenine, satisfy most of the discussed criterions. In addition, neither of them is a product of the direct chemical methylation of DNA or polynucleotides (PEGG, 1983). Due to this feature cells are protected from the interference of accidental chemical methylations with specific signals of biomethylation.

Though recent data show the wide spread of 4mC and 5mC in DNA of bacteria (Ehrlich at al., 1985, 1987), in higher organisms only 5mC is predominant (Hattman, 1981). We tried to elucidate up the reasons of this evolutionary selection by comparing physical and chemical effects of the methylation on DNA structure. On the basis of the present work it can be concluded that 5mC facilitate the B-->Z transition stronger as compared to 4mC, and also produce different effects in respect of chemical stability (spontaneous deamination) and double helix stability (melting). One more difference revealed by the interaction of methylated substrates with restriction endonucleases is a strong interference of N4-methylcytosine with the recognition of methylated sites. The named feature could also be essential if overlapping of genomic regulation regions occured.

REFERENCES

1. V. Butkus, S. Klimašauskas, D. Keršulytė, D. Vaitkevičius, A. Lebionka, and A. Janulaitis, Nucleic Acids Res., 13;5727-5746 (1985).
2. V. Butkus, S. Klimašauskas, L. Petrauskienė, Z. Manelienė, A. Lebionka and A. Janulaitis, Biochim. Biophys. Acta (in press) (1987).
3. M. Ehrlich, M.A. Gama-Sosa, L.H. Carreira, L.G. Ljungdahl, K.C. Kuo, and C.W. Gehrke, Nucleic Acids Res., 13,,1399-1413 (1985).
4. M. Ehrlich, G.G. Wilson, K.C. Kuo, and C.W. Gehrke, J. Bacteriol., 169,939-943 (1987).
5. J. D. Engel and P.H. von Hippel, J. Biol. Chem., 253, 927-934 (1978).
6. S. Hattman, in: "The Enzymes," P.D. Boyer, ed., Academic Press, New York, v. 14, 517-548 (1981).
7. A. Janulaitis, S. Klimašauskas, M. Petrušytė, and V. Butkus, FEBS Lett., 161, 131-134 (1983).
8. A. A. Janulaitis, P.S. Stakėnas, M.P. Petrušytė, J.B. Bitinaitė, S.J. Klimasauskas and V.V. Butkus, Molekulyarnaja Biologiya (Moscow), 18, 115-129. (1984).
9. J. D. Kemp, and D.W. Sutton, Biochim. Biophys. Acta, 425, 148-156 (1976).
10. J. Klysik, S.M. Stirdivant, C.K. Singelton, W. Zacharias, and R.D. Wells, J. Mol. Biol., 168, 51-71 (1983).
11. A. E. Pegg, Rec. Res. Cancer Res., 126, 781-790 (1983)
12. L. J. Petrauskienė, S.J. Klimašauskas, V.V. Butkus and A.A. Janulaitis, Bioorg. Khim. (Moscow), 12, 1597-1603 (1986).
13. J. Stawinski, T. Hozumi, S.A. Narang, C.P. Bahl, and R. Wu, Nucleic Acids Res., 4, 353-371 (1977).
14. S. Tran-Dinh, J. Taboury, J.M. Neumann, T. Huynh-Dinh, B. Genissel, C, Gouyette and J. Igolen, FEBS Lett., 154, 407-410 (1983).

REGULATION OF RNA SYNTHESIS IN YEAST:

THE STRINGENT CONTROL OF RIBOSOMAL RNA SYNTHESIS

L. W. Waltschewa and P. V. Venkov

Institute of Molecular Biology
Bulgarian Academy of Sciences
1113 Sofia, Bulgaria

In procaryotic and yeast cells the restriction of RNA syntheis that occurs on starvation of an auxotrophic strain for required amino acids has been called the stringent control. In response to being deprived of a required amino acid these cells rapidly shut off protein synthesis and synthesis of ribosomal RNA (rRNA) (Shulman et al., 1977). The existence of relaxed mutants relatively unresponsive to amino acid deprivation has led to extensive studies of the molecular mechanism of the stringent control in bacteria. Recently, we isolated and characterized the first relaxed mutant of yeast called Saccharomyces cerevisiae SY15 (Waltschewa et al., 1983). Upon starvation for required tyrosine the SY15 mutant cells synthesize and accumulate rRNA at rates comparable to those in the non-starved controls. In crosses with wild type strains the relaxed phenotype was found to segregate 2:2 demonstrating the existence of one nuclear mutation (designated str) confering the relaxed genotype in the SY15 mutant (Stateva and Venkov, 1984).

During our studies we observed that even after long starvation for required amino acids the SY15 cells incorporate labeled amino acids at about 10% of the control rate (Waltschewa et al., 1983). It is of extreme importance to know if this incorporation reflects synthesis of functionally active protein molecules or is an artefact. In the former case, the synthesis of proteins, even at reduced rate, could determine a phenotype similar to that observed in the SY15 relaxed mutant and could lead to wrong conclusions about its genotype.

In this communication we present data showing that the reduced incorporation of the labeled amino acids in the relaxed SY15 mutant starved for required amino acids does not reflect true protein synthesis and most probably is due to starvation-induced translational errors.

EXPERIMENTAL

Strains

The Saccharomyces cerevisiae strains used in this study are listed in Table 1. The stringent A364 strain is parental to SY15 relaxed mutant (Waltschewa et al., 1983). The VIIA and XVD relaxed strains are members of tetrades obtained for the genetic analysis of the SY15 mutant (Stateva and Venkov, 1984). The relaxed strains 23 and 10 were obtained from the progeny of a cross between SY15 relaxed mutant and a wild type stringent strain.

Cultivation

All strains were grown at 30°C to middle exponential phase in minimal media (Shulman et al., 1977) supplemented with the nutritional requirements of every particular strain. To initiate starvation the cells were washed three times in cold minimal media without amino acids and suspended in minimal media supplemented with the necessary amino acids except the one for which the cells will starve the next 90 minutes.

Protein and rRNA synthesis

Protein synthesis into whole cells from 1 ml culture was measured as incorporation of ^{3}H cpm (^{3}H-Amino Acid Mixture, 37 MBq/ml, Amersham) in hot TCA precipitable material (Waltschewa et al., 1983). For determination of rRNA synthesis, total RNA was extracted from cells labeled for 30 minutes with (5-^{3}H)uracil (880 GBq/mmole, Chemapol, Prague) and the rRNA purified using the LiCl procedure. The synthesis of rRNA was measured by estimating the cpm incorporated into 50 μg rRNA for 30 minutes and expressed as the rate of rRNA synthesis in starved cells relative to the non-starved controls set at 1.00.

Induction of Enzymes

The synthesis of arginase was induced by transferring the cells from media containing 0.02 M $(NH_4)_2SO_4$ to media with 1 mg per ml arginine as a carbon source. The enzyme activity was measured as already described (Messenguy et al., 1971). Invertase synthesis was induced by replacing the media containing repressing (2.0% wt/vol) level of glucose with media supplemented with derepressing (0.1% wt/vol) level of glucose. The amount of cytoplasmatic invertase was measured using described procedure (Perlman et al., 1984). The activity of inducible acid phosphatase was assayed using intact cells as enzyme source and p-nitrophenylphosphate as substrate as described by Toh-E et al. (1973). To induce the synthesis of acid phosphatase the cells cultivated in high phosphate media were harvested, washed and suspended in the same volume of media lacking phosphate.

Table 1. Saccharomyces Cerevisiae Strains

Strain	Genotype [a]	Source
A364	MATa,gal1 ade2 ura1 his7 lys2 tyr1	L. Hartwell
SY15	MATa,gal1 ade2 ura1 his7 lys2 tyr1 Arg⁻ Met⁻ Srb⁻ Ts⁻ Str⁻	P. Venkov
VIIA	MATa,ade2 ura1 lys2 tyr1 His⁻ Met⁻ Str⁻	P. Venkov
XVD	MATα,ura1 his7 Arg⁻ Str⁻	P. Venkov
23	MATα,his3 tyr1 Str⁻	This study
10	MATα,his3 tyr1 lys2 Str⁻	This study

a: Str⁻ is a symbol for the mutation in the SY15 strain conferring its relaxed phenotype (Stateva and Venkov, 1984).

RESULTS AND DISCUSSION

The relaxed SY15 mutant was crossed to different wild type strains, the obtained diploids were sporulated and several haploids with relaxed control were identified from the progeny (Table 1). The genotypes of these relaxed strains were determined by classical genetic methods. The relaxed strains VIIA, XVD, 23, 10 and the originally isolated SY15 relaxed mutant and its parental A364 stringent strain were cultivated in the corresponding minimal media, the cultures were divided and half of the cells were starved for a required amino acid. Protein and rRNA synthesis were measured in starved and non-starved control cells and the results are presented in Table 2. As expected, the deprivation of a required amino acid shut off rRNA synthesis in the stringent A364 cells while in all relaxed strains the synthesis of rRNA is relatively less affected. Under the same culture regimens, protein synthesis in the stringent A364 and the relaxed 10 strains is completely inhibited. However, the incorporation of labeled amino acids in hot TCA precipitable material continues in the remaining relaxed strains at rates representing 10-20% of the non-starved control. Since this property was not found in stringent segregants (data not shown), we are inclined to think that the low level incorporation of labeled amino acids in cells starved for a required amino acid is a property which might appear in relaxed strains.

A similar observation was reported in studies with E. coli (O'Farrell, 1978). It was found that starvation of relaxed (rel⁻) cells for a required amino acid leads to missence errors and premature termination of protein synthesis. As a result, a complex group of new small polypeptides was accumulated in starved rel⁻ cells. However, starvation-induced translational errors were not detected in a stringent (rel⁺) strain. Since

Table 2. Protein and rRNA Synthesis in the Stringent A364 and Strains with Relaxed Control

| Strain[a] | Regimen[b] | Relative Rates[c] of | |
		Protein Synthesis	rRNA Synthesis
A364	Control	1.00	1.00
A364	-Tyrosine	0.03	0.06
SY15	Control	1.00	1.00
SY15	-Tyrosine	0.16	0.81
VIIA	Control	1.00	1.00
VIIA	-Tyrosine	0.16	0.52
XVD	Control	1.00	1.00
XVD	-Arginine	0.21	0.95
23	Control	1.00	1.00
23	-Tyrosine	0.21	0.80
10	Control	1.00	1.00
10	-Tyrosine	0.02	0.66

a: SY15 is a relaxed mutant isolated from A364 stringent
   strain, VIIA, XVD, 23 and 10 are relaxed strains
   isolated from crosses of SY15 to wild type strains.
b: Non-starved controls and cells starved 90 minutes for a
   required amino acid were labeled for 30 minutes with
   5 μCi/ml of either ³H-Amino Acid Mixture or ³H-uracil.
c: Values expressed as a ratio of cpm in the starved
   culture to the controls, set at 1.00.

protein synthesis is a process principally identical in E. coli and S. cerevisiae, one can assume that the incorporation of amino acids on the relaxed yeast strains starved for a required amino acid reflects only a residual elongation of polypeptide chains due to translational errors. Such "protein synthesis" will permit some incorporation of labeled amino acids in hot TCA precipitable material; however, it will never produce functionally active protein molecules.

We verified this suggestion by measuring the synthesis of different inducible enzymes in S. cerevisiae. We chose this approach because the synthesis of an inducible enzyme involves both, expression of the corresponding gene(s) and translation of its mRNA into functionally active protein molecules. Since starvation of S. cerevisiae cells for a required amino acid has little effect on the synthesis of mRNA (Shulman et al., 1977; Waltschewa et al., 1983), the induction of an inducible gene leads to its transciption in mRNA molecules even under starvation conditions. Therefore, the appearance of the corresponding enzyme activity after the induction of starved cells is a measure for the proper translation of mRNA into functionally active protein molecules.

In the experiments described here, stringent and relaxed cells already starved for a required amino acid were induced for the synthesis of either arginase, or invertase, or acid phosphatase. At the start of the induction the required amino acid was added to half of the culture (controls), while the rest of the cells were induced under starvation conditions. In S. cerevisiae arginase, invertase and acid phosphatase are inducible enzymes which synthesis is regulated at transcriptional level (Messemguy et al., 1971; Perlman et al., 1984; Toh-E et al., 1973). Therefore the appearance of enzyme activity after the induction is not due to modulation of pre-existing protein molecules and reflects "de novo" synthesis of mRNA and its proper translation. From the data presented in Tables 3, 4 and 5 it is evident that the induction of the control cells leads to synthesis of arginase, invertase or acid phosphatase. Contrary to this, there is no synthesis of these enzymes after the induction of cells starved for a required amino acid. It should be noted that virtually no enzyme activity was found in starved cells of the relaxed strains SY15, VIIA, XVD and 23 despite the low level incorporation of labeled amino acids detected in these strains in other experiments.

Table 3. Induction of Arginase

Strain	Regimen	Enzyme Activity (Units)[a] Minutes after the induction		
		30	60	90
A364	Control	22	49	70
A364	−Tyrosine	5	4	5
SY15	Control	38	72	112
SY15	−Tyrosine	6	5	4
23	Control	40	68	98
23	−Tyrosine	8	8	10
10	Control	18	53	78
10	−Tyrosine	3	4	3

a: Arginase specific activity is expressed in µmole of urea produced per hour and mg protein.

Table 4. Induction of Invertase

Strain	Regimen	Invertase Activity (Units/10^7 Cells) [a]			
		Minutes After the Induction			
		30	60	90	180
A364	Control	0.2	0.5	0.9	2.2
A364	-Tyrosine	0.0	0.0	0.0	0.0
SY15	Control	0.4	0.6	1.1	2.3
SY15	-Tyrosine	0.0	0.0	0.0	0.0
23	Control	0.1	0.3	0.8	1.8
23	-Tyrosine	0.0	0.0	0.0	0.0
10	Control	0.5	1.1	1.7	2.9
10	-Tyrosine	0.0	0.1	0.0	0.0

a: 1 Unit is the amount of enzyme producing 1 μmole of glucose per minute at 30°C.

These results clearly show that starvation for a required amino acid inhibits completely the synthesis of functionally active protein molecules in yeast. The low level incorporation of labeled amino acids found after starvation in some relaxed strains does not represent true protein synthesis and is most likely due to translational errors. This conclusion is further supported by the fact that the low level incorporation of amino acids is strain-dependent: strain 23 does, and strain 10 does not incorporate amino acids under starvation conditions although both strains derive from the same cross between SY15 relaxed and a stringent strains.

Table 5. Induction of Acid Phosphatase

Strain	Regimen	Acid Phosphatase Activity (Units/10^7 Cells) [a]				
		Hours After the Induction				
		10	12	14	16	18
A364	Control	0.09	0.12	0.20	0.28	0.38
A364	-Tyrosine	0.00	0.00	0.00	0.00	0.00
SY15	Control	0.10	0.14	0.19	0.26	0.35
SY15	-Tyrosine	0.00	0.00	0.00	0.00	0.00
23	Control	0.11	0.18	0.25	0.32	0.41
23	-Tyrosine	0.00	0.00	0.00	0.01	0.01
10	Control	0.05	0.12	0.19	0.25	0.29
10	-Tyrosine	0.00	0.00	0.00	0.00	0.00

a: 1 Unit is the amount of enzyme which liberates 1 μmole of p-nitrophenol per minute.

REFERENCES

O'Farrel, P. H., 1978, Cell, 14:545-557.

Messenguy, F., Pennicky, M., and Wiame, J. M., 1971, Eur. J. Biochem., 22:277-286.

Perlman, D., Raney, P., and Havorson, H. O., 1984, Mol. Cell. Biology, 4: 1682-1688.

Shulman, R. W., Sripati, C. E., and Warne, J. R., 1977, J. Biol. Chem., 252:1344-1349.

Stateva, L. I., and Venkov, P. V., 1984, Mol. Gen. Genet., 195:234-237.

Toh-E, A., Ueda, J., and Oshima, Y., 1973, J. of Bacteriol., 113:727-738.

Waltschewa, L., Georgiev, O., and Venkov, P., 1983, Cell, 33:221-230.

IN VITRO STUDY ON RIBOSOME DEGRADATION IN RAT LIVER

E. N. Nikolov and T. K. Nikolov

Institute of Molecular Biology
Bulgarian Academy of Sciences, 1113
Sofia, Bulgaria

Degradation of ribosomes in mammalian cells is a well regulated process, but surprisingly little is known of how ribosomes are disposed of. We have previously shown that the great augmentation of ribosomes, found in the cytoplasm of rat liver cells after partial hepatectomy (PH) (Dabeva and Dudov, 1982), is partly due to a dramatic decrease in the rate of ribosome degradation (Nikolov et al., 1983; 1987). 18S and 28S rRNA and ribosomal proteins apparently turn over with identical half-lives both in the proliferative and post-proliferative phases of liver regeneration (Nikolov and Dabeva, 1983; Nikolov et al., 1987), suggesting that the ribosome is degraded as a unit, perhaps as an intact ribosome or as 80S monomers out of the ribosome cycle, and that the regulation of ribosome breakdown is on the level of the whole particle. There is no known mechanism to account for altered rates of ribosome degradation. In resting cells a larger proportion of ribosomes is found "out-of-cycle" as single ribosomes (Ruthland et al., 1975) and it has been proposed that these provide a pool for ribosome destruction (Perry, 1973). Alternatively, degradation mechanisms may act on the ribosomal subunits formed by dissociation of the 80S ribosomes (Melvin et al., 1976; Melvin and Keir, 1978). The polysome size-distribution patterns of regenerating liver indicate an increase in the percentage of polyribosomes, with a relative reduction of monomers and small oligomers (Cammarano et al.,1965; Rizzo and Webb, 1968) and hence ribosome degradation might be expected to decrease in growing cells.

We have turned to in vitro experiments, which could suggest possible mechanisms of ribosome destruction. Our specific aim was to establish whether single ribosomes, which are regarded as a reserve pool temporarily not participating in protein synthesis and whose quantity in the cell is modulated by different functional states, are the target for degradation and if they are involved in the control of ribosome destruction.

METHODS

All studies were performed on $190 \pm 10g$ Wistar male albino rats. Free and membrane-bound polyribosomes were prepared according to Ramsey and Steele (1979) and separated into polysomal and monosomal fractions. Auto-degradation of polyribosomes or monoribosomes was assayed in 1ml of a mixture containing 25 to 50 A_{260nm} units of ribosomes and 1mM cycloheximide in Buffer A (25mM Tris-HCl, pH 7.6, 50mM KCl, 5mM $MgCl_2$, 6mM 2-mercap-

toethanol) at 37°C. Aliquots were removed from the incubation mixture at
various times, treated with SDS-urea and the rRNA was analyzed by agar-urea
gel electrophoresis (Dudov et al.,1976). Another portion was further
subjected to sucrose density centrifugation and the changes in the profiles
were evaluated. Single ribosomes not engaged in protein synthesis were
prepared from post-mitochondrial supernatant according to Storb and Martin
(1972). The remaining methods appear in legends.

RESULTS AND DISCUSSION

Figure 1 shows typical profiles of free and membranebound polyribosomes obtained from rat liver. The two ribosomal populations prepared
by the method adopted in this study, have a well preserved structure and
little cross-contamination. The peaks corresponding to monomers and dimers
(fraction 1) and the peaks corresponding to free and membrane-bound polysomes (fractions 2 and 3) were collected separately and pelleted by centrifugation.

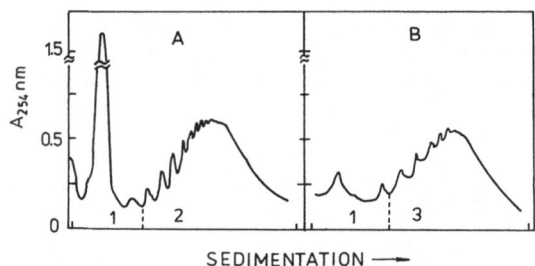

Fig. 1. Sedimentation profiles of free (A) and membrane-bound (B)
polyribosomes on sucrose gradients. Each absorbance profile was
obtained from one liver. Triplicate experiments gave similar
results. Ribosomal suspensions (25 and 15 A_{260nm} units for (A)
and (B) respectively) were layered on 10 to 45% linear sucrose
gradients (wt/v) in gradient buffer (50mM Tris-HCl, pH 7.6, 75mM
KCl, 20mM $MgCl_2$) and centrifuged at 26000 rev/min for 135 min at
2°C in a SW 27 rotor of a Beckman L5 65 ultracentrifuge.
Gradients were monitored at 254 nm using a u.v. spectrophotometer
(LKB Uvicord).

Breakdown of Monoribosomes

When ribosomal particles from fraction 1 (containing approx. 75% of
monomers) were incubated in autodegradation buffer A, no measurable breakdown to subunits occured in 30 min. After 2 hr the proportion of 80S ribosomes had decreased only to 45% with a corresponding increase in 60S subunits (Figure 2A). However, the derived 40S subunits were unstable: they
gave rise to material sedimenting near the top of the gradient. This
"spontaneous" degradation of the 40S particles may reflect only a limited
breakdown (or unfolding) because the RNA component was degraded only to
more rapidly migrating species (Figure 3A). Upon longer incubations the
ribosomes were rapidly converted to subunits, relatively few 40S subunits
been recovered (Figure 2B). The proportion of 80S monoribosomes decreased
from 75 to less than 10% in 4 hours of incubation with a concomitant relatively slow destruction of 60S subunits and further conversion of 40S
subunits to more slowly sedimenting material (Figure 2B). A complete

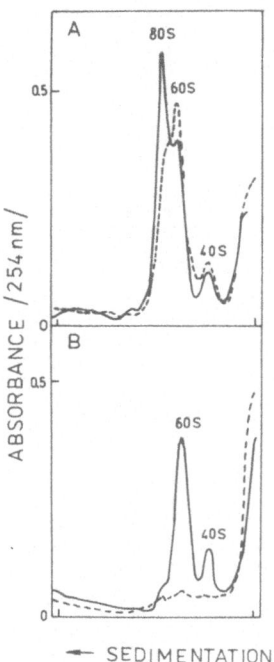

Fig. 2. Absorbance profiles of monoribosomes after incubation. A pellet
of monoribosomes (fraction 1) was suspended at 0°C in buffer A,
containing 1mM cycloheximide, and incubated at 37°C in 1ml final
volume (25 A_{260}/ml). The reaction was stopped by raising the
MgCl$_2$ concentration to 20 mM and chilling. Aliquots (2.5 A_{260}
units) were layered over 12 ml of 10 to 40% linear sucrose
gradient in gradient buffer and centrifuged (150 min, 40000
rev/min, 2°C) in the SW 41 rotor. (A) Unincubated (—) or
incubated for 2 hr (---); (B) Incubated for 4 hr (—) and 8 hr
(---).

degradation of monoribosomes occured after 8 hr of incubation with all the
ribosome material remaining at the top of the gradient (Figure 2B). The
electrophoretic analysis of the partial digests of rRNA (Figures 3A and 3B)
indicates an endonucleolytic cleavage of RNA. The pattern showing a heterogeneous size distribution of cleaved rRNA after 2 hr, is shifted to
smaller sizes after 4 to 8 hr. Moreover, no appreciable ultraviolet-
absorbing PCA-soluble products were detected even after 8 hr of incubation
when all the 28S and 18S rRNA were completely degraded as shown on Figure
3B.

Autodegradation of Polyribosomes

Autodegradation of polyribosomes was analyzed under exactly the same
experimental conditions as for monoribosomes. The distribution patterns on
linear sucrose gradient of polysomes (fraction 2 and 3) after different
incubation times are presented in Figure 4. A partial polysome destruction
is evident on the 4th hour of incubation (Figure 4A), accompanied with a
decrease in the proportion of heavy polysomes, a redistribution of u.v.
absorbing material in the zone of lighter polysomes, forming four-, ztri-
and dimers, and an accumulation of monomers. At the same time the electrophoretic pattern of rRNA extracted from polysomes was unchanged compared
with that of the controls (Figure 5A). Eight hours of autodegradation were
characterized by further increase in the proportion of monomeric ribosomes
(Figure 4B) and some initial signs of rRNA degradation (a slight decrease

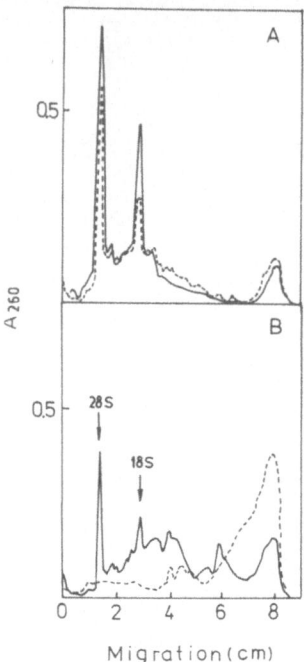

Fig. 3. Electrophoretic analysis of rRNA in agar–urea gels. An aliquot of incubation mixture (see legend to Figure 2) containing 1.8 A_{260} units of RNA was fractionated. (A) Unincubated (—) or incubated for 2 hr (---); (B) Incubated for 4 hr (—) and 8 hr (---).

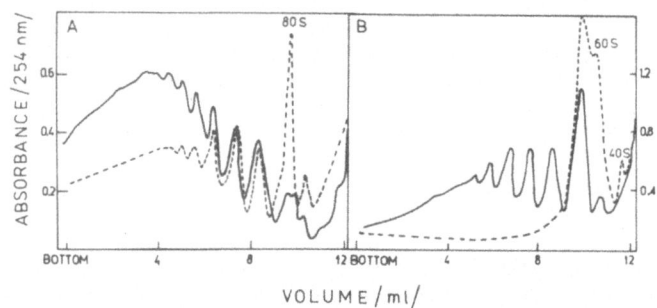

Fig. 4. Absorbance profiles of polyribosomes. A pellet of polyribosomes was suspended in autodegradation buffer to a final concentration of 50 A_{260} units/ml. Approx. 5.5 units were layered on each gradient. Further details are as in the legend to Figure 2, except that gradient centrifugation was for 90 min. (A) Unincubated (—) or incubated for 4 hr (---); (B) Incubated for 8 hr (—) and 16 hr (---).

of the 18S rRNA peak) (Figure 5B) (note that at the same time of incubation monoribosomes had already been completely destructed). The analysis of distribution patterns in sucrose gradients after 16 hr showed an almost complete destruction of polyribosomes, a shift of v.v. absorbing material to the fraction of 80S particles, and a partial dissociation of monosomes to subunits (Figure 4B). A significant degradation of 28S and especially of 18S rRNA, leading to the formation of fragments with sedimentation coef-

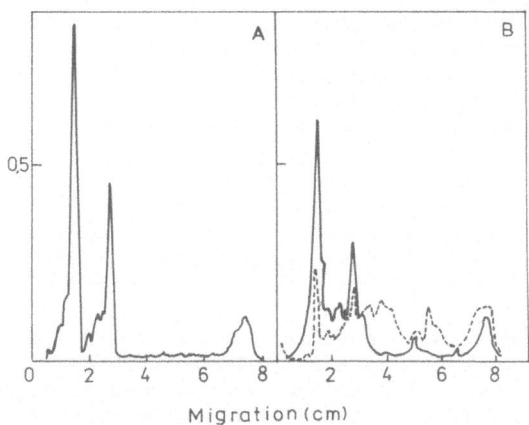

Fig. 5. Electrophoretic fractionation of polyribosomal rRNA. (A)
Unincubated (——) or incubated for 4 hr (---); (B) Incubated for
8 hr (——) and 16 hr (---).

ficient of 18 to 3S, was observed (Figure 5B). Complete degradation of
rRNA was not accomplished even after 24 hours of incubation (results not
shown).

The above results show that polyribosomes possess less autodegradation
activity than monoribosomes, isolated from the same homogenate. A similar
finding has been reported previously by Bransgrove and Cosquer (1978).
However, it was impossible to isolate polyribosomes without any RNase
activity. The increase in the amount of monomers during the incubation of
polyribosomes in vitro (Figure 4A and 4B) is due to a low endogenous RNase
activity, and not to the running-off of ribosomes, which was prevented by
cycloheximide. The derived monomers probably contain a segment of mRNA and
carry peptydil-tRNA. Such complexes are more stable to different influ-
ences than free of mRNA and peptydil-tRNA particles (Naslund and Hultin,
1971; Freedman et al., 1972). This could apparently explain the increase
of the monomer peak during incubation of polysomes and their preserved rRNA
profile. With the purpose of creating a model system, resembling closely
the in vivo conditions, all experiments were carried out at 37°C. pH 7.6,
in the presence of 5mM $MgCl_2$ and 50mM KC1. Under these conditions the
monomeric ribosomes were unstable, while 80S monosomes in polysomal structures were reasonably stable, suggesting that the 80S monemers rather than
polymers are the substrate for degradative enzymes.

Sensitivity to RNase Treatment of Rat Liver Ribosomes and rRNA

The above shown difference in autodegradation activity of polyribosomes and monoribosomes, isolated from the same homogenate, could be
attributed to a qualitative alteration in ribosomes and/or rRNA. So it was
deemed of interest to verify whether fresh ribosomal preparations from
norman rat liver differ by their sensitivity to mild RNase treatment.
Ribosomes or rRNA were incubated in buffer A in the presence of RNase A.
The amount of products soluble in 10% PCA, expressed as percentage of the
total absorbance at 260nm, was taken as a measure for the degradation of
RNA.

Preparations of normal liver monoribosomes incubated with RNase at a
final concentration of 50ng/ml showed a linear rate of RNA degradation that
reached values of 9, 12 and 19% after 20, 30 and 60min, respectively. Both
free and membrane-bound polyribosomes were practically unaffected by the

same treatment, and the enzyme concentration had to be increased 10-fold to obtain rates of hydrolysis of the order of 20% at comparable intervals. Polyribosomes incubated with 500ng/ml RNase actually gave values of 9, 15, 18 and 26% hydrolysis after incubation for 10, 20, 30 and 60 minutes respectively, the results obtained with free and membrane-bound polysomes being practically identical. When expressed per ng of RNase, the rates of degradation of normal liver polyribosomes and monosomes can be more directly compared, and such data presented in Figure 6 make evident that the monribosome preparations are 5 to 6 times as sensitive to RNase A treatment as polyribosomes from the same tissue. The differences noted at the various intervals were always statistically significant (P < 0.001).

Similar experiments with purified rRNA revealed that both monosomal and polysomal rRNAs undergo appreciable hydrolysis during incubation with RNase at concentrations as low as 5ng/ml. In such conditions, the rRNA from polyribosomes showed 2, 4, 8 and 16% degradation after 10, 20, 30 and 60 min, respectively. The rRNA from monoribosomes gave slightly higher values, and reached 12% hydrolysis after 30 min of incubation, but the differences were never statistically significant (P > 0.05).

The present study demonstrates that monoribosomes isolated from normal rat liver are 5-6 times as sensitive to RNase treatment as polyribosomes from the same tissue. This property of liver monoribosomes seems to be mainly due to some modification of the ribosome structure rather than an

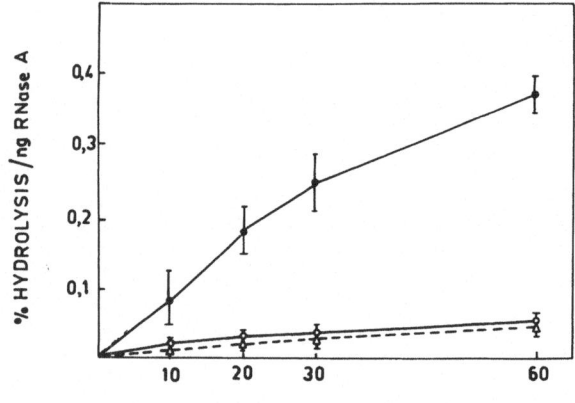

TIME OF INCUBATION(min)

Fig. 6.   Percentage hydrolysis of monoribosomes (●—●), free (o—o) and membrane-bound (△--△) polyribosomes from normal rat liver, per ng of RNase after different period of incubation. Means ± S.D. from 3 determinations. rRNA was extracted by incubating the samples for 5 min at 37°C in a SDS-DEPC mixture (sodium dodecyl sulphate 1%, diethylpyrocarbonate 3%) according to Tsuei and Rouhandeh (1976). The final pellets of ribosomes and rRNA were dissolved in buffer A to a concentration of 25 A_{260nm} units/ml and exact volumes of stock solutions of RHase A (from bovine pancreas, Serva) were added to 1 ml samples. The preparations were incubated at 37°C and aliquots of 0.2 ml were transferred at regular intervals into 5 ml of cold 10% PCA and the absorbance of the 10000 rev/min supernatants at 260 nm was taken as a measure of the amount of hydrolyzed RNA. In all cases, the amounts of RNA hydrolyzed by RNase were expressed as the percentage of total RNA, the latter being determined by complete hydrolysis of 5 A_{260} units of ribosome suspension or rRNA solution after incubation with 5 ml of 10% PCA for 18 min at 70°C.

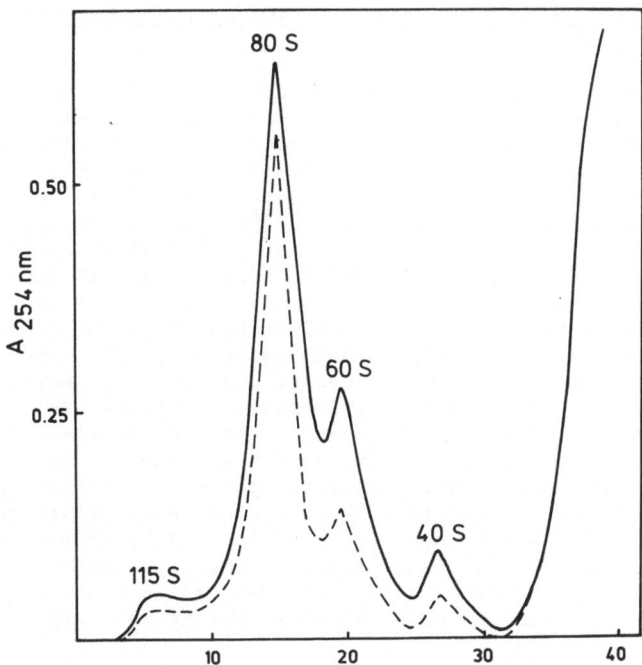

Fig. 7. Typical distribution pattern of active and inactive ribosomes from
normal (—) and 48 hr regenerating (---) rat livers. Aliquots of
post-mitochondrial supernatents were treated with RNase A (2
μg/ml) for 5 min, 0.25 vol of 2.5M KCl was added, the samples were
immediately layered on linear sucrose gradients 10 to 40% (wt/v)
in gradient buffer, containing 0.8M KCl, and centrifuged for 6 hr
at 26000 rev/min (SW 27 rotor) at 4°C.

alteration in rRNA molecles, since purified rRNAs from the same populations
differ little in their sensitivity to RNase. A change in ribosome
structure that would make some RNA more directly accessible to RNase might
result from a difference in protein-RNA interactions, but other factors
should also be considered.

The results taken all together show that the availability of ribosomes
to degradative enzymes is determined by the stage of ribosome function, as
well as by their structure. The pool of monomeric ribosomes would appear
to be most likely source for the coordinate disposal of both subunits. It
may be that in such ribosomes the site at which an initial attack is made
is among those occupied by mRNA, tRNA, and initiation or elongation factors
while the ribosome is participating in protein synthesis. If subunits are
a form in which ribosomes become susceptible to degradation, then splitting
of 80S monomers to subunits may represent the initial steps in ribosome
breakdown. The derived 40S subunits were unstable: they were altered
spontaneously to slowly-sedimenting (but acid-precipitable) material that
probably consisted of unfolded subunits. Thus an endonucleolytic attack
directed against the 3'- terminal portion of the 18S rRNA might be an
initial event which produces a pool of ribosomes incapable of participating
in protein synthesis and thus exposed to further attack (Scott, 1977).
This and the observed simultaneous turnover of the two ribosomal subunits
(Nikolov and Dabeva, 1983) perhaps suggest that 40S determines the fate of
the 60S subunit with which it is associated, or that some other modification selects the 60S subunit destined for degradation.

Number of Ribosomes Active in Protein Synthesis in Normal and Regenerating Rat Liver

To investigate the possible involvement of "out-of-cycle" ribosomes in controlling ribosome degradation, polyribosomal profiles in high-salt media after RNase treatment (in which inactive monomers dissociate into subunits, wheras active ribosomes in polysomal structures do not) of normal and 48 hour regenerating rat liver (a time at which liver cell proliferation is proceeding at a maximal rate [Nikolov et al., 1983]) were obtained (Figure 7). The results show that in normal liver 77.5% of ribosomes are active in protein synthesis, i.e. monomeric ribosomes comprise 22.5% from the total ribosomal population. From the distribution pattern in regenerating liver it is evident that 85.4% of ribosomes are in polysomal structures, the inactive monomers being 14.6%, which shows that the proportion of "out-of-cycle" ribosomes decreased by nearly 35% in regenerating liver., This fractional decrease is however accompanied by a large (30%) increase in the total concentration of ribosomes per gram of liver, that in fact the absolute number of free monomers per unit mass of tissue falls by only about 16%. Thus, the previously estimated 2.4-fold lower rate of ribosome degradation in regeneration rat liver (Nikolov et al., 1987) could not be accounted for by this small decrease in the amount of monomeric ribosomes. This suggests that the proportion of ribosomes which are "out-of cycle" is probably not involved in the control of ribosome breakdown, although such ribosomes may be the target for destruction.

This suggests that some quite different mechanism is triggered during the transition to steady-state conditions to promote the breakdown process. Possible enzymes involved in such a mechanism may be a trypsin-like proteinase (Langner et al., 1982; Bilinkina et al., 1978) and ribonuclease (Bransgrove and Cosquer, 1978; Krechetova et al., 1972).

REFERENCES

Bilinkina, V.S., Levyant, M.I., Gorach, G.G. and Orechovich, V.N. 1978, _Biohimiya_ 43; 100-102.

Bransgrove, A.B. and Cosquer, L.C. 1978, _Biochem. Biophys. Res. Commun._ 81; 504-511.

Cammarano, P., Guidice, G. and Lukes, B. 1965, _Ibid_. 19; 487-493.

Dabeva, M.D. and Dudov, K.P. 1982, _Biochem J._ 204, 179-183.

Dudov, K.P., Dabeva, M.D. and Hadjiolov, A.A. 1976, _Anal. Biochem._ 76; 250-258.

Freedman, M.L., Velez, R. and Mucha, J. 1972, _Exp. Cell Res_. 72; 431-435.

Krechetova, G.D., Chudinova, I.A. and Shapot, V.S. 1972, _Biochim. Biophys. Acta_ 277, 161-178.

Langner, J., Kirschke, H., Bohley, P., Wiederanders, B. and Korant B.D. 1982, _Eur. J. Biochem._ 125; 21-26.

Melvin, W. T. and Keir, H.M. 1978, _Biochem.J._ 176; 933-941.

Melvin, W.T., Kumar, A. and Malt, R.A. 1976, _J. Cell Biol._ 69; 548-566.

Naslund, P.H. and Hulton, T. 1971 _Biophys. Acta_. 254; 104-116.

Nikolov, E.N. and Dabeva, M.D. 1983, _Bioscience Rep_. 3; 781-788.

Nikolov, E.N., Dabeva, M.D. and Nikolov, T.K. 1983, _Int. J. Biochem._ 15; 1255-1260.

Nikolov, E.N., Dineva, B.B., Dabeva, M.D. and Nikolov, T.K. 1987, _Ibid_. 19; 159-163.

Perry, R.P. 1973, _Biochem. Soc. Symp._ 37; 114-135.

Ramsey, J.C. and Steele. W.J. 1979, _Anal. Biochem._ 92; 305-313.

Rizzo, A.J. and Webb, T.E. 1968, _Biochem. Biophys. Acta_ 169; 163-174.

Ruthland, P.S., Weil, S.. and Hunter, A.R. 1975, _J. Mol. Biol._ 96; 745-766.

Scott, J. 1977, _Exp. Cell Res_. 108, 207-219.

Storb, U. and Martin, T.E. 1972, Biochem. Biophys. Acta 281; 406-415.
Tsuei, D. and Rouhandeh, H. 1976, Experientia 32; 749-751.

SPECIFIC LABELLING IN THE FUNCTIONAL DOMAIN OF ELONGATION

FACTOR EF-Tu FROM <u>BACILLUS STEAROTHERMOPHILUS</u> AND FROM

<u>BACILLUS SUBTILIS</u>

J. Jonák, K. Karas and I. Rychlík

Institute of Molecular Genetics
Czechoslovak Academy of Sciences
166 37 Prague 6, Czechoslovakia

INTRODUCTION

Elongation factor Tu plays a pivotal role in the elongation cycle of both prokaryotic and eukaryotic protein biosynthesis. The prokaryotic factor in the form of EF-Tu. GTP forms a complex with aminoacyl-tRNA and promotes its binding to the A-site of mRNA-programmed ribosomes. During the reaction, GTP in the complex is hydrolyzed to GDP, and a stable form of EF-Tu, EF-Tu. GDP, which has a low affinity for aminoacyl-tRNA, is released from the ribosome. The recognition of aminoacyl-tRNA, is released from the ribosome. The recognition of aminoacyl-tRNA by EF-Tu represents a natural model for the study of complex interaction processes between nucleic acids and proteins.

The available date indicate that the 3' terminal region of aminoacyl-tRNA is very probably the most important part of the polynucleotide for the binding to EF-Tu and, in EF-Tu from <u>E.coli</u>, the binding site for the 3' terminus of aminoacyl-tRNA is situated in the region around cysteine 81, spanned from histidine 66 to histidine 118 (for review see e.g. Jonák et al., 1980, Jonák et al., 1984), even though other regions of EF-Tu are probably also involved (Parmeggiani et al., 1987).

Recently, a model of the tertiary structure of the GDP binding domain of EF-Tu from <u>E.coli</u> based on high resolution crystal X-ray diffraction studies was presented (la Cour et al., 1985, Jurnak, 1985, McCormick et al., 1985). An analysis of the model reveals that the GDP-binding domain of EF-Tu has a structure similar to that of other nucleotide-binding proteins. The protein elements that take part in the nucleotide binding are located in four loops connecting parallel β-strands with α-helices presenting a structural feature called the Rossman fold. The GDP ligand is linked to the protein via a Mg^{2+} ion, which forms a salt bridge with the side chain of aspartic acid 80, the amino acid residue located next to Cys-81. Thus, these results suggest that the binding sites for GDP and aminoacyl-tRNA in EF-Tu from <u>E.coli</u> are in close vicinity or might overlap. Moreover, the GTP-bound form of EF-Tu differs from the GDP-bound form in its ability to interact with aminoacyl-tRNA, and the basis of this difference may be conformational changes that accompany GTP and GDP binding. La Cour and co-workers (1985) have predicted that the conformation of the guanine nucleotide binding domain of EF-Tu is likely to be different when GTP is bound rather than GDP. Inspection of the model indicates that the pocket for GDP does not seem large enough to accommodate an additional γ phosphate without moving the loop consisting of residues Asp-80 to His-84. Thus structural properties of this region of EF-Tu appear to have key importance for the function of the protein.

To elucidate the question whether the same or similar molecular groups and interactions are involved in the recognition of GDP/GTP and aminoacyl-tRNA also by other EF-Tu's, we isolated elongation factors Tu from two taxonomically distant prokaryotes, B.subtilis and B.stearothermophilus, and looked for structural homologies in this region between both of these factors and EF-Tu from E.coli.

RESULTS AND DISCUSSION

Elongation factors Tu from B.stearothermophilus and from B.subtilis were purified from frozen cells by modified classical procedures (Arai et al., 1972, Jonåk et al., 1986) and were obtained in virtually homogeneous form. Figure 1 shows the results of an electrophoretic experiment on SDS/polyacrylamide slab gel with EF-Tu's from E.coli B,Streptomyces aureofaciens, B.stearothermophilus and B.subtilis. Relative to the standards on the gel, which are now shown here, the mobility of EF-Tu both from B. stearothermophilus and B.subtilis indicated the same Mr of about 50 000.

The amino acid composition of the factors, obtained after 20 h hydrolysis and based on Mr = 50 000, is shown in Table 1. There is a great

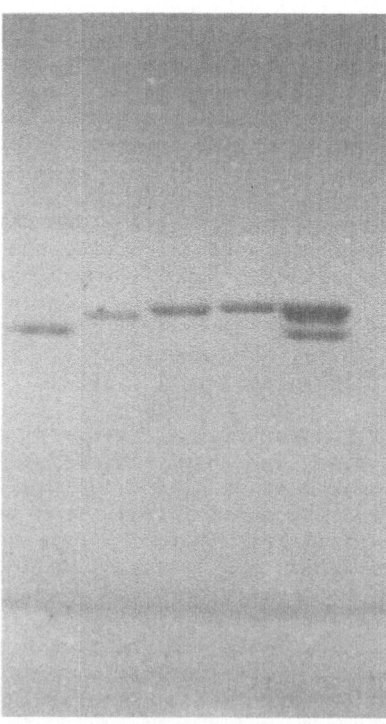

Fig. 1. SDS-polyacrylamide gel electrophoresis of elongation factors Tu from different bacteria. Left to right: E.coli, S.Aureofaciens, B.stearothermophilus, B.subtilis, a mixture of all four elongation factors.

Table 1. Amino acid composition of EF-Tu from B.stearothermophilus,
B.subtilis and E.coli (Jones et al., 1980). B.ST.,
B.stearothermophilus, B.SUB., B.subtilis, n.d., not determined.

Amino acid	B.ST.	B.SUB.	E. coli
	(residues per molecule)		
Lys	25-26	26	23
His	12	14	11
Arg	25	23	23
Cys	3	3	3
Asx	36-37	42	31
Thr	34-35	37-38	30
Ser	15-16	21-22	11/10
Glx	60-63	61-64	45
Pro	25-27	17-18	20
Gly	41-43	43-44	40/41
Ala	37	32-33	27
Val	42-43	43	37
Met	13	13	10
Ile	26-27	26	29
Leu	27-29	28	28
Tyr	11	12-13	10
Phe	13-14	14-15	14
Trp	1-2	n.d.	1

deal of similarity between EF-Tu from E.coli and EF-Tu's of bacillary
origin with respect to their amino acid composition. However, the absolute
amounts of Glx, Ser, Pro and Ala residues are substantially higher in EF-Tu
from the thermophilic organism, and absolute amounts of His, Asx, Glx and
Ser are higher in EF-Tu from B.subtilis, than in E.coli EF-Tu. On the
other hand, the bacillary EF-Tu's substantially differ from each other in
the content of Ser and Pro residues, whereas the content of cysteine
residues is the same as in E.coli EF-Tu.

To isolate and characterize from the bacillary EF-Tu's a peptide which
would be analogous to the Cys-81-containing peptide of the GDP-binding
domain of E.coli EF-Tu, we used the same simple strategy as was applied in
the case of E.coli EF-Tu (Jonák et al.,1982). We treated the factors with
radioactive N-tosyl-L-phenylalanyl-chloromethane (TPCK), the reagent shown
previously to label specifically Cys-81 in E.coli EF-Tu. Both bacillary
factors were affected by the reagent in a similar manner as E.coli EF-Tu
(Table 2).

The treatment by TPCK destroyed their ability to promote protein
synthesis, the loss of about one cysteine residue per molecule of either
factor was detected by amino acid analysis, and at the same time radioactivity measurements revealed that approximately one molecule of the
reagent was incorporated per each molecule of either factor. To localize
the site of TPCK attachment, the EF-Tu proteins inactivated by treatment
with the radioactive inhibitor were carboxymethylated, digested with trypsin, and the resulting peptides were separated by HPLC on a CGC Separon Six
C_{18} reverse-phase column with a gradient of methanol (Jonák et al., 1986).
Figure 2 and Figure 3 show that a new distinct peak (indicated by arrow)
not detected in the hydrolysate of control factors, appeared in the
chromatographic profile of the digest of TPCK-treated EF-Tu both from B.
stearothermophilus and B.subtilis.

Table 2. Effect of TPCK on EF-Tu from B.stearothermophilus and B.subtilis. Cysteine was determined as cysteic acid, TPCK incorporation was measured as in Jonák et al. (1980), polyphenylalanine formation as in Jonák et al. (1986).

Elongation factor Tu	Cysteine residues (per mole EF-Tu)	TPCK incorporated (per mole EF-Tu)	Phenylalanine polymerized (pmol)
I. Control			
B.stearothermophilus	2.84	–	5.5
B.subtilis	2.99	–	5.1
II. TPCK – treated			
B.stearothermophilus	2.30	0.90	0.8
B.subtilis	2.15	0.99	0.2

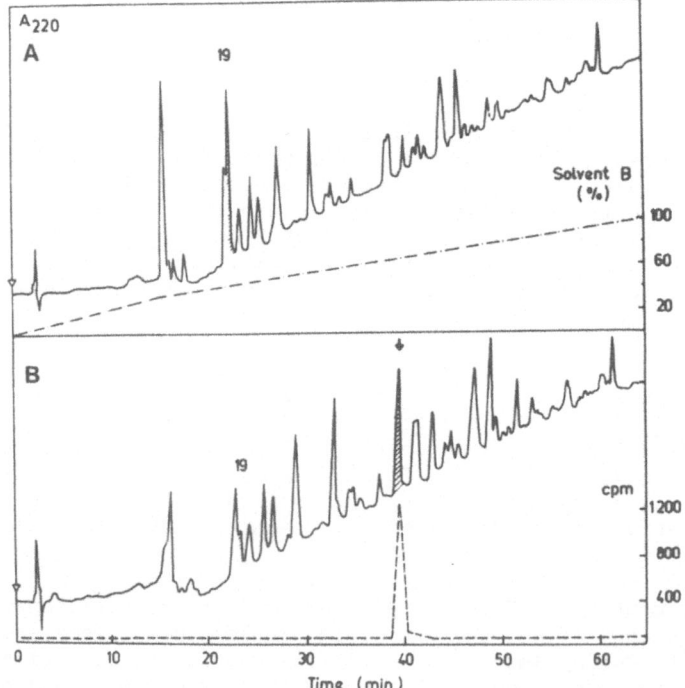

Fig. 2. HPLC of tryptic digest of EF-Tu from B.stearothermophilus. Absorbance at 220 nm was recorded and radioactivity (---) in a 100 μl aliquot of each fraction was measured: (A) control EF-Tu; (B) TPCK-labelled EF-Tu.

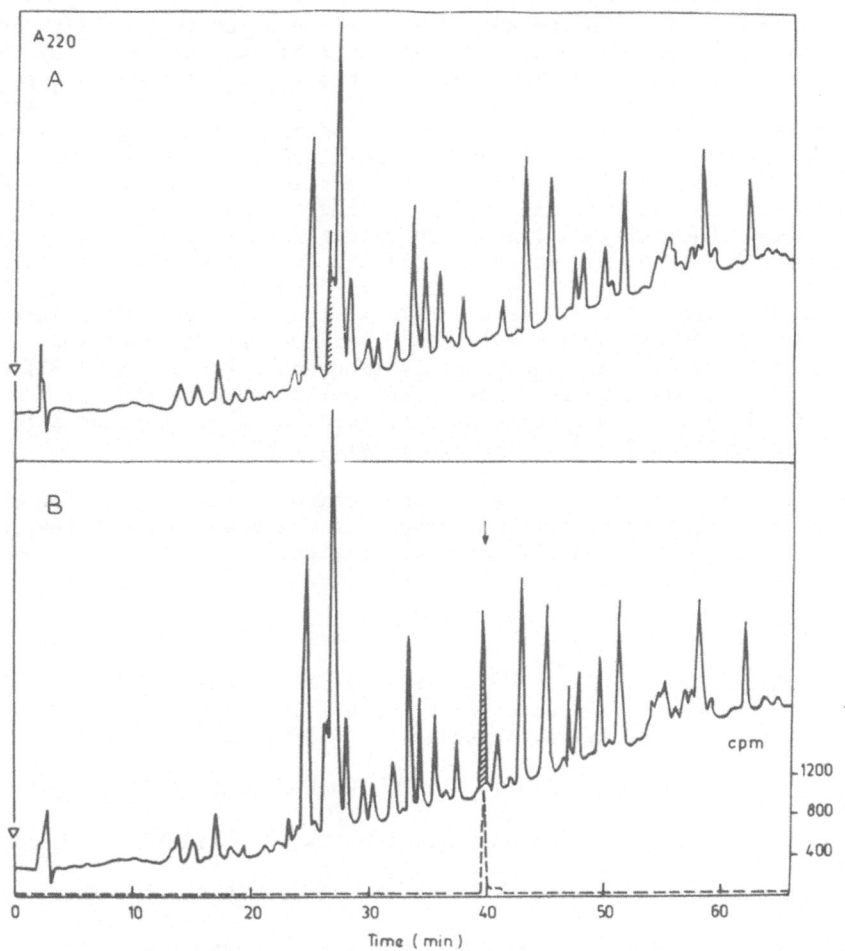

Fig. 3. HPLC of tryptic digest of EF-Tu from B.subtilis. Absorbance at
220 nm was recorded, and radioactivity (---) in a 100 μl aliquot
of each fraction was measured: (A) control EF-Tu; (B)
TPCK-labelled EF-Tu.

These peaks were the only radioactive fractions from all collected
during the chromatography and were eluted at the same concentration of
methanol. This suggested that both peptides might be identical, and the
results of amino acid analyses supported this assumption. Amino acid composition of TPCK-labelled peptide from B.stearothermophilus: Asx 2.05, Pro
0.83, Gly 1.45, Ala 2.0, Val 176, Try 1.69, Lys 1.05, His 2.55. Amino acid
composition of TPCK-labelled peptide from B.subtilis: Asx 2.26, Pro 0.74,
Gly 1.57, Ala 2.04, Val 1.81, Tyr 1.57, Lys 1.0, His 2,76. Finally, the
TPCK-labelled peptide from B.stearothermophilus was sequenced by automated
Edman degradation and the sequence of the TPCK-labelled peptide from B.
subtilis was determined manually by the 4-N,N-dimethylaminoazobenzene-
4'-isothiocyanate/phenylisothiocyanate metod (Chang et al., 1978). Thus,
it was found that the target site for TPCK attachment to both bacillary
EF-Tu's had the same sequence: His-Tyr-Ala-His-Val-Asp-Cys-Pro-Gly-His-
Ala-Asp-Tyr-Val-Lys. The position of the cysteine residue was determined
indirectly from the results of radioactivity measurements of products in
each degradation step.

The present report shows that EF-Tu from B.Stearothermophilus and EF-Tu from B.subtilis share a region of complete sequence homology 15 amino acid residues long. It is identical with the E.coli peptide that forms one of the four loops of the GDP-binding domain and contains the important residues Asp-80 and Cys-81. The integrity of this peptide appears to be essential also for the function of the bacillary factors, because alkylation of the cysteine residue of the peptide results in complete loss of the polymerization activity of both factors. The above data suggest that although the bacillary factors Tu differ from E.coli EF-Tu in the amino acid composition and molecular weight, evolutionary constraints on function have maintained this structure from organism to organism. The same molecular groups in different EF-Tu's might be involved in interaction processes with GDP/GTP, aminoacyl-tRNA or ribosomes. In addition, chloroplast EF-Tu from Euglena gracilis (Montandon and Stutz, 1983) bears the same peptide, and the sequence of the analogueous peptide in mitochondrial EF-Tu from S.cerevisiae differs in only two amino acid residues at position 3 and 4 (Nagata et al., 1983).

Thus it is reasonable to assume that common properties shared by prokaryotic elongation factors may have a strong basis also in the primary structure of this functionally important region.

REFERENCES

Arai, K., Kawakita, M., Kaziro, Y. (1972) J. Biol. Chem. 247, 7029-7037.
Chang, J.Y. Brauer, D., Wittmann-Liebold, B. (1978) FEBS Lett. 93, 205-214.
Jonák, J., Smrt, J., Holy, A., Rychlík, I. (1980) Eur. J. Biochem. 105, 315-320.
Jonák, J., Petersen, T.E., Clark, B.F.C., Rychlík, I. (1982) FEBS Lett. 150, 485-488.
Jonák, J., Petersen, T.E., Meloun, B., Rychlík, I. (1984) Eur. J. Biochem. 144, 295-303.
Jonák, J., Pokorná, K., Meloun, B., Karas, K. (1986) Eur. J. Biochem. 154, 355-362.
Jones, M.D., Petersen, T.E., Nielsen, K.M., Magnusson, S., Sottrup-Jensen, L., Gausing, K., Clark, B.F.C. (1980) Eur. J. Biochem. 108, 507-526.
La Cour, T.F.M., Nyborg, J., Thirup, S., Clark, B.F.C. (1985) The EMBO J. 4, 2385-2388.
McCormick, F., Clark, B.F.C., La Cour, T.F.M., Kjeldgaard, M., Norskov-Lauritsen, L., Nyborg, J. (1985) Science 230, 78-82.
Montandon, P. -E., Stutz, E. (1983) Nucleic Aids Res. 11, 5877-5892.
Nagata, S., Tsunetsugu-Yokota, Y., Naito, A., Kaziro, Y. (1983) Proc. Natl. Acad. Sci. USA 80, 6192-6196
Parmeggiani, A., Swart, G.W.M., Mortensen, K.K., Jensen, M., Clark, B.F.C., Dente, L., Cortese, R. (1987) Proc. Natl. Acad. Sci. USA, in press.

GTP EXPENSE AT CODON-SPECIFIC TRANSLATION AND DURING MISREADING

L. P. Gavrilova and D. G. Kakhniashvili*

Institute of Protein Research, Academy of Sciences of
the USSR, 142292 Pushchino, Moscow Region, USSR

*Institute of Plant Biochemistry, Georgian Academy of
Sciences, Tbilisi, USSR

For a long time it was practically impossible to measure the stoichiometry of GTP hydrolysis during peptide synthesis proceeding on ribosomes,
as ribosome preparations usually consist of no more than 10-50% active
ribosomes. The other 50-90% of ribosomes in ribosome preparations are
incapable of synthesizing peptide in vitro, but can carry out uncoupled
with protein synthesis factor-dependent GTP hydrolysis. As yet there are
no means of measuring correctly this uncoupled factor-dependent GTP
hydrolysis.

Here, the GTP expense during peptide synthesis was determined under
conditions in which all the ribosomes were active and engaged in peptide
synthesis. Active ribosomes were separated from inactive ones using poly
(U) covalently bound to Sepharose (Belitsina et al. 1973; Baranov et al.,
1979; Ustav et al., 1979). Peptide synthesis proceeded on ribosomes
carrying oligophenylalanine presynthesized on poly(U)-Sepharose
(Kakhniashvili et al., 1983). The GTP expense was measured during peptide
elongation both in the process of codon-specific translation (poly(U)-
directed polyphenylalanine synthesis) and during misreading (poly(U)-
directed polyleucine or polyisoleucine synthesis) (Gavrilova and
Rutkevitch, 1980).

Figure 1 demonstrates the dependences on MG^{2+} concentration of the
rates of elongation of poly(Phe), poly(Leu) and poly(Ile) using
poly(U)-Sepharose as a template. It is seen that the elongation rate of
each polypeptide has a fairly narrow optimum. It is noteworthy that the
Mg^{2+} optima for the elongation rates of the 3 polypeptides synthesized are
very different; the lowest, about 6 mM Mg^{2+}, characterizes codon-specific
elongation, whereas maximal rates of miselongation require higher Mg^{2+}
concentrations. The difference in Mg^{2+} optima between poly(Phe), poly(leu)
and poly(Ile) synthesis may be consistent with the assumption that the
lower the affinity of anticodon to codon the higher is the Mg^{2+}
concentration required for aminoacyl-tRNA uptake by the ribosome.
Stoichiometry of GTP hydrolysis during elongation or miselongation was
studied at three different Mg^{2+} concentrations: at the Mg^{2+} optima and at
lower and higher Mg^{2+} concentrations where the elongation rate was half of
that in the optima.

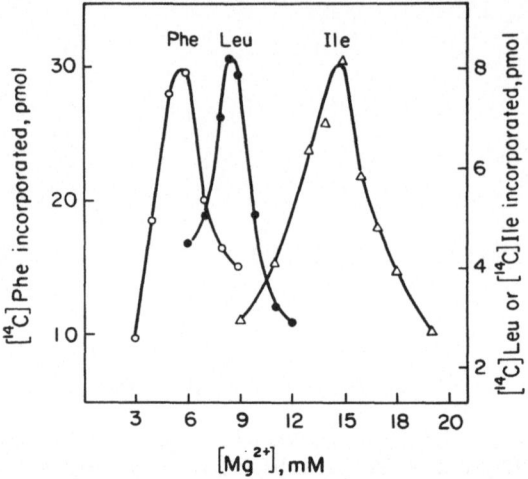

Fig. 1. Mg^{2+} dependence of Phe, Leu or Ile incorporation in peptides
elongating in the [oligo(Phe)-tRNA-ribosome-poly(U)-Sepharose]
system at 37°C. 'Incubation: Phe, 1.5 min; Leu or Ile, 10 min.

To determine the stoichiometry of GTP hydrolysis during peptide
elongation on ribosomes in codon-specific translation or misreading of
polyuridylic acid, the following were measured in each experiment: (1)
kinetics of poly(U)-directed peptide elongation; (2) kinetics of GTP
hydrolysis accompanying peptide elongation, and (3) kinetics of background
GTP hydrolysis in the mixture where all components expect ribosomes (i.e.,
EF-T, EF-G, poly(U)-Sepharose, total tRNA acylated with [^{14}C]leucine,
[^{14}C]isoleucine or [^{14}C]phenylalanine, pyruvate kinase and phosphoeol-
.pyruvate) were present.

Figure 2 shows typical experimental kinetic curves at 37°C and 6mM
MgCl$_2$ (Mg^{2+} optimum for polyphenylalanine elongation on poly(U)-Sepharose).
In all cases elonga'tion started from oligophenylalanine presynthesized in
the [oligo(Phe)-tRNA-ribosome-poly(U)-Sepharose] complex. The rates of
polyphenylalanine or oligophenylalanine-initiated polyleucine (or polyisoleucine) elongation and GTP hydrolysis were calculated from the linear
parts of the kinetic curves.

The results of GTP/peptide bond ratio determination during codon-
specific elongation are summarized in table 1. It is seen that the GTP/Phe
ratio was about 2 for polyphenylalanine elongation on poly(U)-Sepharose at
the Mg^{2+} optimum of the system (6mM). This value did not practically
change at 3 and 9 mM Mg^{2+} when the polyphenylalanine elongation rate was
about two times lower than that at the Mg^{2+} optimum (Kakhniashvili et al.,
1983; Gavrilova el al., 1984).

The results of GTP/peptide bond ratio determination during misreading
of poly(U) are summarized in table 2 (see data without streptomycin). The
data showed that the GTP/peptide bond ratio determination during misreading
of poly(U) are summarized in table 2 (see data without streptomycin). The
data showed that the GTP/peptide bond ratio during miselongation of poly-
(Leu) or poly(Ile) (GTP/Leu or GTP/Ile) was higher than during codon-
specific elongation (GTP/Phe). During misreading of poly(U) and elongation
of poly(Leu) or poly(Ile) 6 to 20 GTP molecules were spent on one peptide
bond instead of only two GTP ones in the process of codon-specific elongation. It was found that in contrast to codon-specific elongation during
miselongation on poly(U), the GTP/Leu or GTP/Ile ratio changed with a

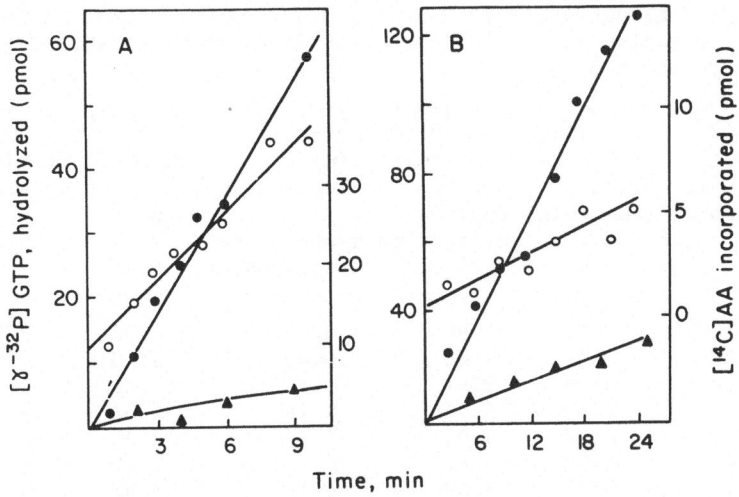

Fig. 2. Typical kinetics of peptide elongation and GTP hydrolysis used for
calculation of stoichiometry of GTP expense per peptide bond in
the cell-free system with poly(U)-Sepharose at 6 mM $MgCl_2$, 37°C.
Peptide elongation (o—o); GTP hydrolysis during peptide synthesis
(●—●); GTP hydrolysis with all components of the system except
ribosomes (▲—▲). (A) Polyphenylalanine; (B) Polyleucine.

change of Mg^{2+} concentration. Moreover our data showed that during misreading the GTP expense is determined not by the absolute Mg^{2+} concentration, but rather by the relative position of the Mg^{2+} point on the Mg^{2+}
dependence curve of a given elongation system: the same GTP expense is
attained at 9mM Mg^{2+} in the case of poly(Ile) elongation and at 6 mM Mg^{2+}
in the case of poly(Leu) elongation, both points being on the left slope of
their Mg^{2+} dependence curve (see Table 2 and Figure 1) (Smailov et al.,
1984; Gavrilova, 1985). The left slope of the curve seems to provide the
conditions where aminoacyl-tRNA binding is the rate-limiting step of the
elongation cycle, whereas the Mg^{2+} range above the Mg^{2+} optima makes translocation a rate-limiting step (Gavrilova et al., 1976).

In accordance with the accepted point of view, GTP hydrolysis during
ribosomal protein synthesis is coupled both with EF-Tu and with EF-G functioning. However, the experiments performed do now allow to determine
which of the GTP-dependent stages of the elongation cycle, i.e. either
EF-Tu-promoted aminoacyl-tRNA binding or EF-G-promoted translocation, is
responsible for the excess of GTP expenditure. We have found previously by
using systems of poly(U) translation with different sets of elongation
factors, such as a complete one, including both EF-Tu and EF-G, the
EF-Tu-promoted one without EF-G, and the EF-G-promoted one without EF-Tu,
as well as the factor-free translation system, that EF-Tu can reduce
translation errors in poly(U)-directed cell-free systems (Gavrilova and
Perminova, 1981; Gavrilova et al., 1981).

Determination of the stoichiometry of the EF-Tu-dependent GTP hydrolysis during elongation was done in the following way. Before the elongation
was started, [γ-^{32}P]GTP, EF-Tu and [^{14}C]aminoacyl-tRNA were incubated for 5
min at 37° to form the ternary complex. Then the [oligo (Phe)-tRNA·
ribosome·poly(U)-Sepharose] complex was added to the mixture. The elongation was initiated by addition of EF-G and unlabeled GTP. The results are
summarized in table 3. It was found that only one GTP molecule per amino
acid residue was expended in the EF-Tu-promoted binding of Phe-tRNA during

103

Table 1. Stoichiometry of GTP Hydrolysis during Polyphenylalanine Elongation in the [Oligo(Phe)-tRNA-Ribosome-Poly(U)-Sepharose] System at 37°C.

[MgCl$_2$] (mM)	EF-Tu in a sample (pmol)	[^{14}C]Phe polymer-ized (pmol/min)	[γ^{-32}P]GTP hydrolyzed (pmol/min)		Total GTP cleaved per peptide bond	GTP cleaved minus back-ground per peptide bond
			During peptide elongation	Background of the system with-out ribosome		
3	40	0.75	1.90	0.40	2.52	2.00
	40	0.65	2.00	0.45	3.10	2.38
	130	0.78	2.91	1.23	3.78	2.18
6	40	2.58	5.97	0.78	2.31	2.01
	40	0.92	3.07	1.13	3.33	2.11
	40	2.80	5.81	0.59	2.09	1.88
	40	2.93	5.50	–	1.90	–
	130	3.48	8.15	0.87	2.34	2.09
	130	1.33	2.90	0.30	2.37	2.12
	130	2.28	6.51	2.20	2.86	1.89
	130	1.87	3.80	–	2.03	–
8	130	2.53	6.54	0.64	2.58	2.33
9	40	0.95	1.64	0.00	1.72	1.72
	130	1.29	2.49	0.48	1.93	1.60

codon-specific elongation. This means that during codon-specific elongation practically every Phe-tRNA bound to [oligo(Phe)-tRNA·ribosome· poly(U)-Sepharose] as a [Phe-tRNA·EF-Tu·GTP] complex does not dissociate from the ribosome after GTP hydrolysis and almost inevitably forms a peptide bond with the presynthesized oligophenylalanine. At the same time, the expenditure of GTP in the EF-Tu-promoted binding of Leu-tRNA during misreading is about 20-times higher (Kakhniashvili et al., 1986). From the presented results it can be deduced that under our experimental conditions only one out of 20 Leu-tRNA molecules interacting with the translating ribosomes is finally accepted by the ribosome, and thus is incorporated into a peptide. The other Leu-tRNA molecules are withdrawn from the ribosome after EF-Tu-mediated GTP hydrolysis. Such a mechanism is called proof-reading. Hopfield was the first to propose that the error correction during elongation can proceed at the stage of EF-Tu-promoted binding of aminoacyl-tRNA and is coupled with additional expenditure of GTP (Hopfield, 1974). Our data supports Hopfield's hypothesis on the EF-Tu-correcting function during peptide elongation.

Our results demonstrate that all GTP excessively hydrolyzed in the course of mistranslation (as least, under our conditions) is expended at the stage of EF-Tu-dependent binding of non-cognate aminoacyl-tRNA. Never-theless, it is not unlikely that a second correction may be realized at the translocation step of the elongation cycle with the participation of EF-G. At this step EF-G with GTP could promote dissociation of non-cognate tRNA carrying peptidyl from the ribosome. There is some data in favour of the appearance of abortive peptidyl-tRNA's in the process of protein synthesis (Menninger, 1976; Menninger, 1978; Caplan and Menninger, 1979).

It has long been known that streptomycin stimulates misreading of the template during translation (Davies et al., 1964). It was clear that streptomycin results in a decrease in the selectivity of codon-dependent aminoacyl-tRNA binding to the ribosome. Selectivity of a substrate with a

Table 2. Stoichiometry of GTP Hydrolysis during Peptide Elongation in the [Oligo(Phe)-tRNA-Ribosome-Poly(U)-Sepharose] System in the Absence and Presence of 5×10^{-6}M Strptomycin (-SM and +SM) at 37°C (average values).

^{14}C-labeled amino acid	[Mg^{2+}] (nM)	Amino acid, polymerized (pmol/min)		[γ^{-32}P]GTP, hydrolyzed (pmol/min)		GTP cleaved per peptide bond	
		-SM	+SM	-SM	+SM	-SM	+SM
Phe	3	8.5	7.7	18.8	16.2	2.2	2.1
	6(opt)		22.5		51.7		2.3
	9	7.0	6.3	15.8	13.3	2.2	2.1
Leu	6	0.4	0.8	7.2	3.3	20.4	4.0
	9(opt)	0.8	2.8	7.3	8.0	9.1	2.9
	11	0.6	1.3	4.0	5.1	6.4	4.0
Ile	9	0.5	1.5	8.5	4.8	18.9	3.3
	15(opt)	0.8	2.0	5.7	5.7	7.3	2.8
	20	0.6	1.1	3.2	3.9	5.9	3.5

GTP hydrolysis background without ribosomes was 0.63±0.39 pmol GTP hydrolyzed per min in 54 experiments summarized. Data are without background subtraction.

Table 3. Stoichiometry of GTP Hydrolysis of the Preformed [Aminoacyl-tRNA-EF-Tu [γ^{-32}P]GTP] Complex during Elongation in the Sepharose-bound Poly(U) Translation System (6 mM MgCl$_2$, 37°C).

Expt no.	^{14}C-labeled amino acid (Aa)	Special conditions	Aa polymerized (pmol/min)	[γ^{-32}P]GTP hydrolyzed (pmol/min)[a]	GTP cleaved per peptide bond
1	Phe	-	29.91	35.14	1.2
		-	27.75	23.84	0.9
2	Leu	-	0.54	11.70	21.7
		+[γ^{-32}P]GTP[b]	0.51	10.75	21.1
3	Leu	-	0.48	10.61	22.1
		+[γ^{-32}P]GTP[b]	0.44	9.25	21.0
4	Leu	-	0.64	9.82	15.3
		+SM[c]	1.60	2.59	1.6
5	Leu	-	0.32	5.58	17.4
		+SM[c]	1.29	1.51	1.2

[a] Background of [γ^{-32}P]GTP cleavage (in the mixture without ribosomes) was usually not higher than 0.5 pmol GTP/min; values in this column are given without its subtraction.
[b] To eliminate the contribution from EF-G-catalyzed cleavage of free [γ^{-32}P]GTP present in the incubation mixture, excess unlabeled GTP (400 μM) was added to all incubation assays, except in this special case, where 50 μM [γ^{-32}P]GTP was added instead of the unlabeled GTP (see text).
[c] Streptomycin (SM) at 5×10^{-6} M was present.

ribosome is a result of at least two successive discrimination steps in aminoacyl-tRNA binding: the initial codon-anticodon interaction of the [aminoacyl-tRNA-EF-Tu-GTP] complex with the template on the ribosome (before GTP-hydrolysis), and the retention/rejection of the codon-bound aminoacyl-tRNA on the ribosome after GTP hydrolysis (the correction or

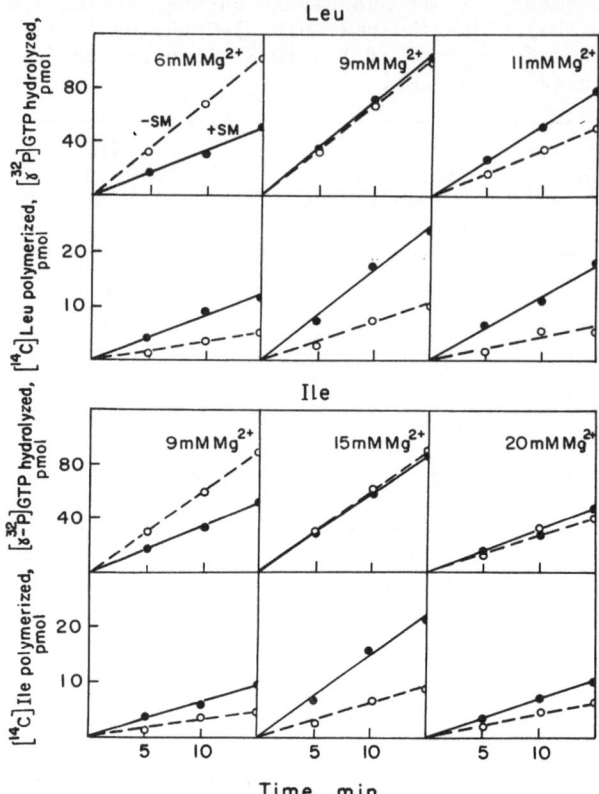

Fig. 3. Kinetics of peptide elongation and GTP hydrolysis during misreading in the [oligo(Phe)–tRNA·ribosome·poly(U)–Sepharose] system at 37°C. Miselongation for poly(Leu) at 6, 9 and 11 mM MgCl$_2$; for poly(Ile) at 9, 15 and 20 mM MgCl$_2$. Without streptomycin (o–o); with 5 x 10^{-6} M streptomycin (●–●). Experimental series with leu or Ile were done simultaneously using the same sample of the [oligo(Phe)–tRNA·ribosome·poly(U)–Sepharose] complex.

proof-reading step). There is some data that streptomycin decreases the GTP/peptide bond ratio during misreading (Yates, 1979; Thompson et al., 1981; Ruusala and Kurland, 1984). Hence the correction step of aminoacyl-tRNA binding has been suggested to be inhibited by streptomycin. In this case, however, the difficulty of direct measurement of peptide bound-coupled GTP hydrolysis against a background of uncoupled EF-Tu and EF-G-catalyzed GTP hydrolysis on the non-translating ribosomes was unavoidable, and the authors used an indirect method (Ruusala et al., 1982; Andersson and Kurland, 1983; Ruusala et al., 1984).

Here we present the results of the study of the streptomycin effect on GTP stoichiometry in the system where all ribosomes were active and engaged in peptide synthesis during codon-specific elongation or misreading (see Figure 3 and Table 2). It is seen that in the case of codon-specific elongation the presence of 5 x 10^{-6} M streptomycin had little effect either on the elongation or on the GTP hydrolysis rates. Just as under normal conditions (in the absence of streptomycin) the GTP/Phe ratio was about 2 in the presence of streptomycin.

In the cases of misreading, however, the effect of streptomycin was obvious. Figure 3 and the data in Table 2 show that the rates of elongation of both poly(leu) and poly(Ile) on poly(U) were stimulated 2-4 fold, an increase in the synthesis being observed in the presence of streptomycin at all Mg^{2+} concentrations. At the same time, the rate of GTP hydrolysis was not affected significantly by streptomycin at the Mg^{2+} optima for elongation (9 and 15 mM for Leu and Ile, respectively) and at higher Mg^{2+} concentrations, or it was even decreased at Mg^{2+} concentrations below the optimum (see Figure 3). Hence, the augmentation of misreading in the presence of streptomycin was not due to a correspondingly increased consumption of non-cognate ternary [Aa-tRNA·EF-Tu·GTP] complexes, but to an increased economy of their use, i.e. a higher efficiency of the final binding and incorporation of non-cognate aminoacyl-tRNAs. As has been shown above, during misreading in the absence of streptomycin the number of GTP molecules cleaved per peptide bond was about 20 below the Mg^{2+} optimum for elongation (the Mg^{2+} points for the highest proof-reading), about 10 at the Mg^{2+} optimum, and about 6 above the Mg^{2+} optimum (see Tables 2 and 3); this means that only one of the respective 20, 10 or 6 non-cognate aminoacyl-tRNA molecules entering the ribosome as ternary complexes (with GTP) is incorporated into the peptide bond, whereas the rest are <u>rejected</u> after GTP cleavage. In the presence of streptomycin the level of rejection is strongly reduced; as a result of incorporation of non-cognate amino acid residues increases, and the number of the GTP cleaved per peptide bond reaches 2-4, thus approaching that of about 2 in the case of codon-specific elongation. Naturally the maximum effect of streptomycin is exerted under conditions where the EF-Tu-dependent proof-reading is manifested the most, i.e. at an $Mg2+$ point lower than the Mg^{2+} optima (see Figure 1, Tables 2 and 3).

At present work is proceeding on the effect of other antibiotics (inhibitors of peptide-bond formation and translocation step of translation) on the stoichiometry of GTP hydrolysis during peptide elongation.

REFERENCES

Baranov, V.I., Belitsina, N.V., Spirin, A.S. 1979, <u>Methods Enzymol</u>, 59, 382-397.
Belitsina, N.V., Girshovich, A.S., Spirin, A.S. 1973, <u>Dokl. Akad. Nauk SSSR</u>, 210, 224-227.
Caplan, A.B., Menninger, J.R. 1984, <u>Mol. Gen. Genet</u>. 194, 534-538.
Davies, J., Gilbert, W., Gorini, L. 1964, <u>Proc. Natl. Acad. Sci</u>. USA, 51, 883-890.
Gavrilova, L.P., Kakhniashvili, D.G., Smailov, S.K. 1984, <u>FEBS Lett</u>. 178, 283-287.
Gavrilova, L.P., Kostiashkina, O.E., Koteliansky, V.E., Rutkevitch, N.M., Spirin, A.S. 1976, <u>J.Mol. Biol</u>., 101, 537-552.
Gavrilova, L.P., Perminova, I.N. (1981) Proc. 4th Int. Symp. Metabolism and Enzymology of Nucleic Acids, Smolenice, June 8-11, Publ. House of the Slovak Acad. Sci., Bratislava, 1982.
Gavrilova, L.P., Perminova, I.N., Spirin, A.S. 1981, <u>J. Mol. Biol</u>., 149, 69-78.
Gavrilova, L.P., Rutkevitch, N.M. 1980, <u>FEBS Lett</u>., 120, 135-140.
Hopfield, J.J. 1974, <u>Proc. Natl. Acad. Sci</u>. USA, 71, 4135-4139.
Kakhniashvili, D.G., Smailov, S.K., Gavrilova, L.P. 1986, <u>FEBS Lett</u>, 196, 103-107.
Kakhniashvili, D.G., Smailov, S., Gogia, I.N., Gavrilova, L.P. 1983, <u>Biokhimiya</u>, 48, 959-969.
Menninger, J.R. 1976, <u>J. Biol. Chem</u>. 251, 3392-3398.
Menninger, J.R. 1978, <u>J. Biol. Chem</u>. 253, 6808-6813.

Ruusala, T., Anderson, D.I., Ehrenberg, M., Kurland, C.G. 1984, EMBO J., 3, 2575-2580.

Ruusala, T., Ehrenberg, M., Kurland, C.G. 1982, EMBO J. 1, 741-745.

Ruusala, T., Kurland, C.G. 1984, Mol. Gen. Genet. 198, 100-104.

Smailov, S.K., Gavrilova, L.P. 1985, FEBS Lett. 192, 165-169.

Smailov, S.K., Kakhniashvili, D.G., Gavrilova, L.P. 1984, Biokhimiya, 49, 1868-1873.

Thompson, R.C. Dix, D.B., Gerson, R.B., Karim, A.M. 1981, J.Biol. Chem. 256, 6676-6681.

Ustav, M.B., Remme, Ya.L., Lind, A.Ya., Villems R.L.E., 1979, Bioorgan. Khimiya, 5, 365-369.

Yates, J.L. 1979, J. Biol. Chem. 254, 11550-11554.

SITE-DIRECTED MUTAGENESIS OF ELONGATION FACTOR Tu

A. Parmeggiani, E. Jacquet, M. Jensen**, P.H. Anborgh,
R.H. Cool, J. Jonák* and G.W.M. Swart**

Laboratoire de Biochimie, Laboratoire Associé du CNRS n°240
Ecole Polytechnique, F-91128 Palaiseau Cedex, France
*Institute of Molecular Genetics, Czechoslovak Academy of
Sciences, CS-166 37 Praha 6, Czechoslovakia

Elongation factor Tu (EF-Tu), a monomeric protein of 393 amino acid residues (M.W. 43,000), is the most abundant protein in E.coli and one of the best studied guanine nucleotide binding proteins, a family of enzymes involved in signal transduction in higher and lower organisms (for references see Bosch et al., 1984; Gilman, 1984; Parmeggiani and Swart, 1985; Bourne, 1986). These proteins bind GTP and GDP, are able to hydrolyze GTP and show typical homologies in their primary structures, especially in the N-terminal 150-200 amino acids. The finding that the ras p21 protein, a mutant variant of which is responsible for oncogenic transformation, is a guanine nucleotide binding protein (Scolnick et al., 1979) further emphasizes the importance of this family. EF-Tu is an essential component of protein biosynthesis, acting as the carrier of aa-tRNA to the ribosome (for references, see Miller & Weissbach, 1977). As with the other guanine nucleotide binding proteins, GTP induces the active form of the factor: only EF-Tu·GTP is capable of interacting with aa-tRNA, forming a ternary complex. The GTPase activity of EF-Tu, associated with the binding of the ternary complex to the ribosomal A-site, converts the factor into the inactive form EF-Tu·GDP, which dissociates from the ribosome. This allows the formation of a new peptide bond between aa-tRNA in the ribosomal A-site and peptidyl-tRNA in the P-site. EF-Tu is the specific target of the antibiotic kirromycin that keeps EF-Tu in the active form even in the presence of GDP, blocking the release of the factor from the ribosome (for references, see Parmeggiani & Swart, 1985), which inhibits the elongation of the polypeptide chain.

In E.coli EF-Tu is encoded by two almost identical genes, tufA and tufB, the two respective gene products (EF-TuA and EF-TuB) having an identical primary structure but for the C-terminus (for references, see Bosch et al., 1984). Crystallographic studies of the EF-Tu-GDP complex have revealed three domains, of which the N-terminal domain, in a model refined to 2.9Å, shows an alternation of six α-helices and six β-strands, a situation typical for the class of nucleotide binding proteins (Jurnak,

---

** Present address: Division of Biostructural Chemistry, Department of
Chemistry, Aarhus University, DK-8000 Aarhus C, Denmark (M.J.); EMBL,
Postfach 1022.09, Mayerhofstrasse, D-6900 Heidelberg, F.R.G. (G.W.M.S).

1985, La Cour et al., 1985). The GDP binding pocket is constituted by four loops connecting β-strands with α-helices.

The present knowledge of EF-Tu makes it a suitable model to investigate function-structure relationships (Parmeggiani et al., 1986). In this article we report the properties of EF-Tu mutants, obtained by site-directed mutagenesis of tufA (for details Swart, 1987; Parmeggiani et al., 1987).

STRATEGY FOR CONSTRUCTION AND OVERPRODUCTION OF EF-Tu MUTANTS

We have cloned a 2kb DNA fragment containing the structural tufA gene into the unique SmaI site of pEMBL9, a vector susceptible to be secreted as single stranded DNA upon phage F1 superinfection (Dente et al., 1983). Point substitutions or deletion of the tufA portion coding for the Middle and C-terminal domains (570 bp) from Glu-203 to the penultimate residue Leu-392 have been introduced by means of synthetic oligonucleotides, using the gapped duplex method (Kramer et al., 1982). Screening of the transformed colonies was performed by cell colony hybridisation. Mutations were confirmed by DNA sequencing. The mutated tufA was cloned into the EcoRI-HindIII polylinker site of the expression vector pCP40, in which it is under temperature-inducible λP_L control. (Remaut et al., 1983). Shift to 42°C activates P_L-expression and runaway replication. The host cells are either the kirromycin-sensitive E.coli 71/18 (EFTuAs and EF-TuBs) or the kirromycin resistant E.coli PM 455 (EF-TuAr, see also below) and its recA variant PM1455. GDP binding assay is carried out on nitrocellulose filters and GTP hydrolysis measured as release of P_i. For further details see legends to figures and previous publications (Parmeggiani & Sander, 1981; Parmeggiani & Swart, 1985; Swart, 1987).

ISOLATION OF MUTANT FACTORS

Figure 1 illustrates the model of the N-terminal domain of EF-Tu derived from X-ray diffraction studies of crystals of mildly trypsinized EF-TuGDP complex (from La Cour et al., 1985, s'ightly modified). The tufA mutants obtained sofar and the presumed functional involvement of each mutation are resumed in Table 1.

Overexpression of the mutated gents in the runaway system yields variable results. The induction conditions must be adapted to each mutant in order to obtain optimum production of protein associated with optimum solubility. In our system the optimum induction temperature, 42°C in the original publication (Remaut et al., 1983), varies between 38 and 42°C and optimum induction time, 3 hrs in the original system, lies between 1 and 4 hrs. Under optimum conditions we have obtained for instance a solubility of our mutants varying from 5 to 80%. For comparison, the solubility of overproduced wild type EF-Tu (reaching 50% the total cell proteins) was found to be at least 90%.

For the isolation of the mutant factors we have applied the following two procedures (see also Parmeggiani et al., 1987):

A) In the case of point substitutions in the EF-Tu molecule the presence of chromosomic EF-Tu constitutes a major problem. To distinguish between chromosomic and plasmid-borne EF-Tu we have applied the "runaway expression" system in the host cell E.coli PM 455 or PM1455 (recA⁻), in which the only active tuf gene (tufA) confers kirromycin resistance while tufB is inactivated by Mu insertion (Van der Meide et al., 1983). The sonicated cell extract was centrifuged for 3 hrs at 100,000g and the super-

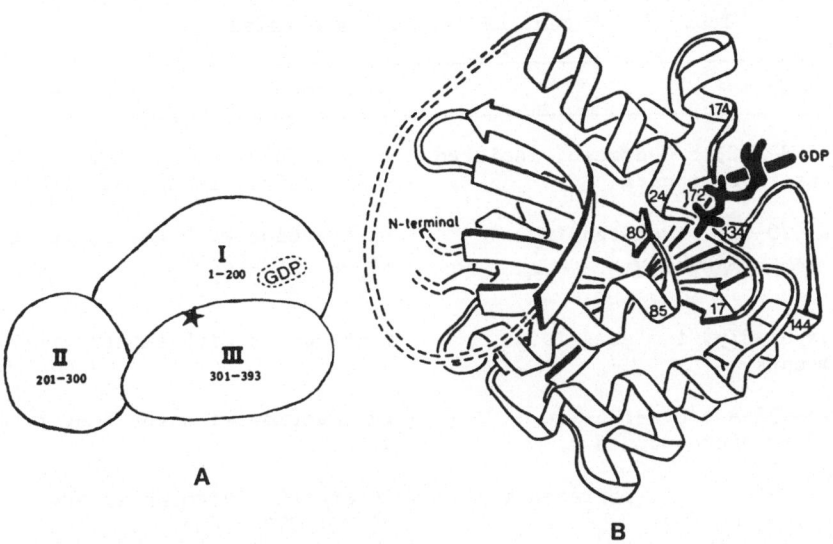

A

B

Fig. 1. Three-dimensional model of mildly trypsinized EF-Tu-GDP complex
derived from X-ray diffraction analysis (from La COur et al. 1985;
Parmeggiani et al. 1986, modified). A: Representation of the
overall shapes of the three EF-Tu domains with the GDP binding
side in I (N-terminal domain). II and III represent the Middle
and C-terminal domains, respectively. The numbers indicate the
polypeptide chain residues participating in the formation of the
respective domains. The star indicates position 375, responsible
for kirromycin resistance, situated near the cleft between domain
I and II. B: Cartoon of the N-terminal domain at 2.9 Å
resolution. The numbers indicate the amino acid residues
delimiting the four loops involved in the formation of the GDP
binding pocket.

natant (S-100) fractionated on a DEAE-Sepharose column. To separate
kirromycin resistant (EF-TuAr) and kirromycin sensitive EF-Tu (EF-TuAs) we
have followed the method already described (Swart et al., 1984, and Swart
et al., 1987). This method is based on the different affinity of EF-TuAr
and EF-TuAs towards kirromycin, and the different behaviour of EF-TuAr and
EF-TuAs on DEAE-Sepharose or DEAE-Sephadex chromatography in the presence
or in the absence of EF-Ts. It permits a complete separation of the two
populations of EF-Tu factors present in the cell.

B) In the case of G-domain (the produce of the deletion eliminating
the Middle and C-terminal) and the derived mutants the "runaway expression"
has been applied either in the host cell E.coli 71/18 or in E.coli PM455.
After chromatography on DEAE-Sepharose (pharmacia) the active fractions
(40-50% pure) are concentrated and filtrated on Ultrogel AcA44(IBF). The
last purification step of the 80% pure G-domain is carried out with the
FPLC system (Pharmacia) using a MonoQ column. After this step, the
G-domain is 95-100% pure on SDS-PAGE (not shown).

PROPERTIES OF MUTATED EF-Tu FACTORS

In this section we describe the properties of A) the G-domain, the
first constructed EF-Tu mutant isolated in preparative amounts (Parmeggiani

Table 1. Mutant EF-Tu factors obtained so far.

EF-Tu mutants	Functions possibly involved
EF-Tu(Δ203-392) = G-domain, isolated in preparative amounts	guanine nucleotide binding GTP hydrolysis interaction with ribosomes
EF-TuVal-20-Gly, isolated in preparative amounts	interaction with the phosphoryl groups of GDP/GTP GTPase activity
EF-TuCys-81-Gly, isolated in microamounts	interaction with aminoacyl-tRNA?
EF-TuAsn-135-Asp/Lys-136-Ile, isolated in microamounts	interaction with the base ring of GDP/GTP
G-domain Val-20-Gly, isolated in preparative amounts	interaction with phosphoryl groups of GDP/GTP GTPase activity
G-domainPro 82-Thr, isolated in preparative amounts	interaction with phosphoryl groups of GDP/GTP GTPase activity
G-domainHis-84-Gly, isolated in preparative amounts	interaction with phosphoryl groups of GDP/GTP GTPase activity
G-domainAsp-138-Asn, isolated in microamounts, insoluble	interaction with the base ring of GDP/GTP

et al., 1986; Parmeggiani et al.,1987) and B) other EF-Tu mutants that are presently under characterization.

A) The G-domain of EF-Tu: Binding of Guanine Nucleotides.- The isolated G-domain (M.W. 21,000), the homologue of Ha-ras p21 (Halliday, 1983), is able to bind GDP/GTP confirming the evidence from X-ray diffraction analysis that the binding site for GDP is located in the N-terminal domain of EF-Tu. GDP and GTP are bound in a stoichiometry approaching the molar ratio 1:1, as found for EF-Tu. The K'_d values of the G-domain complexes with GDP and GTP are similar, in contrast to the situation found with EF-Tu, whose GDP complex shows a K'_d smaller than that of EF-TuGTP by two orders of magnitude. In this regard we like to mention that the K'_d-values of the GDP and GTP complex of the Ha-ras protein p21 are also similar, though they are smaller than in the case of the G-domain (Scolnick et al., 1979). The higher affinity of EF-Tu towards GDP is characterized by an extremely slow dissociation rate of GDP from the EF-TuGDP complex, whereas GTP is released from EF-TuGTP at a much faster rate (Fasano et al., 1979). With the G-domain the dissociation rates of the two complexes are again approximately similar, emphasizing the radical difference between its behaviour and that of the intact molecule. Table 2 resumes these properties and for comparison some of the properties of EF-Tu and p21.

The GTPase Activity.- The G-domain can hydrolyze GTP. The hydrolysis rate has a linear course (Figure 2), displaying in experiments lasting several hours the characteristics of a multiple turnover reaction (not

Table 2. Dissociation constants (A) and dissociation half time (B) of GTP and GDP complexes with G-domain relative to the corresponding complexes of EF-Tu and p21.

A

complexes	dissociation constants (K'_d), mM
G-domain GTP[a]	4-8
G-domain GDP[a]	2-3
EF-Tu GTP[b]	0.3-0.8
EF-Tu GDP[b]	0.001-0.007
p21 GTP[c]	0.02-0.05
p21 GDP[c]	0.02-0.05

B

complexes	half-time, sec
G-domain GTP[a]	118
G-domain GDP[a]	166
EF-Tu GTP[d]	117
EF-Tu GDP[d]	3,120

[a] from Parmeggiani et al. 1987; [b] from Arai et al. 1974, Miller and Weissbach 1977, and Fasano et al. 1978; [c] from Scolnick et al. 1979; [d] from Fasano et al. 1978.

shown). By contrast the hydrolysis of EF-TuGTP is essentially a single turnover reaction, since the very slow dissociation rate (Table 2) of the bound GDP inhibits the regeneration of the EF-TuGTP complex (Fasano et al., 1982) The turnover rate of the G-domain GTPase corresponds to 0.1 mmol·sec^{-1}·mol^{-1} G-domain, a value that is well within the range of the GTPase activity of ras p21 protein and other members of the guanine nucleotide binding proteins. The turnover rate of the G-domain GTPase is close to the initial velocity of the single turnover hydrolysis of the EF-Tu·GTP complex (0.06 mmol·sec^{-1}·mol^{-1} EF-Tu).

The EF-Tu Effectors and the G-domain Activities.- The GTPase activity of the G-domain is strongly enhanced by the ribosome. This is the only clear effect induced on the G-domain activities by the EF-Tu ligands tested (ribosomes, aa-tRNA, elongation factor Ts, EF-Ts, and kirromycin). Tight couples 70S ribosomes are particularly effective, increasing the G-domain GTPase activity up to 15 times (Figure 2). The other EF-Tu ligands can slightly affect the GTPase activity of the G-domain as well as its interaction with GDP and GTP but probably in an unspecific way, since most of their effects are not of the same kind as those observed with the intact molecule. EF-Ts and kirromycin are known to strongly affect both the exit and entry rate of the EF-Tu interaction with GDP and GTP while aa-tRNA strongly retards the dissociation of EF-Tu·GTP; moreover, kirromycin activates the GTPase activity of EF-Tu (Parmeggiani & Swart, 1985).

B) Properties of Other EF-Tu Mutants.- We like to summarize the properties of the EF-Tu mutants sofar isolated and presently under characterization.

1) The point substitutions introduced in the N-terminal domain of EF-Tu most probably concern residues that do now affect the kirromycin resistance. Since kirromycin sensitivity is dominant over kirromycin

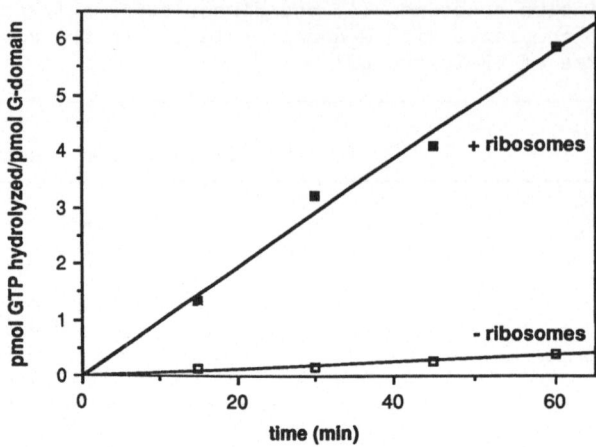

Fig. 2. GTPase of the G-domain in the absence (□) and presence of ribosomes (■). The reaction mixture contains, in a final volume of 180 µL, 50 mM imidazol-acetate, pH 7.6, 10 mM $MgCl_2$, 50 mM NH_4Cl, 1 mM dithiothreitol, 15.8 µM $[\gamma-^{32}P]GTP$ (specific activity 1175 cpm/pmol), 0.8 µM G-domain and 0.8 µM "tight couples" ribosomes. Incubation is carried out at 30°C. Aliquots of 40 µl are withdrawn at the given times and the GTPase activity determined as inorganic phosphate liberation as described (Parmeggiana and Sander, 1981). Blanks values (15% in the presence of ribosomes and 1-2% in their absence) have been subtracted.

resistance (Swart et al., 1984), disappearance of kirromycin resistance in the cell extract of the E.coli PM455 containing a plasmid with a mutated kirromycin sensitive EF-Tu allows to determine whether the mutant factor is still active in protein synthesis. All the point substitution mutants of the whole molecule sofar tested, EF-TuVal-20→Gly, EF-TuCys-81→GLy and EF-TuAsn-135→Asp;Lys-136→Ile appear to be able to sustain poly(U)-directed poly(Phenylalanine) synthesis at least to some extent.

2) Substitution in EF-Tu of Val20 by Gly enhances the GTPase activity. This is in agreement with the results obtained with the ras p21, where the wild-type protein (Gly in position 12) shows much higher GTPase activity than the oncogenic variant (Val in position 12).

3) Substitution in the G-domain of Val-20 by Gly destabilizes the G-domain·GDP complex, increasing the K'_d value by two orders of magnitude (from 2 to about 100µM).

4) Substitution of Pro-82 by Thr stabilizes the G-domain complexes with GDP and GTP by one order of magnitude (K'_d=0.3-0.4µM).

5) With mutation His-84→Gly, the K'_d of the G-domain complexes with GDP and GTP remains essentially unchanged.

6) Substitution of Pro-82 by Thr or His-84 by Gly inhibits the GTPase activity. This effect is particularly pronounced with the latter substitution inducing a >90% inhibition. This points to the important role of His-84 in GTP hydrolysis.

CONCLUSIONS

A number of common structural and functional features appear to relate the nucleotide binding domains of the different guanine nucleotide binding proteins (Halliday 1983; Lebermann & Egner 1984; Master et al., 1986). On this basis, a tentative model for the tertiary structure of the p21 ras protein has been recently proposed from the crystallographic model of EF-Tu·GDP (McCormick et al., 1985). Site-specific mutations of the G-domain will be therefore a useful tool for investigating the structure-function relationships not only of EF-Tu but also of other members of this class of proteins.

In this article we show that it is possible to isolate a functionally active, stable G-domain, which represents the N-terminal domain of EF-Tu as determined by X-ray diffraction. This domain, when isolated, still conserves several of the functions observed with the intact molecule of EF-Tu, suggesting that removal of the Middle- and C-domain of EF-Tu does not substantially effect the tertiary structure of the N-terminal domain. The isolation of the G-domain allows therefore the study of both the basic activity of EF-Tu and the functional role of the two other domains. The most striking difference between the intact molecule and the G-domain is the failure of the latter to be specifically and differentially affected by GDP and GTP. This indicates that the Middle and C-terminal domain play an essential role in the mechanism of the allosteric controls regulating the activities of EF-Tu.

Acknowledgements

The work reported in this article is supported by the grant BAP-0066-F from the Commission of the European Community. R.H.C. has been supported by a Fellowship of the Ecole Polytechnique; G.W.M.S. by a long-term EMBO Fellowship and J.J. by a short-term EMBO Fellowship.

REFERENCES

Bosch, L., Kraal B., Van der Meide, P.H., Duisterwinkel, F.J., Van Noort, 1983, Prog. Nucleic Acid Res. Mol. Biol., 20; 91-126.
Bourne, H.R. 1986, Nature 321; 814-816.
Dente, L., Cesareni, G., Cortese, R. 1983, Nucl. Acids Res. 11; 1645-1655.
Fasano, O., Bruns, W., Crechet, J.B., Sander, G., Parmeggiani, A., 1978, Eur. J. Biochem., 124; 53-58.
Fasano, O., De Vendittis, E., Parmeggiani, A., 1982, J. Biol. Chem., 257; 3145-3150.
Gilman, A.G. 1984, Cell, 36; 577-579.
Halliday, K.R., 1983, J. Cyclic Nucl. Phosph. Res. 9; 435-448.
Jurnak, F., 1985; Science 230; 32-36.
Kramer, W., Shughart, K., Fritz, H.-J. 1982; Nucl. Acids Res., 10; 6475-6485.
La Cour, T.F.M., Nyborg, J., Thirup, S., Clark, B.F.C., 1985, Embo J. 4; 2385-23886.
Lebermann, R., Egner, U. 1984, Embo J., 3; 339-341.
Master, S.B., Stroud, R.M., Bourne, H.R. 1986, Protein Engineering, 1; 47-54.
McCormick, F., Clark, B.F.C., La Cour, T.F.M., Kjeldgaard, M., Norskov-Lauritsen, L., Nyborg. 1985, Science, 230; 78-82.
Miller, D.L., Weissbach H. 1977, in: "Molecular Mechanisms in Protein Biosynthesis," Weissbach, H., Pestka, S., eds.,pp. 323-373, Academic Press, New York.
Parmeggiani, A., Sander, G. 1981, Molec. Cell Biochem., 35; 129-158.
Parmeggiani, A., Swart, G.W.M., 1985, Annu. Rev. Microbiol. 39; 557-577.

Parmeggiani, A., Anborgh, P.H., Canceill, D., Jacquet, J., Jonak, J.,
    Merola, M., Mortensen, K.K., Swart, G.W.M. 1986, in "Structure, Func-
    tion and Genetics of Ribosomes," Hardesty, B. & Kramer, G., eds., pp.
    672-685, Springer Verlag, New York.
Permeggiani, A., Swart, G.W.M., Mortensen, K.K., Jensen, M., Dente, L.,
    Cortese, R., 1987, Proc. Natl. Acad. Sci., in print, U.S.A.
Remaut, E., Tsao, H., Fiers W. 1983, Gene, 22; 103-113.
Scolnick, E.M., Papageorge A.G., Shih, T.Y. 1979, Proc. Natl. Acad. Sci.,
    U.S.A., 76; 5335-5339.
Swart, G.W.M., Merola, M., Guesnet, J., Parmeggiani, A. 1984, in "Metabo-
    lism and Enzymology of Nucleic Acids Including Gene Manipulations 5,
    Zelinka, J. & Balan,J., eds., pp. 277-288, Slovak Academy of Sciences,
    Bratislava.
Swart, G.W.M., Parmeggiani, A., Kraal, B., Bosch, L., 1987, Biochemistry.
    26; 2047-2054.
Swart, G.W.M., 1987, in "The Polypeptide Chain Elongation Factor Tu from
    E.coli. Characterization of Mutants and Protein Engineering," Ph.D.
    Thesis, Leiden, pp. 83-119.
Van der Meide, P.H., Vijgenboom, E., Talens, A., Bosch, L. 1983, Eur. J.
    Biochem.,130; 397-407.

A SITE-MUTATED rrsD GENE EXPRESSION CONSTRUCT:

IMPAIRMENT OF RIBOSOMAL FUNCTION

J. Sedláček, *J. Smrt, **J. Lakota,
I. Rychlík, M. Fábry and O. Melzochová

Institute of Molecular Genetics, 166 37 Prague
*Institute of Organic Chemistry and
Biochemistry, 166 10 Prague
**Cancer Research Institute
812 32 Bratislava, Czechoslovakia

Ribosomal RNA serves as the backbone in ribosomal assembly and also, at least to some degree, as an active component of the mature ribosome. The introduction of nucleotide changes (e.g., by site-directed mutations) clearly offers the most promising tool for elucidating the function of ribosomal RNA.

The complexity of the in vivo pathway from primary rRNA transcript to the functional ribosome (RNA processing, methylations and binding of 52 different ribosomal proteins in E. coli) does not allow to set up a system for in vitro biosynthesis of ribosomes. Reliance on the in vivo expression of cloned rRNA genes encounters a major obstacle: ribosomes containing plasmid coded rRNA can only form in host cells already containing background wild-type ribosomes (with their rRNAs encoded by chromosomal operons). Specific labeling of plasmid-coded transcripts has been described (Stark et al., 1982) and separation of the "new" ribosomal population from the background one has also been undertaken (J. Ofengand, personal communication). In an attempt to get a well-defined system facilitating separation of the two ribosomal populations, we designed and constructed a novel mutation in the 16 S rRNA, with this underlying rationale: the $GGm_2^6Am_2^6A$ - loop sequence in positions 1516 – 1519 of E. coli 16 S rRNA is known to give the 30 S ribosomal subunit special binding properties for some ligands (e.g., the antibiotic kasugamycin), but the presence or absence of dimethylations has not been found to bear on any vital function of the ribosome (Knippenberg et al., 1985). The prevention of dimethylations by substituting target adenosines was thus expected to provide a system which would in vivo express ribosomes altered enough for subsequent in vitro separation but still active as to their vital functions.

We show here that the expectations were not justified by the first chosen substitution, GGAA → CCTT, in the cloned rrnD operon: the expression of rRNA from the mutated plasmid substantially retarded the growth of host cells, thus signalizing impairment of ribosomal function.

The source of the wild-type rrnD-operon DNA was the plasmid pBK8 of Boros et al. (1979). The mutation was initially introduced into plasmid pNR225 (Sedláček et al., 1984) containing the 3' portion of the gene for

16 S RNA. The design of oligonucleotides for bridging the unique BstEII and BamHI sites of pNR225

```
            15-mer  ─────
            GTAACCGTAGGCCTT     10-mer
                          CCTGCGGTTG
            GCATCCGGAAGGACGCCAACCTAG
                      24-mer
```

avoided extensive self-complementarities and provided new restriction sites, StuI or BspRI (overlined), for easy identification of recombinants. Oligonucleotides were synthesized by a modified chlorophosphite method and purified electrophoretically using 20% polyacrylamide gel containing 7.5 M urea. The 10-mer and 15-mer were phosphorylated. The synthetic duplex was inserted by a two-step procedure: the 24-mer and the 15-mer annealed at 56°C (above the melting temperature of the self-complementarity of the 24-mer, 54°C) and the annealed mix was used in large excess for ligation, following addition of the 10-mer, yielded the circular construct. The site-mutated plasmids (pNR1248) were identified by GGCC (BspRI) spans within the EcoRI-BamHI fragment, which are different from those of pNR225.

The scheme of the attempt to build up the mutant rrnD operon including its P2 promoter is shown in Figure 1. To extend the mutant insert of pNR1248, several (wild-type) fragments of pBK8 were used. The final step of the scheme, insertion of the mutant DNA fragment comprising the 3' portion of the rrsD gene and the entire downstream sequence of the operon into the vector comprising the upstream sequence (P2 promoter and 5' portion of rrsD), produced no recombinants.

To confirm the "nonclonability" of the mutant gene comprising the endogenous promoter (i.e., to show that the negative result was not caused by technicalities), a comparative experiment was done (Figure 2). The mutant insert was extended somewhat further (pNR341) to allow extraction of a DNA fragment having identical ends (BglII) and also including the 3' portion of the rrsD gene and all the downstream sequences. The wild-type counterpart to this BglII fragment was extracted from plasmid pBK8; it was found to recombine with the pBK8-derived vector equally well in both orientations, without bias against the completion of the operon. In contrast, all the fourteen characterized recombinants of the mutant DNA fragment with the same vector had the insert in the "wrong" orientation and a correctly completed operon was not found.

The failure in completion of the mutant operon led us to yet other cloning schemes (Figure 3). The mutant operon was built up anew without any endogenous promoters, using the AluI (1/2 PvuII) site closely down-stream of P2 for the required deletion. This construct, pNR722, with no promoter in front of the structural part of the mutated rrnD operon gave viable transformants. In order to get regulated expression, the structural part of the mutated rrnD was put under the control of the P_L promoter in plasmid pNR2683 constructed with a vector derived from plasmid pPLa2411 (Remaut et al., 1981) and containing a polylinker of pUR222 (Rüther et al., 1981). In the chosen host strain, E. coli M5219 (Remaut et al., 1981) repression is provided by a thermosensitive lambda repressor encoded by the bacterial chromsome. pNR2683 gave transformants viable at permissive temperature. An analogous wild-type rrnD expression construct (pNR1191) was obtained from pNR2683 by substitution of a segment harboring the mutation; its identification was based on the absence of the extra StuI site which had been included in the design of the mutant oligonucleotides.

The results of the cloning, seen in an overview, might suggest a lethality of the mutation introduced. The failure of the first two schemes

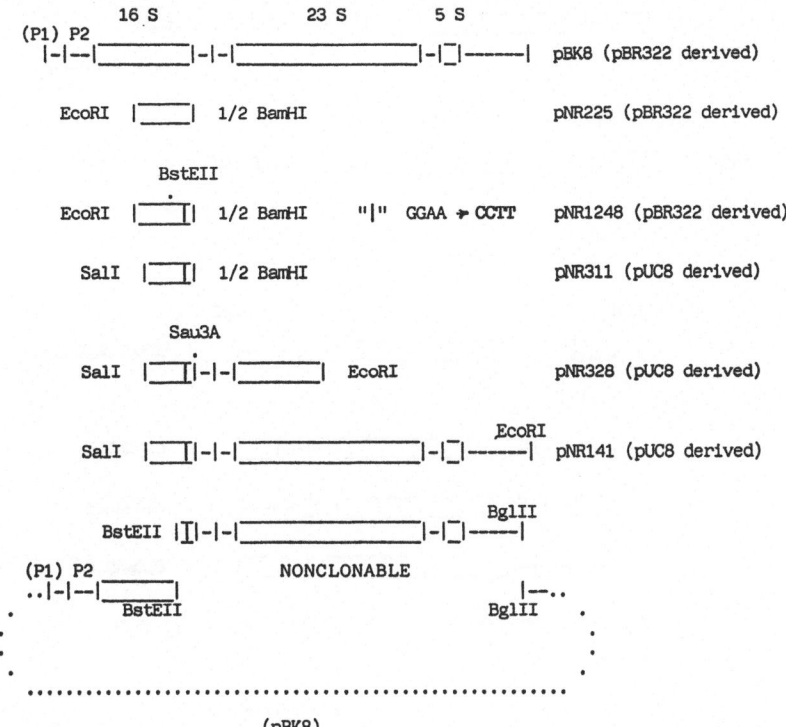

Fig. 1. Scheme of the attempt to build up the site-mutated rrnD operon including its endogenous P2 promoter. Inserts of the respective plasmids are shown; the designation of each plasmid as well as the origin of its vector part (in brackets) are given in the right-hand column. Structural parts of the rrsD and rrlD genes, or portions of them, are identified by their respective 16 S, 23 S and 5 S colinear products (top) and drawn as rectangles. The position of the mutation is marked by the bold vertical bar specified in the line of pNR1248. Brackets around P1 indicate its incompleteness. Restriction sites shown at thin vertical bars are limits of inserts and fragments; internal restriction sites (shown by overwritten points) are those involved in the cloning procedures.

of constructing a mutant version of pBK8 indicates an unescapable detrimental impact of a constitutive or endogenously controlled expression of the putative mutant construct. As to the P_L promoter- and lambda repressor-controlled expression system (pNR2683 or pNR1191 in M5219 cells), the above picture fits perfectly with observations of growth upon solid medium: at nonpermissive temperature (above 42°C), streaks of the pNR2683 clone were not found to form visible colonies, while the reference pNR1191 and pPLa2411 clones displayed confluent growth; at the permissive 28°C confluent growth was found with all three clones. Growth in liquid media indicated that the notion of the "lethality" was possibly an exaggeration, but confirmed beyond question a strong detrimental effect. A shift of cultivation temperature from 28°C to 42.5°C caused a different response. Cultures of the reference clones grew at a doubling time of 28 min. (pNR1191)

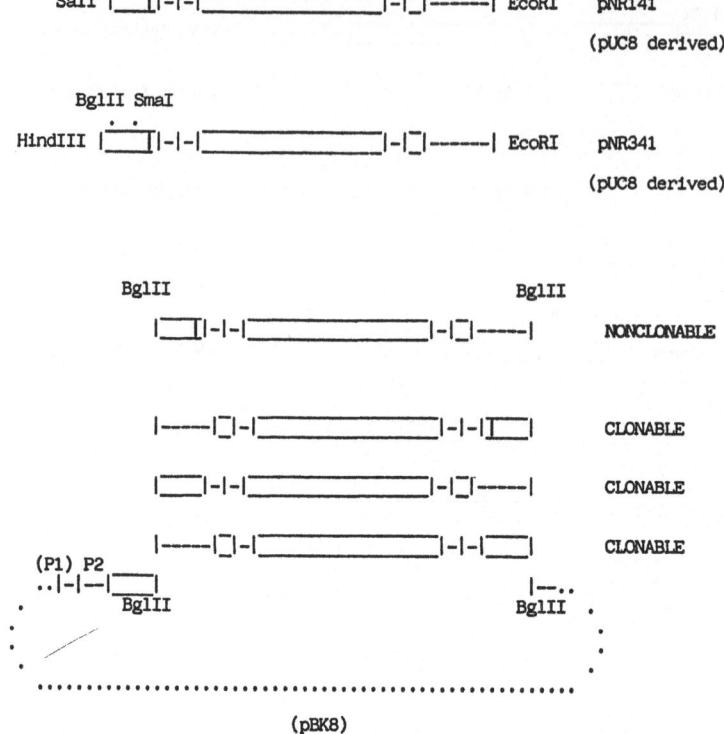

Fig. 2. Scheme of experiment confirming nonclonability of the site-
mutated rrnD operon with its endogenous P2 promoter. The
explanation of symbols is given in the legend to Figure 1.

or 30 min. (pPLa2411) during the exponential phase and their growth
levelled off at A_{550} close to 2.5; the pNR2683 culture had a doubling time
of 47 min. and its plateau was reached at 0.83 (curves not shown).

Knowledge of the time sequence of ribosomal assembly as well as
analogy with other expression systems for rrn operons (Gourse et al.,
1985) rather imply that the four adjacent point mutations close to the
3' end of 16 S RNA cannot conceivably prevent the formation of ribosomal
particles. The observed strong detrimental effects of the mutation have
thus to be connected with an impairment of ribosomal function. On the
other hand, we are aware of the need for in vitro analysis of the expression
product and understand that the in vivo determination of detriment allows
explanation by a variety of defects (ranging from formation of incomplete
particles to a block in some subtle step of ribosomal performance).

The main conclusion which can up to now be drawn from the results is
that dispensability of dimethylations of adenosines in positions 1518,
1519 does not automatically imply a tolerance for nucleotide substitutions.
The design of the mutation described otherwise tried to preserve the local
pattern of RNA folding (four looping purines were substituted by pyrimidines) but a harmful effect was found nevertheless. It is premature to
evaluate the apparent intolerance, but the strict evolutionary conservation
of the purine loop in question (Knippenberg et al., 1984) can be called to
mind in this respect.

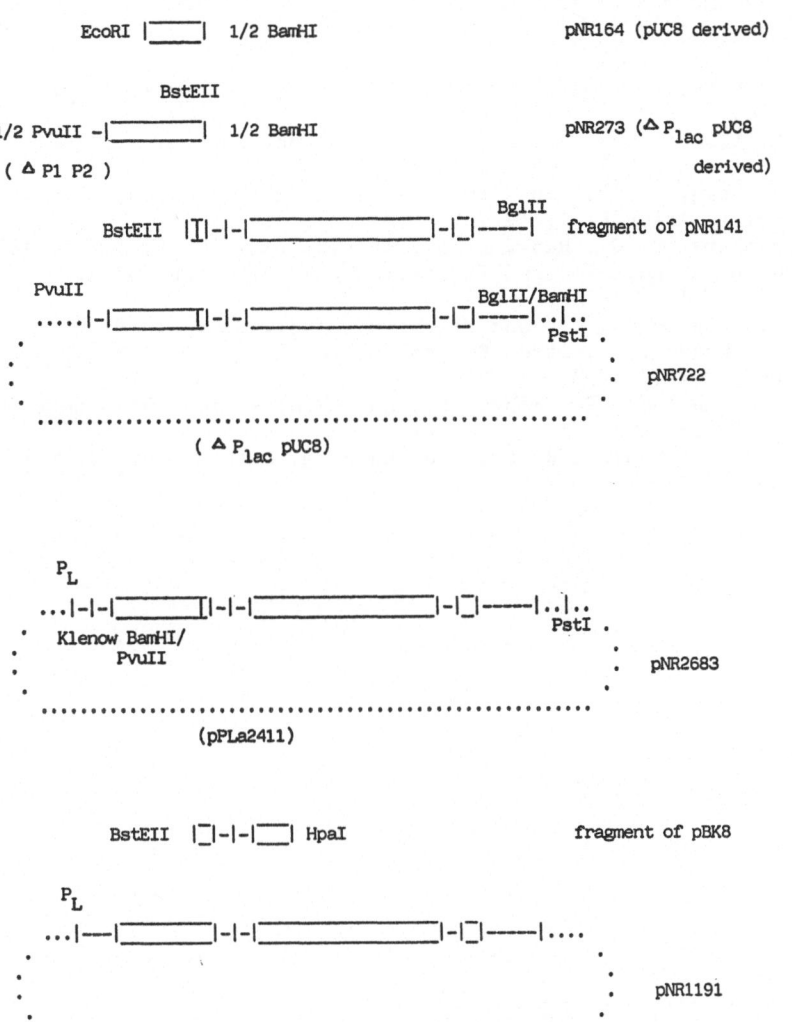

Fig. 3. Schemes of completing the site-mutated <u>rrnD</u> operon with
deletion of endogenous promoters and of the cloning for
expression. Explanation of symbols is given in the legend
to Figure 1.

The results themselves do not invalidate the idea of access to a
separable plasmid-coded population of ribosomes via nucleotide alterations,
but they make search for such changes as would provide also reasonably
active particles much more difficult.

We are grateful to Dr. Z. Hostomský for help with the purification
of oligonucleotides, to Dr. F. Rozkot for determining the melting temperature of the 24-mer oligonucleotide, to Dr. B. Sain for providing the
plasmid pBK8 and to Dr. Malti Adhin for plasmid pPLa2411 with polylinker.
To Prof. P. H. van Knippenberg and Dr. H. A. Heus we owe thanks for discussions and to Prof. L. Bosch for support of part of this work.

REFERENCES

Boros, I., Kiss, A., and Venetianer, P., 1979, Nucleic Acids Res., 9:1817–1830.

Gourse, R. L., Takebe, Y., Sharrock, R. A., and Nomura, M., 1985, Proc. Natl. Acad. Sci., USA, 82:1069–1073.

van Knippenberg, P. H., van Kimmenade, J. M. A., and Heus, H. A., 1984, Nucleic Acids Res., 12:2595–2604.

van Knippenberg, P. H., Heus, H. A., and van Buul, C. P. J. J., 1985, in: "Gene Manipulation and Expression", R. E. Glass and J. Spížek, eds., Croom Helm, London, pp 478–496.

Remaut, E., Stanssens, P., and Fiers, W., 1981, Gene, 15:81–93.

Rüther, U., Koenen, M., Otto, K., and Müller-Hill, B., 1981, Nucleic Acids Res., 9:4087–4098.

Sedláček, J., Holoda, E., Fábry, M., and Rychlík, I., 1984, this series 5, pp 327–332.

Stark, M. J. R., Gourse, R. L., and Dahlberg, A. E., 1982, J. Mol. Biol., 159:417–439.

EXPRESSION OF PROCHYMOSIN cDNA IN E.COLI

M. Fábry, P. Kašpar, S. Zadražil, F. Kaprálek
and J. Sedláček

Institute of Molecular Genetics
166 37 Prague, Czechoslovakia

Calf chymosin is an aspartyl protease whose specific activity in inducing the coagulation of milk plays an important role in cheese manufacture. Cloning and expression of calf prochymosin in Escherichia coli has been reported from several laboratories (e.g. Nishimori et al., 1982, Emtage et al., 1983, Liebscher et al., 1985, Sedláček et al., 1987).

Our first recombinant plasmid for the expression of calf prochymosin in E.coli, pMG424, had prochymosin cDNA inserted into the pUC9 vector and utilized the lac promoter. The 5' -end of prochymosin cDNA beginning with the BamHI site (which corresponds to the 5th amino acid of prochymosin) was fused with the lacZ segment and a portion of the pUC9 polylinker. In this fusion, the 5th amino acid of prochymosin is thus preceded by seven amino acids of the polylinker translate and five amino acids of β-galactosidase (see Figure 1). The host strain for the expression of pMG424-coded prochymosin was E.coli JM 105; induction was with IPTG. This new type of fusion proved satisfactory: the produce was activable in the usual way (acid activation), the pro- part apparently tolerating the structural change. The yield of some thousand molecules of enzymatically active chymosin per cell correlates well with the strength of the lac regulatory elements present in plasmid pMG424 (Sedláček et al., 1987).

In order to increase the expression to levels suitable for further experimental and production purposes, "stronger" transcriptional control signal was introduced. The weak lac promoter was replaced by the strong tac promoter while leaving the other "proven" structural elements of the plasmid constructs unchanged. The tac-promoter DNA segment originated from the pKK223-3 expression vector obtained from Pharmacia P-L Biochemicals. A usable restriction fragment that contained the tac promoter followed by the same sequence as the lac promoter in pMG424 was obtained via several cloning steps. Briefly, plasmid pKK223-3 had the unique EcoRI site converted into PvuII by cutting with EcoRI, removing the overhand by S1 nuclease treatment, and ligating with a stuffer fragment that brought in the 3' half of the PvuII recognition sequence. Between this newly created PvuII site and an Eco RI site (in the stuffer fragment), a 0.5 kb AluI-(HindIII)-EcoRI fragment (representing the N-terminal part of the fusion prochymosin) was inserted. The resulting intermediary construct already contains the required fragment that spans the BamHI site upstream of the tac promoter to the BamHI site of the polylinker, the sequence 3' to the promoter being identical to that in pMG424. To replace the lac promoter region of the pMG424, the above BamHI fragment was blunt ended with Klenow polymerase, cut at the HindIII site, gel purified, and inserted into the vector molecule obtained by digesting pMG424 with PvuII (partially) and HindIII. The final construct, pMG225, is thus a <u>tac</u>-promoter version of the pUC9-derived plasmid pMG424 (Figure 1). Both the cloning and the expression experiments with pMG225 were done in <u>E.coli</u> strain MT, a hybrid obtained by mating the HB101 recipient strain with the donor of the F' factor carrying the <u>lac</u>I^Q repressor gene. Inside the induced cells harboring pMG225, inclusion bodies of the prochymosin product (Schoemaker <u>et al</u>., 1985) were visible on phase-contrast micrographs.

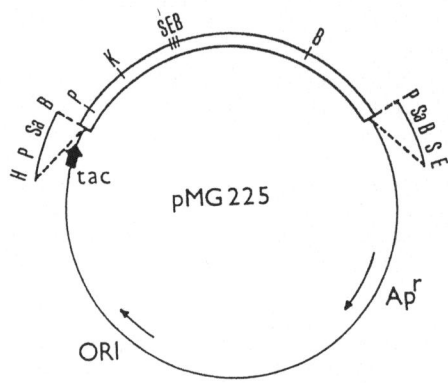

```
                     met thr met ile thr pro ser leu ala ala gly arg arg ile pro ..
tac promoter ..
     ..CAGGAAACAGCT ATG ACC ATG ATT ACG CCA AGC TTG GCT GCA GGT CGA CGG ATC CCT ..
        SD    AluI                         HindIII  PstI         BamHI
                                                        SalI
              β-galactosidase |           polylinker        |5th amino acid
                                                            |of prochymosin
```

Fig. 1. Recombinant plasmid pMG225 for the expression of calf prochymosin in <u>E.coli</u>. The fusion of prochymosin to polylinker and β-galactosidase is below the plasmid map. Plasmid pMG225 is identical with pMG424 except for <u>tac</u> substituting <u>lac</u>. For details see text. The symbols of restriction sites are: B, BamHI; E, EcoRI; H, HindIII; K, KpnI; P, PstI; S, SmaI; Sa, SalI.

Processing 0.2 g portions of wet cell mass using the procedure of Marston <u>et al</u>. (1984) adapted (scaled down, omitting ionex batch step, and applying a stepwise salt gradient) gave yields of about 1.2 mg of activable, i.e. renatured, prochymosin product per gram of wet cells. The estimate was based on measuring the enzymatic activity of the activated product in the adapted kinetic turbidimetric milk-clotting assay of McPhie (1974).

To further increase the yield we systematically varied the cultivation conditions so as to obtain high growth of biomass simultaneously with a high content of prochymosin. This is an inherently contradictory demand and optimization is necessary. Typically, cultivations were done in 8

Table 1. Culture parameters in a typical laboratory fermentation run.

Cultivation time (hours)	3.5	11	23
Biomass dry weight concentration, x (mg/ml)	0.75	3.51	4.68
Biomass wet weight concentration, $\bar{x}$ (mg/ml)	2.7	12.5	16.6
Biomass volume concentration, V_b (μm^3/ml)	23×10^8	108×10^8	143×10^8
Concentration of prochymosin inclusion bodies, V_i (μm^3/ml)	0.3×10^8	25.2×10^8	40.7×10^8
Ratio V_i/V_b (%)	1.3	23.3	28.5
Chymosin enzymic activity A (mg enzyme per g wet biomass)	0.05	0.82	1.00
Concentration of extractable active enzyme in culture (mg/l)	0.1	10.3	16.6

liters of a complex medium in a Magnaferm MA114 laboratory fermentor (New Brunswick, USA) agitated and aerated with air at 37°C. We measured the dry-weight biomass concentration, x (mg/ml), and determined under the microscope: the cell concentration, n (number of cells/ml), length and width of the average cell (μm), from which the volume v_b (μm^3) of the average cell was calculated, the mean diameter of the prochymosin inclusion body (μm), from which the mean volume of the inclusion body, v_i (μm^3), was calculated, and the ratio n_i/n, i.e. the relative number of cells containing an inclusion body. From these data the volume concentration of biomass, $V_b = n.v_b$ (μm^3/ml of culture), and the volume concentration of prochymosin inclusion bodies, $v_i = n.v_i.n_i/n$ (μm^3/ml of culture), were calculated. The parameter V_i indicates the absolute content of prochymosin in unit volume of culture and the ratio V_i/V_b expresses the relative amount of prochymosin in biomass.

For the first 3 hours the bacteria grew exponentially and no inclusion bodies were observable. Then slower, linear growth started and synthesis of prochymosin continued during the stationary phase and reached a maximum after 20 hours of cultivation. During fermentation, samples of biomass were withdrawn and chymosin enzymatic activity A (mg enzymatically active chymosin per g wet weight of biomass) was determined using the above-mentioned adapted isolation and renaturation procedure of Marston et al. (1984) and the turbidimetric milk-clotting assay of McPhie (1974). The amount of enzymatically active chymosin correlated with the content of prochymosin inclusion bodies in the biomass.

The results of a typical laboratory fermentation experiment are summarized in Table 1. By the end of fermentation the concentration of extractable enzymatically active chymosin in the culture reached approximately 17 mg/liter. In other words, approximately 30% of the volume as well as of the dry and wet weight of biomass was prochymosin. Prochymosin in the form of inclusion bodies represented 50% of the total protein of biomass. However, only 1.2% of the prochymosin protein was convertible to enzymatically active chymosin by the standard procedure.

The optimalized system for the overproduction and cytoplasmic accumulation of the prochymosin product just described is comparable to if not better than, the systems based on alternative control elements developed by others (Emtage et al., 1983, Nishimori et al., 1984). Its potential is best shown by our ability to produce and isolate sufficient amounts of active enzyme to have made experimental batches of cheese with recombinant chymosin.

The system is also well suited for the preparation of site-directed chymosin mutants in amounts and purity required for kinetic measurements and X-ray crystallography. The first mutant of our mutational program carried the 111 Val → Phe substitution in the P2 subsite of calf chymosin. The mutation site maps between the SmaI and EcoRI restriction sites of the coding sequence. To introduce the mutation, a duplex of 22 + 26 nucleotides with a single nucleotide change from the natural sequence (Harris et al., 1982) was used to bridge the SmaI and EcoRI cloning sites. In the duplexes, two additional nucleotide changes were included, both producing synonymous codons and at the same time creating a new RsaI site, which facilitated simple and straightforward positive identification of the mutant plasmids. The impact of the codon substitutions on the expression level was not significant: the wild-type pMG225 clone and its mutated pMG13195 counterpart were found to accumulate their respective products at the same rates.

Acknowledgements

The experimental batches of cheese, using recombinant chymosin, were prepared by Drs. R. Špaček, K. Zelený et al. at JZD AK Slušovice.

REFERENCES

Emtage, J.S., Angal, S., Doel, M.T., Harris, T.J.R., Jenkins, B., Lilley, G. and Lowe, P.A. 1983, Proc. Natl. Acad. Sci. USA, 80; 3671-3675.
Harris, T.J.R., Lowe, P.A., Lyons, A., Thomas, P.G., Eaton, M.A.W., Millican, T.A., Patel, T.P., Bose, C.C., Carey, N.H. and Doel, M.T. 1982, Nucleic Acids Res. 10; 2177-2187.
Liebscher, D.-H., Lipoldová, M., Smrt, J., Schwertner, S., Wambutt, R., Speter, W., Pohlmann, Ch., Pfeifer, M., Hahn, V., Rapoport, T.A., Rosenthal, S. and Zadrazil, S. 1985, Folia biol. (Praha) 31; 81-92.
Marston, F.A.O., Lowe, P.A., Doel, M.T., Schoemaker, J.M., White, S. and Angal, S. 1984; Biotechnology 2; 800-804.
McPhie, P. 1976, Anal. Biochem 73; 258-261.
Nishimori, K., Kawaguchi, Y., Hidaka, M., Uozumi, T. and Beppu, T. 1982, Gene 19; 337-344.
Nishimori, K., Shimizu, N., Kawaguchi, Y., Hidaka, M., Uozumi, T. and Beppu, T. 1984; Gene 29; 41-49.
Schoemaker, J.M., Brasnett, A.H. and Marston, F.A.O. 1985; EMBO Journal, 4; 775-780.
Sedláček, J., Fábry, M., Kašpar, P., Zadražil, S. and Liebscher, D.-H. 1987, Folia biol. (Praha) 33; 145-153.

CONTROL OF THE CELL CYCLE START BY PROTEIN

KINASE GENES IN SACCHAROMYCES CEREVISIAE

H. Küntzel, J. Lisziewicz, A. Godány*, E. Hostinová*
H. H. Förster, M. Trauzold and H. Sternbach

Abteilung Chemie, Max-Planck-Institut für
experimentelle Medizin, D-3400 Göttingen, GFR

*Permanent address: Slovak Academy of Sciences
Institute of Molecular Biology
84251 Bratislava, Czechoslovakia

The cell division cycle of the budding yeast Saccharomyces cerevisiae is controlled during the G1 phase by a complex series of events designated as START. During START the yeast cell monitors its size and integrates nutritional and/or pheromonal signals in order to decide whether to stay in a resting state (G$_0$), to sporulate (in a diploid state), to conjugate or to start a new round of division. START is controlled by several cell division cycle (CDC) genes, and two groups of temperature-sensitive cdc mutants arrest upon shift to restrictive temperature (37°) either prior to the pheromone-sensitive step (group A, including cdc25 and cdc35) or after this step (group B, including cdc28) as stationary-like, round, unbudded and uninuclear cells (Pringle and Hartwell, 1981). Group A START mutants are unable to grow at 37°, do not form a spindle pole body satellite, and are unable to conjugate, and at least two mutants of this group (cdc25 and cdc35) can enter the meiotic pathway in a fully complemented medium, if in a diploid state, whereas start of mitosis is blocked (Shilo et al., 1978).

The CDC35 gene codes for adenylate cyclase (Boutelet et al., 1985; Casperson et al., 1985), whereas the CDC25 gene product is involved in the positive control of adenylate cyclase (Camonis et al., 1986) by acting upstream the GTP-binding RAS2 protein (Robinson et al., 1987).

We have transformed a temperature-sensitive, uracil-requiring cdc25, ura3 mutant (ts⁻, ura⁻) with a yeast wild type genomic library in the shuttle vector pFL1, and isolated from ts⁺, ura⁺ transformants two groups of plasmids carrying two different genomic regions. By targeted integration of these cloned genes into the yeast genome, we could demonstrate that one plasmid contains the wild type allele CDC25, whereas the other plasmid acts as an extragenic suppressor of the cdc25 mutation. Both genes were sequenced, and the suppressor gene could be identified as a protein kinase closely related to the beef heart catalytic subunit (C$_A$) of cyclic AMP-dependent protein kinase (Lisziewicz et al., 1987). The cdc25-suppressing protein kinase gene is targeted into a chromosomal site unlinked to the CYR2 gene reported to encode C$_A$ (Uno et al., 1984), and the suppressor kinase is functionally unrelated to the CYR2 product, since a temperature-sensitive cyr2 mutant is not complemented by the suppressor plasmid (Lisziewicz et al., 1987).

In order to functionally characterize the cdc25 suppressor kinase we have developed an assay system to study the endogenous protein phosphorylation activity of cellular lysates, by using the $[^{35}S]$-labeled phosphorothioate analogue ATPγS (Eckstein, 1985). This substance is a good substrate for purified yeast cAMP-dependent protein kinase (Sternbach and Küntzel, manuscript in preparation), and the thiophosphoproteins produced by this enzyme (which phosphorylates serine and threonine residues) are resistant against yeast phosphatases, since the incorporation of thiophosphate into crude extract protein proceeds for at least two hours, and the labeled products are completely stable.

The thiophosphorylation assay demonstrated that crude extracts of cdc25 strains containing the plasmid-amplified protein kinase gene exhibit a six-fold increased endogenous protein kinase activity, as compared to extracts from plasmid-less mutants and allelic revertants. The plasmid-dependent protein kinase activity could be suppressed by added cAMP-free regulatory subunit isolated from purified yeast cAMP-dependent protein kinase, and restored in the presence of 10^{-5} M adenosine-3', 5'-cyclic phosphorothioate (Sp stereoisomer), a phosphodiesterase-resistant analogue of cAMP (Eckstein, 1985).

We conclude from these observations that the cdc25 suppressor gene encodes the catalytic subunit of cAMP-dependent protein kinase (Lisziewicz, Agoston and Küntzel, manuscript submitted), implying that the CYR2 gene cannot encode this enzyme, as previously stated (Uno et al., 1984). Recent sequence data indicate that the suppressor gene is identical to one of three related genes coding for C_A (T. Toda, personal communication).

Two of these C_A genes (termed TPK1, TPK2 and TPK3) can be replaced by two cloned gene variants interrupted by the marker genes HIS3 and URA3, respectively, without affecting viability, whereas the third gene copy appears to be essential for growth and division. Furthermore, deletion experiments with the suppressor plasmid containing the C_A gene TPK1 indicate that the N-terminal 37 amino acids of the yeast C_A protein (which are missing in the beef heart enzyme) as well as the 45 C-terminal amino acids are not required for complementation of the cdc25 lesion, whereas a more drastic deletion of 110 C-terminal amino acids abolishes the complementing function of C_A (Hostinova and Küntzel, manuscript in preparation).

The Figure summarizes the functions of genes involved in the cAMP-mediated control of the yeast cell cycle START, and suggests how an amplified C_A gene can suppress the cdc25 lesion. The production of the effector molecule adenosine-3', 5'-cyclic monophosphate (cAMP) is controlled by three genes, CDC35 (coding for the catalytic subunit of adenylate cyclase, AC), RAS2 (coding for a GTP-binding AC-activating protein homologous to the mammalian oncogene ras protein, Toda et al., 1985) and CDC25 (coding for a 178 kDalton protein activating adenylate cyclase via RAS2, Camonis et al., 1986; Robinson et al., 1987).

The effector cAMP activates the holoenzyme $[C_A R_A]$ by binding to the regulatory subunit R_A (encoded by BCY1, Toda et al., 1987), thus releasing the free active C_A subunit. The arrest of the START mutant cdc25 at restrictive temperature is caused by a deficiency of C_A and its phosphoprotein products required to initiate nuclear division, but is triggered by a far upstream function: the mutated cdc25 protein cannot sufficiently activate adenylate cyclase (via RAS2) to produce the level of cAMP required to activate the holoenzyme. This C_A-deficiency is bypassed by the plasmid-amplified suppressor gene, resulting in an overexpression of free C_A. Interestingly, an increase of the C_A gene dosage from three to four copies by integrative duplication of TPK1 does not bypass the cdc25 lesion (Lisziewicz et al., 1987).

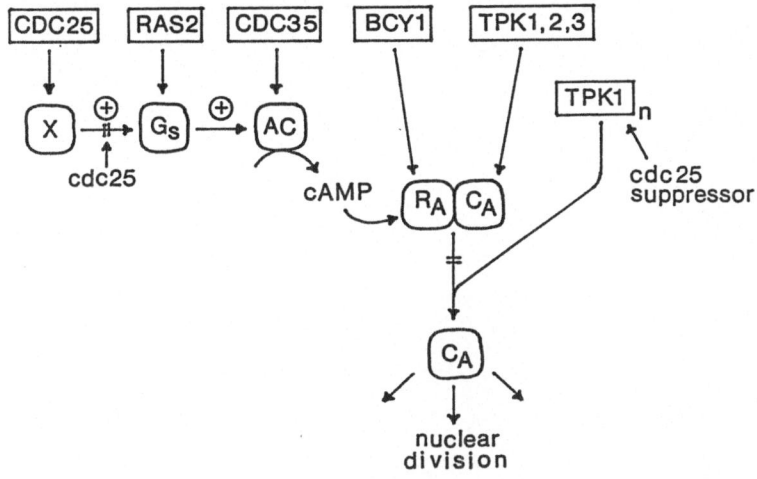

Figure 1

Thus our data directly demonstrate the central role of cAMP-dependent protein kinase activity for the control of START. This enzyme triggers a phosphorylation cascade involving other protein kinases, which amplifies nutritional signals transduced by the effector-producing START genes CDC25, RAS2 and CDC35. We are now studying amplifying protein phosphorylation cascades by purifying cAMP-independent protein kinases from yeast (Sternbach and Küntzel, 1987), and by isolating yeast protein kinase genes using synthetic DNA probes (Trauzold, Krengel and Küntzel, unpublished data).

REFERENCES

Boutelet, F., Petitjean, A., and Hilger, F., 1985, EMBO J., 4:2635-2641.
Camonis, J. H., Kalekine, M., Gondré, B., Garreau, H., Boy-marcotte, E., and Jacquet, M., 1986, EMBO J., 5:375-380.
Casperson, G. F., Walker, N., and Bourne, H. R., 1985, Proc. Natl. Acad. Sci., USA, 82:5060-5063.
Eckstein, F., 1985, Ann. Rev. Biochem., 54:367-402.
Lisziewicz, J., Godany, A., Förster, H. H., and Küntzel, H., 1987, J. Biol. Chem., 262:2549-2553.
Pringle, J., and Hartwell, L., 1981, in: "The Molecular Biology of the Yeast Saccharomyces: Life Cycle and Inheritance", J. Strathern, E. Jones and J. Broach, eds., Cold Spring Harbor, New York, pp 97-142.
Robinson, L. C., Gibbs, J. B., Marshall, M. S., Sigal, I. S., and Tatchell, K., 1987, Science, 235:1218-1221.
Shilo, V., Simchen, G., and Shilo, B., 1978, Exp. Cell Res., 112:241-248.
Sternbach, H., and Küntzel, H., 1987, Biochemistry, in press.
Toda, T., Cameron, S., Sass, P., Zoller, M., Scott, J. D., McMullen, B., Hurwitz, M., Krebs, E. G., and Wigler, M., 1987, Mol. Cell. Biol., 7:1371-1377.
Toda, T., Uno, I., Ishikawa, T., Powers, S., Kataoka, T., Broek, D., Cameron, S., Broach, J., Matsumoto, K., and Wigler, M., 1985, Cell, 40:27-36.
Uno, I., Matsumoto, K., Adachi, K., and Ishikawa, T., 1984, J. Biol. Chem., 259:12508-12513.

MOLECULAR ANALYSIS OF MITOCHONDRIAL DNA AND CLONING

OF ITS rDNA FRAGMENTS FROM TOMATO CELL SUSPENSION CULTURE

B. Hause, F. Baldauf*, K. Stock*
C. Wasternack and M. Metzlaff*

Division of Plant Biochemistry and
*Division of Genetics, Section of Biosciences
University of Halle, GDR

INTRODUCTION

In the last years investigation on mtDNA of higher plants has been intensified. In connection with problems of cytoplasmic male sterility, organelle inheritance and genetic recombination in the organellar genome after somatic hybridization there has been an increasing body of information about plant mtDNA.

The molecular weight of mtDNA ranges from approximately 200 kb in Brassica species (Chetrit et al., 1984; Palmer and Shields, 1984) to about 2.500 kb in Cucumis melo (Ward et al., 1981). In contrast to other mitochondrial genomes the restriction patterns of plant mtDNA are very complex. Circular and linear molecules have been observed in varying proportions and length distributions (Leaver and Gray, 1982) and small plasmid-like molecules have been found in different species.

Among the structural models of plant mtDNA proposed in recent years the idea of physically distinct circular molecules arising by recombination is that most favored by experimental data (Lonsdale et al., 1983a; Palmer and Shields, 1984; Stern and Palmer, 1986).

Further, it has been shown that in plant genomes analyzed large (26S) and small (18S) rRNA genes appear to be far apart and that 5S rRNA genes, undetectable so far in other mitochondrial systems, are closely linked with the 18S rRNA genes (Gray et al., 1982; Stern et al., 1982; Chao et al., 1983; Brennicke et al., 1985).

Among the limited number of species, in which the structure of mtDNA have been studied, no data exist on tomato cell suspension cultures. For this reason, we used a cell suspension culture of Lycopersicon esculentum cv. "Lukullus" to analyze its mtDNA.

MATERIALS AND METHODS

Plant Material and Isolation of mtDNA

The cell line Lycopersicon esculentum cv. "Lukullus" was originally established in 1976, and cultivation was performed as described by Tewes et al. (1984). The isolation of mtDNA followed the method described by Hause et al. (1986). Using this method the crude mitochondrial fraction was purified by two successive sucrose density gradient centrifugation steps and lysed with sodium-sarcosyl and pronase. This DNA preparation was used directly for restriction enzyme analysis. For molecular cloning the mtDNA was further purified by CsCl-density gradient centrifugation.

Isolation of Plastid (pt-) DNA

ptDNA was isolated from leaves of Lycopersicon esculentum cv. "Lukullus" following a method described by Atchison et al. (1976) modified by Metzlaff et al. (1981).

Restriction Analysis of mtDNA and Identification of the rDNA

The isolated mtDNA was digested by Bam HI (Pharmacia), Bgl I, Bgl II and Eco RI (Boehringer GmbH, Mannheim) under the conditions recommended by the suppliers. The separation of the restriction fragments, their transfer to nitrocellulose, the hybridization with ^{32}P-dATP nick-translated plasmids and the autoradiography were carried out following the standard methods (see Southern, 1975 and Maniatis et al., 1982). The plasmids used for the hybridization analyses consist of the vector pBR322 and 16S rDNA, 23S, 4.5S and 5S rDNA (Barbier, 1980) and the gene of the large subunit of ribulose-1,5-bisphosphate carboxylase (pSocB149/LS) (Zurawski et al., 1981), all from spinach chloroplasts.

Molecular Cloning of the rDNA

"Shot-gun-cloning" of the Bam HI-digested mtDNA using the vector pBR322 (Bolivar et al., 1977) was carried out as described by Maniatis et al. (1982). After selection of positive colonies of transformed E. coli HB101 by tetracycline sensitivity these were characterized by Bam HI restriction enzyme analysis and hybridization with a radioactively labeled insert from heterologous gene probes.

RESULTS AND DISCUSSION

We used cell fractionation and two subsequent sucrose density gradient centrifugations to isolate mtDNA from cell suspension cultures of Lycopersicon esculentum. In our hand a DNase treatment of isolated mitochondria often described in the literature (Kolodner and Tewari, 1972; Kemble et al., 1980; Hanson et al., 1986) was unsuccessful because of difficulties in removing the DNase. The prepared mtDNA was analyzed by digestion with restriction endonucleases (Figure 1A). The low background smear indicates that mtDNA does not seem to be contaminated with nuclear DNA. To exclude the possibility of contaminating DNA from proplastids we compared the Eco RI cleavage pattern of mtDNA with that of ptDNA from tomato leaves (Figure 2A). In both patterns bands of different and similar molecular weights exist. Therefore, SOUTHERN blots of comparative gels were hybridized with nick-translated genes of the rRNA of spinach plastids. For these cross-hybridization experiments we used the relatively high evolutionary conservation of sequences of the rDNA from various plant organelles (see Bohnert et al., 1980; Chao et al., 1983; Spencer et al., 1983 and 1984). As shown in the Figure 2B the fragments hybridizing with the 23S rRNA gene probe are only partially identical in the two patterns. Thus, contamination with ptDNA was excluded.

In the restriction patterns of the tomato mtDNA which are complex some bands were found in bimolar amounts as shown by the relative intensity of ethidium bromide fluorescence (Figure 1B). Taking into account these

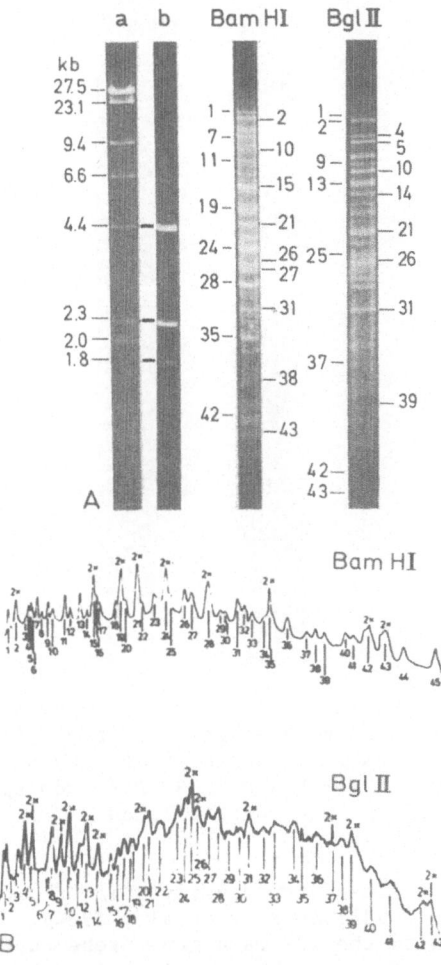

Fig. 1. Bam HI and Bgl II restriction patterns of mtDNA from tomato.
A: Restriction patterns in 1% agarose; (a) λ-DNA digested with
Hind III, (b) pBR322 linearized and digested with Bgl I. B:
Densitometric tracings of the photographic negatives of the gels.
The relative intensity of the bands is marked above the peaks.

fragment stoichiometries the total size of <u>Lycopersicon</u> mtDNA was calculated by addition of all fragments after co-electrophoresis with marker DNA
(Bgl I-fragments of pBR322 and Hind III-fragments of λ-DNA). The resulting
size of about 270 kb is in the same range of mtDNAs from other species.
The difference of this value to the determination of molecular weight of
the mitochondrial genome of whole <u>Lycopersicon</u> plants by McClean and Hanson
(1986) (300 kb without including the fragment stoichiometries) can be
explained: (i) by the relatively inexact method of determination of the
molecular weight and (ii) by the possibility that the mitochondrial genome
is changed by the long cultivation time of the suspension culture, which
was established in 1976. Similar results are shown and discussed by Negruk
et al. (1986) by the comparison of mtDNA restriction patterns of cell
suspension cultures (established in 1979) and whole plants of <u>Vicia faba</u>.

By hybridizing Bam HI-restricted mtDNA with rDNA probes of spinach
plastids the fragments containing mt-rDNA were identified (Figure 3, lanes
a and b). For both used plasmids one specific band of cross-hybridization
was detectable. The sequence homology for both large rRNAs between plas-

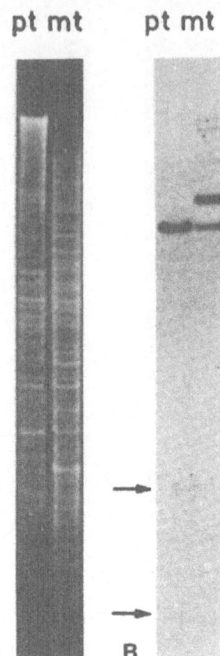

pt mt   pt mt

A   B

Fig. 2.   Comparison between ptDNA and mtDNA from tomato. A: Gel
electrophoresis of Eco RI-digests of ptDNA from leaves and mtDNA
from cell cultures in 1% agarose. B: SOUTHERN blots of A
hybridized with ^{32}P-labelled 23S rDNA from spinach plastids.
Arrows indicate the pt-specific bands.

tids and mitochondria was used to identify rDNA of organelles (Bohnert et
al., 1980; Chao et al., 1983; Spencer et al., 1984).   Therefore, the 5.7 kb
fragment hybridizing with the 23S rRNA gene probe contains the gene for the
26S rRNA and the 3.3 kb fragment hybridizing with the 16S rRNA gene probe
contains the gene for the 18S rRNA of the mitochondria.   Both mtDNA fragments were cloned in E. coli HB101 using the vector pBR322.   The resulting
recombinants pLBmt26 and pLBmt18 were further characterized by plasmid
electrophoresis and cross-hybridization with rRNA gene probes from spinach
plastids (results not shown).

Both recombinant plasmids pLBmt18 and pLBmt26 were rehybridized to
total Bam HI-digested mtDNA (Figure 3, lanes d and e).   In addition to
self-hybridization of the cloned fragments two hybridization bands were
visible in the case of pLBmt26 hybridization (Figure 3, lane e).   From this
result we conclude that a certain sequence in the 5.7 kb Bam HI-fragment is
homologous to some other sequence in the mitochondrial genome.   Such homologous sequences are the basis for the structural model of recombination.
For various angiosperms it has been shown that recombination repeats are
closely linked to the genes for subunit II of cytochrome c oxidase, and 26S
rRNA, respectively (Stern and Palmer, 1984a).   In addition Falconet et al.
(1984) and Stern and Palmer (1986) have shown that the 18S rRNA gene in
wheat mtDNA and the 26S rRNA gene in spinach mtDNA are located within a
recombination repeat.

In various plants homology between mtDNA and ptDNA has been described
(Stern and Palmer, 1984b).   In maize mtDNA a sequence was found to be
homologous to plastid genes for the tRNA^{ile}, tRNA^{val} and 16S rRNA (Stern
and Lonsdale, 1982) as well as for the ribulose-1,5-bisphosphate carboxylase large subunit (LS) (Lonsdale et al., 1983b).   We used pSocB149/LS, a

134

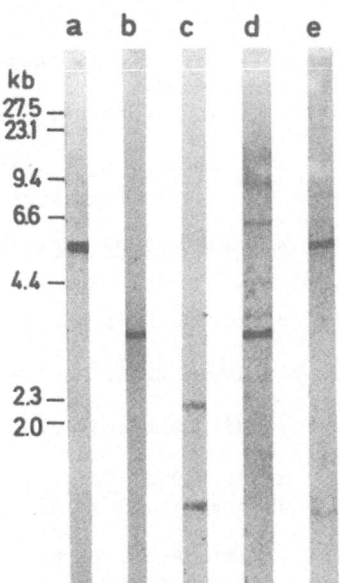

Fig. 3. Autoradiographs of SOUTHERN blots of Bam HI-digested mtDNA of
tomato hybridized with (a) 23S, 4.5S and 5S rDNA from spinach
plastids; (b) 16S rDNA from spinach plastids; (c) pSocB149/LS; (d)
pLBmt18; (e) pLBmt26.

plasmid containing a gene of the LS from spinach ptDNA to probe a Bam
HI-digest of Lycopersicon mtDNA. As shown in Figure 3, lane c, we found
cross-hybridization between mtDNA and pSocB149/LS. Two cross-hybridizing
fragments (2.3 kb and 1.3 kb) were visible.

These results support the hypothesis of recombination between DNA
molecules of different compartments in eukaryotic cells (Stern and
Lonsdale, 1982) and leads to the assumption that organelle DNA transfer is
a general phenomenon in plants and could explain the large size of mitochondrial genomes of higher plants.

REFERENCES

Atchison, B. A., Whitfeld, P. R., and Bottomley, W., 1976, Mol. Gen.
    Genet., 148:263-269.
Barbier, H. A., 1980, Dissertation, University of Science and Medicine,
    Grenoble.
Bohnert, H. J., Gordon, K. H. J., and Crouse, C. J., 1980, Mol. Gen.
    Genet., 179:539-545.
Bolivar, F., Rodriguez, R., Greene, P., Betlach, M., Heynecker, H., Boyer,
    H., Crosa, J., and Falkow, S., 1977, Gene, 2:95-113.
Brennicke, A., Möller, S., and Blanz, P. A., 1985, Mol. Gen. Genet.,
    198:404-410.
Chao, S., Sederoff, R. R., and Levings, C. S. III, 1983, Plant Physiol.,
    71:190-193.
Chetrit, P., Mathieu, D., Muller, J. P., and Vedel, F., 1984, Curr. Genet.,
    8:413-421.
Falconet, D., Lejeune, B., Quetier, F., and Gray, M. W., 1984, EMBO J.,
    3:297-302.
Gray, M. W., Bonen, L., Falconet, D., Huh, T. Y., Schnare, M. N., and
    Spencer, D. F., 1982, in: "Mitochondrial Genes", P. Slonimski, P.
    Borst and G. Attardi, eds., Cold Spring Harbor, NY., pp 483-488.

Hanson, M. R., Boeshore, M. L., McClean, P. E., O'Connell, M. A., and
    Nivison, H. T., 1986, in: "Methods in Enzymology", Vol. CXVIII,
    Academic Press, NY., pp 437-453.
Hause, B., Baldauf, F., Stock, K., Wasternack, C., and Metzlaff, M., 1986,
    Curr. Genet., 10:785-790.
Kemble, R. J., Gunn, R. E., and Flavell, R. B., 1980, Genetics, 95:451-458.
Kolodner, R., and Tewari, K. K., 1972, Proc. Natl. Acad. Sci., USA,
    69:1830-1834.
Leaver, C. J., and Gray, M. W., 1982, Annu. Rev. Plant Physiol.,
    33:373-402.
Lonsdale, D. M., Hodge, T. P., Fauron, C. M. R., and Flavell, R. B., 1983a,
    in: "Plant Molecular Biology", R. B. Goldberg, ed., Liss, NY., pp
    445-456.
Lonsdale, D. M., Hodge, T. P., Howe, L. J., and Stern, D. B., 1983b, Cell,
    34:1007-1014.
Maniatis, T., Fritsch, E. F., and Sambrock, J., 1982, "Molecular Cloning",
    Cold Spring Harbor, NY.
McClean, P. E., and Hanson, M. R., 1986, Genetics, 112:649-667.
Metzlaff, M., Börner, T., and Hagemann, R., 1981, Theor. Appl. Genet.,
    60:37-41.
Negruk, V. I., Eisner, G. I., Redichkina, T. D., Dumanskaya, N. N., Cherny,
    D. I., Alexandrov, A. A., Shemiakin, M. F., and Butenko, R. G.,
    1986, Theor. Appl. Genet., 72:541-547.
Palmer, J. D., and Shields, C. R., 1984, Nature (London), 307:437-440.
Southern, E. M., 1975, J. Mol. Biol., 98:503-518.
Spencer, D. F., Williamson, S., Schnare, M. N., Gray, M. W., Doolittle, W.
    F., and Bonen, L., 1983, Endocytobiol., II:871-879.
Spencer, D. F., Schnare, M. N., and Gray, M. W., 1984, Proc. Natl. Acad.
    Sci., USA, 81:493-497.
Stern, D. B., and Lonsdale, D. M., 1982, Nature, 299:698-702.
Stern, D. B., and Palmer, J. D., 1984a, Nucl. Acid. Res., 12:6141-6157.
Stern, D. B., and Palmer, J. D., 1984b, Proc. Natl. Acad. Sci., USA, 81:-
    1946-1950.
Stern, D. B., and Palmer, J. D., 1986, Nucl. Acid. Res., 14:5651-5666.
Stern, D. B., Dyer, T. A., and Lonsdale, D. M., 1982, Nucl. Acid. Res.,
    10:3333-3340.
Tewes, A., Glund, K., Walther, R., and Reinbothe, H., 1984, Z.
    Pflanzenphys., 113:141-150.
Ward, B. L., Anderson, R. S., and Bendich, A. J., 1981, Cell, 25:793-803.
Zurawski, G., Perrot, B., Bottomley, W., and Whitfeld, P. R., 1981, Nucl.
    Acid. Res., 9:3251-3270.

NUCLEOTIDE COMPARTMENTATION DURING SYNTHESIS OF CYTOPLASMIC rRNA
AND tRNA AS WELL AS CYTOPLASMIC rRNA AND MITOCHONDRIAL rRNA IN
TOMATO CELL SUSPENSION CULTURE

B. Hause and C. Wasternack

Division of Plant Biochemistry
Department of Biosciences, University of Halle
4020 Halle, Neuwerk 1, GDR

INTRODUCTION

The incorporation into RNA of ^{3}H—uridine (or ^{14}C—orotic acid) is
frequently used to estimate quantitatively absolute rate of RNA synthesis
or even of different RNA species through label amount in isolated RNAs or
different RNA-DNA-hybrids. All these procedures assume a common nucleotide
pool. But, with respect to the spatial pattern of pyrimidine metabolism as
well as different RNA species several aspects of compartmentation has to be
taken into account (Wasternack and Benndorf, 1983, Moyer and Henderson,
1985): Compartmentation (i) between de novo- and salvage-derived nucleotides (Genchev et al., 1980), (ii) by metabolic and storage pools of RNA
precursors (Wasternack 1983), (iii) by growth-phase dependent fluctuation
of UTP content (Piper 1981), (iv) by different UTP consumption in rRNA and
mRNA synthesis (Wiegers et al., 1976), (v) by species-dependent exchange
between nuclear and cytoplasmic located UTP (Khym et al., 1978), and (vi)
by different UTP pools used in mitochondrial (mt-) rRNA and cytoplasmic
(cyt-) rRNA and cytoplasmic (cyt-) rRNA synthesis (Soeiro and Ehrenfeld,
1973) or by different deoxyribonucleotide pools used in nuclear and mtDNA
synthesis (Bestwick and Mathews, 1982).

Among various approaches to determine function of UTP pools in the
synthesis of RNA species, comparison of spec. rad. act. of UTP and RNAs
under steady state labeling as well as constant precursor supply (Wiegers
et al., 1976) seems to be confinient by excluding rate-dependent effects.
Using Wiegers approach we question here wether or not the nucleoplas-
matically synthesized tRNA and the nucleolarly synthesized rRNA as well as
the cyt-rRNA and mt-rRNA are built up from a common UTP pool.

MATERIALS AND METHODS

Cells of tomato (Lycopersicon esculentum cv. "Lukullus") were cultivated in suspension and counted as described by Tewes et al. (1984). Cells
were fed under sterile conditions with [5-^{3}H]-uridine during growth indicated by arrows. At given times aliquotes were drawn to determine label
and spec. rad. act., respectively in medium, in cells (uptake), in uridine
and uridine nucleotides derived by different TLC methods after methanolic
extraction, and in different RNA species as described by Hause and
Wasternack (1987a). Isolation of 18S and 25S cyt-rRNA and tRNA by cell

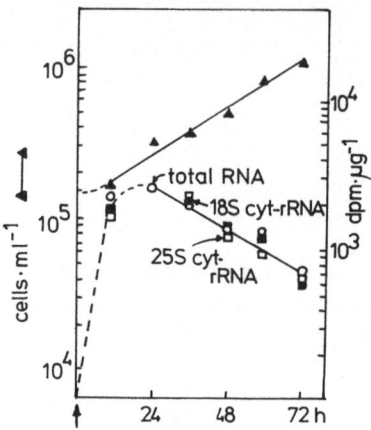

Fig. 1. Spec. rad. act. of 18S and 25S cyt-rRNA and cell number during
growth after one dossage ($\downarrow$) of 25 nmoles.ml^{-1} ^{3}H-uridine (spec.
rad. act.: 4.4 x 10^{7} dpm.nmole^{-1}.

fractionation followed by phenol procedure, rRNA precipitation with LiCL,
Paage and electroelution are described by Hause and Wasternack (1987a).
mt-rRNA-containing fractions were isolated by a 10.000xg-centrifugation,
sucrose density gradient centrifugation, sedimentation of a mt-containing
interphase followed by resuspension, lysis and phenol procedure as
described by Hause and Wasternack (1987b). Quantitation of mt-RNA was
performed by hybridization on NC-filter (Hause and Wasternack, 1987b)
containing known amounts of cloned mt-rDNA fragments of tomato (prepared by
Hause et al., 1986) coding for 18S and 26S mt-rRNA, respectively.

RESULTS AND DISCUSSION

In using Wiegers approach some necessary prerequisites are fulfilled for
tomato cells (Hause and Wasternack, 1987a): (i) cyt-RNAs and tRNA could be
separated intactly without cross-contamination. mt-rRNAs from a fraction
still containing cyt-rRNAs were quantified by hybridization with specific
mt-rDNA clones taking into account efficiency of hybridization. (ii) As
shown in Figure 1, after one dosage of ^{3}H-uridine cell doubling time and
rate of decline of spec. rad. act. of 18S and 26S cyt-rRNA differed by only
1.4 h suggesting minimal rRNA turnover. (iii) Steady state of spec. rad.
act. of RNAs and UTP was achieved between 30 and 54 h of growth by
repeated label addition. (iv) There was no label randomization by cleavage
of ^{3}H-uridine and reductive degradation of its pyrimidine moiety (Tintemann
et al., 1985).

Spec. rad. act. were compared to answer possible UTP compartmentation
during synthesis of cyt-rRNA and tRNA. As shown in Figure 2 A_1 cyt-rRNAs
and tRNA reached an equal spec. rad. act. of $3.2.10^{4}$ dpm per µg RNA under
low concentration of exogenous ^{3}H-uridine suggesting that no UTP compartmentation took place. By comparing the same under high concentration of
^{3}H-uridine the chance is given to exclude concentration-dependent differences in precursor consumption. As shown in Figure 2 B_1, cyt-rRNA and tRNA
reached a lower but again equal spec. rad. act., although the exogenously
given spec. rad. act. of ^{3}H-uridine was lower by 20-fold. In accordance

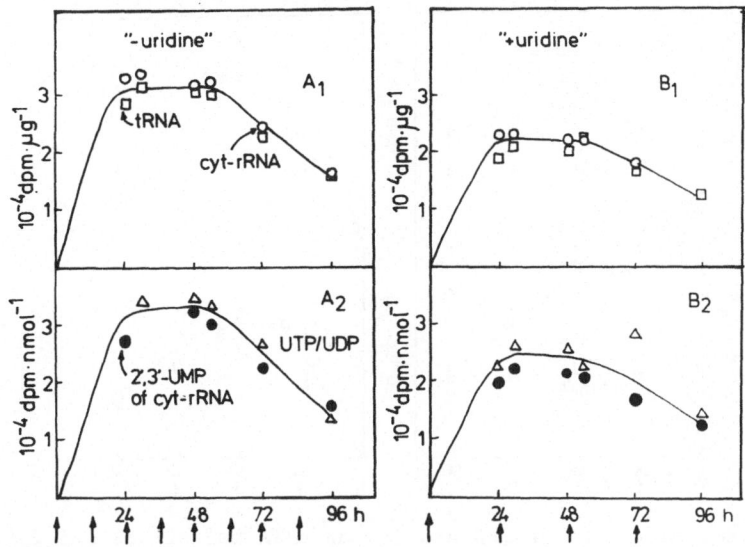

Fig. 2. Spec. rad. act. of 18S and 25S cyt-rRNA and of tRNA after repeated
addition ($\downarrow$) of 50 pmoles.ml^{-1} (A_1,A_2) or 1050 pmoles.ml^{-1} (B_1,B_2)
^{3}H-uridine compared to spec. rad. act. of UTP/UDP and 2',3'-UMP of
cyt-rRNA (A_2,B_2).

with this spec. rad. act. of 2',3'-UMP-residues of cyt-tRNA was found equal
to that of UTP under both spec. rad. act. (Figure 2 A_2 and B_2).

This strongly supports that cyt-rRNA and tRNA were built up from a
common UTP pool. This is conclusively only, if the amount per μg RNA of
uracil (and cytosin) residues is in the same range for both RNA species.
As determined for various plant species 25.9 % UMP and 22.6 % CMP residues
were found in cyt-rRNA (Eckenrode et al., 1985), which is in the same range
as for tRNA calculated from various plant tRNA sequences (Sharp et al.,
1985). This excluded that the determined equal spec. rad. act. of
cyt-rRNA and tRNA was given by chance through differences in mass ratio of
RNA residues.

To study UTP compartmentation during synthesis of cyt-rRNA and mt-rRNA
the same approach was used under equal feeding conditions. At low concentration of exogenous ^{3}H-uridine $3.2.10^4$ dpm per μg RNA were found for
cyt-rRNA at steady state (Figure 3 A_2). On the other hand, both mt-rRNAs
were labeled to $1.8.10^4$ dpm per μg RNA suggesting UTP compartmentation.
This was strengthened by results at high ^{3}H-uridine concentration. Here
spec. rad. act. of cyt-rRNA dropped to $2.5.10^4$ dpm per μg RNA, whereas 18S
and 26S mt-rRNA reached an equal but higher value as under low ^{3}H-uridine
concentration (Figure 3 B_2). The same tendency is expressed of labelling
of mt-rRNA between both exogenous ^{3}H-uridine concentrations by determining
dpm per NC-filter (Figure 3 A_1,B_1).

Beside the above mentioned UTP compartmentation there were some more
facts related to this problem (Table 1). Under low exogenous ^{3}H-uridine
concentration this was diluted intracellularly 1000-fold, whereas a dilution took place 100-fold only at high concentration. With respect to
intracellular amount of uridine in tomato cells, partially located in
vacuoles, of 50-150 pmoles/10^5 cells during different growth stages
(Leinhos et al., 1986) dilution may be caused mainly by de novo-synthesis
of pyrimidine nucleotides. The remarkable lower spec. rad. act. of UMP
compared to UTP and uridine under both feeding conditions suggest compartmentation of UMP at least partially.

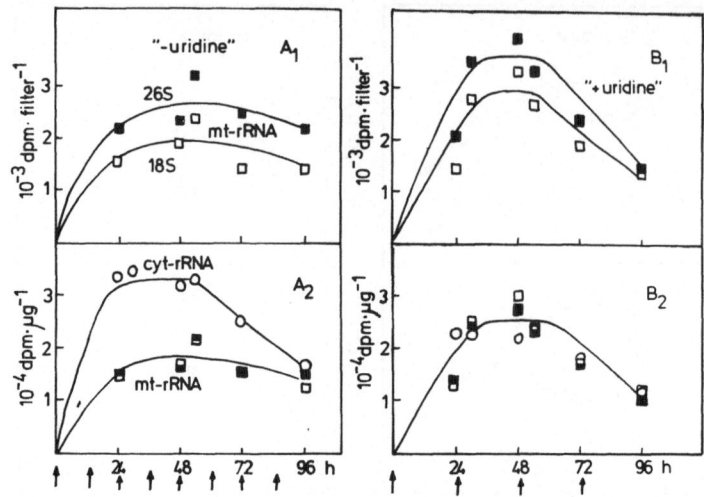

Fig. 3.   Spec. rad. act. of 18S and 26S mt-rRNA and of 18S and 25S cyt-rRNA after repeated addition (↓) of 50 pmoles.ml^{-1} (A_1,A_2) or 1050 pmoles.ml^{-1} (B_1,B_2) ^{3}H-uridine compared to spec. rad. act. of UTP/UDP (A_2,B_2).

Table 1. Steady state spec. rad. act. of uridine and uridine nucleotides as well as cyt-rRNA (dpm per µg RNA or nmole).

Substances	^{3}H–uridine fed	
	50 pmoles.ml^{-1}	1050 pmoles.ml^{-1}
^{3}H–uridine (exogenously)	$4.4 \cdot 10^7$	$2.1 \cdot 10^6$
^{3}H–uridine (intracellularly)	$2.7 \cdot 10^4$	$2.0 \cdot 10^4$
UMP	$4.0 \cdot 10^3$	$4.0 \cdot 10^3$
UTP/UDP	$3.4 \cdot 10^4$	$2.4 \cdot 10^4$
2',3'–UMP of cyt-rRNA	$3.3 \cdot 10^4$	$2.2 \cdot 10^4$
cyt–rRNA	$3.2 \cdot 10^4$	$2.2 \cdot 10^4$

The presented facts allow for tomato cells detection of absolute rate of cyt-rRNA and tRNA synthesis neglecting compartmentation. For mt-rRNA synthesis this has to be taken into account. Compared to controversial results on nucleotide compartmentation during synthesis of different RNA species for animal cells (cf. reviews of Wasternack and Benndorf, 1983, Moyer and Henderson, 1985) by our study nucleotide compartmentation seems to be of importance mainly for RNA synthesis of different organelles.

REFERENCES

Bestwick, R.K., Mathews, C.K. 1982, J. Biol. Chem. 257; 9305-9308.
Eckenrode, V.K., Arnold, J., Meagher, R.B. 1985, J.Mol. Evol. 21; 259-269
Genchev, D.D., Kermekchiev, M.B. Hadjiolov, A.A., 1980, Biochem. J. 188; 85-90.

Hause, B., Baldauf, F., Stock, K., Wasternack, C., Metzlaff, M. 1986, Curr. Genet. 10; 785-790.

Hause, B., Wasternack, C. 1987a, (in preparation).

Hause, B., Wasternack, C. 1987b, (in preparation).

Khym, J.X., Jones, M.H., Lee, W.H., Volkin, E. 1978, J.Biol. Chem. 253; 8741-8746.

Leinhos, V., Krauss, G.-J., Glund, K. 1986, Plant Sci. 47; 15-20

Moyer, J.D., Henderson, J.F., 1985, CRC Crit. Rev. Biochem. 19; 45-61.

Piper, P.W. 1981, FEBS Lett. 131; 373-376.

Sharp, S.J., Schaack, J., Cooley, L., Burke, D.J., Söll, D. 1985, CRC Crit. Rev. Biochem. 19; 107-114.

Soeiro, R., Ehrenfeld, E. 1973, J.Mol. Biol. 77; 177-187.

Tewes, A., Glund, K., Walther, R., Reinbothe, H. 1984, Z. Pflanzenphys. 113; 141-150

Tintemann, H., Wasternack, C., Benndorf, R., Reinbothe, H. 1985; Comp. Biochem. Physiol. 82B; 787-792.

Wasternack, C. 1983, Mol. Cell. Biol. 3; 613-622.

Wasternack, C., Benndorf, R. 1983, Biol. Zbl. 103; 1-16.

Wiegers, U., Kramer, G., Klapproth, K., Hilz, H. 1976, Eur. J. Biochem. 64; 535-540.

INTRACELLULAR LOCATION OF SOME PURINE- AND
PYRIMIDINE-METABOLIZING ENZYMES IN PLANT CELLS

C. Wasternack, R. Walther, K. Glund
*A. Guranowski, H. Tintemann and J. Miersch

Division of Plant Biochemistry, Section of
Biosciences, University of Halle
4020 Halle, Neuwerk 1, GDR
*Katedra Biochimii, Akademia Rolnicza
ul Wolynska 35, PL-60-637 Poznan, Poland

INTRODUCTION

In plant cells purines and pyrimidines are synthesized and metabolized by sequences also known for bacterial and animal systems (Figure 1) (cf. reviews of Wasternack, 1982; Keppler and Holstege, 1982). Although an important role in understanding intracellular availability of nucleotides, location of relevant enzymes is poorly understood for plant cells, and controversial results exist.

CPSase was found in plastids of soybean cells cultured in suspension (Shargool et al., 1978), and OPRTase, ODCase and ATCase were localized in the cytosol (Kanamori et al., 1980) of cultured Vicia rosea cells. On the other hand recent studies of Doremus and Jagendorf (1985) suggest the location of ATCase, OPRTase, ODCase, and DHOase in plastids and of DHO-DH in mitochondria of pea leaves. APRTase was found in mitochondria of cultured Catharanthus roseus cells (Ukaji et al., 1986), but also in the chloroplasts of spinach leaves (Ashihara and Ukaji, 1985).

We tried to localize some purine- and pyrimidine-metabolizing enzymes in tomato cell suspension culture using different procedures such as cell fractionation, sucrose density gradient centrifugation, time-dependent lysis of protoplasts and isolation of vacuoles.

Abbreviations

ATcase – aspartate transcarbamylase; APRTase – adenine phosphoribosyl-transferase; cat – catalase; CPSase – carbamylphosphate synthetase; CS – citrate synthase; DHOase – dihydroorotic acid hydrolase; DHO-DH – dihydro-orotic acid dehydrogenase; EDH – ethanol dehydrogenase; MDH – malate dehydrogenase; MTA – 5'methyltioadenosine; NCβA – N-carbamyl-β-alanine; ODCase – orotidine-5'-phosphate decarboxylase; OPRTase – orotidine-5'-phosphate phosphoribosyltransferase; SAH – S-adenosylhomocysteine; SDH – succinate dehydrogenase.

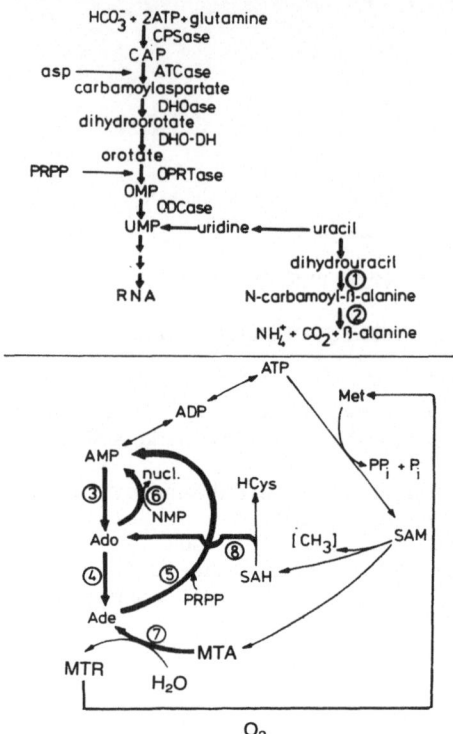

Fig. 1. Pathways of pyrimidine and purine metabolism
and enzymes studied in this paper:
1 - dihydropyrimidinase; 2 - NCβA-amidohydrolase;
3 - 5'-nucleotide phosphatase; 4 - adenosine
nucleosidase; 5 - APRTase; 6 - nucleoside
phosphotransferase; 7 - MTA nucleosidase; and
8 - SAH hydrolase.

MATERIALS AND METHODS

Cells of tomato (Lycopersicon esculentum cv. "Lukullus") were cultured
and counted as described by Tewes et al. (1984). Preparation of crude
extracts, cell fractionation, sucrose density gradient centrifugation,
time-dependent lysis of protoplasts, isolation of vauoles and mitochondria
were described by Tewes et al. (1984), Glund et al. (1984), Wasternack et
al. (1985), Miersch et al. (1986) and Tinteman et al. (1987). Enzyme
assays were as follows or mentioned therein: APRTase, 5'-nucleotidase, MTA
nucleosidase, SAH hydrolase according to Guranowski and Wasternack (1982),
MDH, SDH, EDH, α-mannosidase, fumarase, dihydropyrimidinase and NCβA-
amidohydrolase according to Tintemann et al. (1987), DHO-DH according to
Miersch et al. (1986), CPSase, ATCase, DHOase according to Walther (1984),
ODCase, OPRTase according to Walther et al. (1984). Electron microscopic
localization was used as described in Walther (1984).

RESULTS

First information was drawn from differential centrifugation. Related
to clearly different occurrence of marker enzymes as EDH, MDH and SDH in
different fractions (not shown), cytosolic location was suggested for most
enzymes under study, whereas DHO-DH occurred <10% in the supernatant
(Table 3). As exemplified in Figure 2, enzymes under study coincide with
the cytosolic marker EDH in a 20-60% and a 17.5-30% sucrose gradient
(Wasternack et al., 1985; Tintemann et al., 1987). Since the mitochondrial

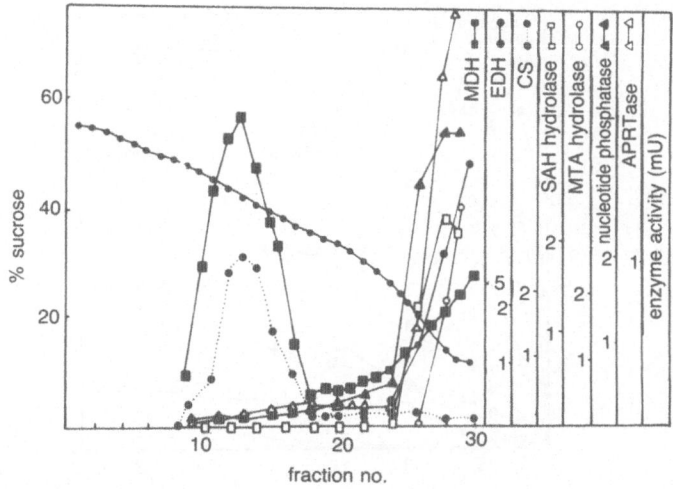

Fig. 2.   Sucrose density gradient centrifugation of a 5 ml homogenate
of 3-d-old tomato cells.

matrix marker MDH and CS were not released or only partially, cytosolic
location cannot be caused by enzyme release from mitochondria. On the
other hand, DHO-DH was found to be mitochondrial using sucrose density
gradients and cell fractionation (Table 1) (Miersch et al., 1986). Further
support on cytosolic location came from time-dependent enzyme release from
protoplasts being ruptured by osmotic lysis (Figure 3). Similar to EDH,
most of the enzymes were found to be released into the supernatant with
almost 100% yield as compared with sonified protoplasts. On the other
hand, organellar enzymes were clearly sedimentable. To exclude rupture of
vacuoles which occurred in all three procedures mentioned above, enzymes
were tested in isolated vacuoles purified from protoplasts (Table 2).
Related to the vacuolar marker α-mannosidase, vacuolar location for all
enzymes under study could be excluded. But, so far in most approaches
used, enzyme release from the nucleus cannot be ruled out as a cause of
their occurrence in cytosolic fractions. In tomato cells, ATCase occurred
in cytosolic fractions using sucrose gradients (Table 3), but its real
nuclear location was evidenced by a histochemical approach. Here P_i,
released by ATCase reaction, was precipitated with Pb and visualized
electron microscopically (Figure not shown).

Table 1.   Activities of DHO-DH in Different Subcellular
Fractions of Tomator Cells (Data from Miersch et
al., 1986)

Fraction	Total Volume (ml)	Total Protein (mg)	Total Act. (nkat)	Spec. Act. (pkat. protein^{-1})	Recovery %
Homogenate	800	1900	3.2	1.7	100
10.000 x g Pellet	17	316	2.6	8.0	81
Mitochondrial Fraction	5	21	1.55	74.0	48

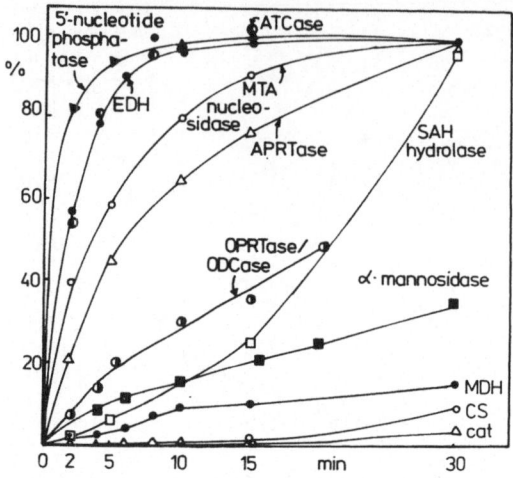

Fig. 3. Time dependent lysis of protoplasts of tomato cells. Protoplasts were resuspended in a mannitol-free medium at a density of 10^6 per ml (= 0 min of lysis). At indicated times aliquotes were separated by rapid silicone oil bromododecane centrifugation and enzyme activities were tested in the supernatant and were expressed in % of the corresponding activity in sonified protoplasts.

As presented in the present paper, most of the enzymes under study were found to be cytosolically, whereas DHO-DH and ATCase were found in mitochondria and nucleus, respectively. It is also shown that care has to be taken with respect to their possible release from organelles during time of the experiment. In this respect divergent procedures as described in this paper are of advantage.

Table 2. Comparison of Activities of Pyrimidine- and Purine-Metabolizing Enzymes (nmoles.h^{-1} at 30°C) in Vacuoles Prepared from Protoplasts according to Tewes et al. (1984) and Glund et al. (1984). Activities of Marker Enzymes are given in nmoles.min.$^{-1}$ at 30°C.

Enzymes (Activity per 10^6 Protoplasts)	Protoplasts	Vacuoles
SAH Hydrolase	28	n.d.
APRTase	200	n.d.
5'-nucleotide Phosphatase	20	n.d.
MTA Nucleosidase	56.9	n.d.
NCβA-amidohydrolase	10.4	n.d.
Dihydropyrimidinase	6.8	n.d.
α-mannosidase	1.2	0.48
EDH	72.6	n.d.
MDH	2120	10.6
Acid Phosphatase	20.8	1.2

n.d. - not detectable

Table 3. Summary of Intracellular Location of Purine- and Pyrimidine-metabolizing Enzymes in Tomato Cells using Different Procedures (Quantitative Data are Expressed in % of Occurrence in the Homogenate or 10^6 Protoplasts in the Case of Vacuoles)

Enzyme	100.00xg (by Cell Fractionation)	Sucrose Density Gradient	Time-Dependent Lysis of Protoplasts	Isolated Vacuoles	Electron-Microscopic Studies
SAH Hydrolase	99	cyt.	cyt.	n.d.	–
MTA Nucleosidase	99	cyt.	cyt.	n.d.	–
APRTase	93	cyt.	cyt.	n.d.	–
5'-nucleotide phosphatase	48	cyt.	cyt.	n.d.	–
NCβA-amidohydrolase	78[b]	cyt.	(cyt.)	n.d.	–
Dihydropyrimidinase	66[b]	cyt.	(cyt.)	n.d.	–
CPSase	–	cyt.	–	–	–
ATCase	–	cyt.	cyt.	–	nucleus
DHOase	–	cyt.	–	–	–
DHO-DH	8[a]	mit.	–	–	–
OPRTase	–	}cyt.	}(cyt.)	–	–
ODCase	–			–	–

a: 10.000xg supernatant; b; 75.000xg supernatant; n.d. – not detectable

Different location of pyrimidine-metabolizing enzymes has to be regarded for plant cells of different origin. Whereas CPSase, DHOase, OPRTase, and ODCase were found in the cytosol and ATCase in the nucleus of proplastid-containing tomato cells grown in suspension, or in the apoplastic mutant W$_3$BUL of the flagellate Euglena gracilis (OPRTase/ODCase) (Walther et al., 1980), these enzymes were localized exclusively in plastids of pea leaves (Doremus and Jagendorf, 1985). This suggests species-specific differences in intracellular location of nucleic acid precursor enzymes.

REFERENCES

Ashihara, H., and Ukaji, T., 1985, J. Biochem., 17:1275-1277.
Doremus, H. D., and Jagendorf, A. T., 1985, Plant Physiol., 79:856-861.
Glund, K., Tewes, A., Abel, S., Leinhos, V., Walther, R., and Reinbothe, H., 1984, Z. Pflanzenphysiol., 113:151-161.
Guranowski, A., and Wasternack, C., 1982, Comp. Biochem. Physiol., 71B: 483-488.
Kanamori, I., Ashihara, H., and Komamine, A., 1980, Z. Pflanzenphysiol., 96:7-16.
Keppler, D., and Holstege, A., 1982, in: "Metabolic Compartmentation", H. Sies, ed., Acad. Press, London, NY, pp 147-203.
Miersch, J., Krauss, G. -J., and Metzger, U., 1986, J. Plant Physiol., 122:55-66.
Shargool, P. D., Steeves, T., Weaver, M., and Russell, M., 1978, Can. J. Biochem., 56:273-279.
Tewes, A., Glund, K., Walther, R., and Reinbothe, H., 1984, Z. Pflanzenphysiol., 113:141-151.
Tintemann, H., Wasternack, C., Helbing, D., Glund, K., and Hause, B., 1987, Comp. Biochem. Physiol. (B), in press.

Ukaji, T., Hirose, F., and Ashihara, H., 1986, J. Plant Physiol., 125: 191-197.

Walther, R., 1984, Ph. Thesis, manuscript, University of Halle.

Walther, R., Wald, K., Glund, K., and Tewes, A., 1984, J. Plant Physiol., 116:301-311.

Walther, R., Wasternack, C., Helbing, D., and Lippmann, G., 1980, Biochem. Physiol. Pflanzen., 175:764-771.

Wasternack, C., 1982, in: "Nucleic Acids and Proteins in Plants", Encycl. of Plant Physiol., B. Partheir and D. Boulter, eds., Springer-Verlag, Heidelberg, 14b:263-301.

Wasternack, C., Guranowski, A., Glund, K., Tewes, A., and Walther, R., 1985, J. Plant Physiol., 120:19-28.

SOME PROPERTIES OF DINUCLEOSIDE OLIGOPHOSPHATE

PHOSPHORYLASE FROM EUGLENA GRACILIS

A. Guranowski, E. Starzynska
and *C. Wasternack

Katedra Biochimii, Akademia Rolnicza
ul. Wolynska 35, PL-60-637 Poznan, Poland
*Division of Plant Biochemistry
Section of Biosciences, University of
Halle, 4020 Halle, Neuwerk 1, GDR

INTRODUCTION

In 1976, ten years after the discovery of Ap_4A (Figure 1A), a by-product of aminoacylation by some tRNA synthetases, its intracellular concentration was found to increase from 10^{-8} M to 10^{-6} M during transition from G_0/G_1-phase to S-phase of animal cells (Rapaport and Zamecnik, 1976). Because of that and the observation that Ap_4A triggers DNA-replication (Grummt, 1978), studies expanded on synthesis, degradation, intracellular concentration and function of Ap_4A.

The presence of Ap_4A was documented for bacterial (Lee et al., 1983), animal and fungi systems (Rapaport and Zamecnik, 1976; Garrison and Barnes, 1984) (for further ref. cf. Wasternack, 1987). In vitro, some aminoacyl-tRNA synthetases purified from such different phyla as archebacteria (Rauhut et al., 1985), eubacteria (Goerlich et al., 1982), fungi (Plateau et al., 1981), plants (Jakubowski, 1983) and mammals (Brevet et al., 1982) can catalize synthesis of Ap_4A (Figure 1B and 1C). Zn^{2+} greatly stimulates Ap_4A synthesis (Mayaux and Blanquet, 1981). The mentioned proliferation-dependent accumulation of Ap_4A may be regulated by synthesis and/or its degradation. Degradation does not proceed according to the same pattern in all organisms. In higher eukaryotes, animals and plants, Ap_4A is cleaved to ATP and AMP by a specific dinulceoside tetraphosphate (asymmetrical) hydrolase (E.C. 3.6.1.17.) (Lobaton et al., 1975). In Physarum polycephalum (Barnes and Culver, 1982) and bacteria Ap_4A is hydrolized to ADP by a specific dinucleoside tetraphosphate (symmetrical) hydrolase (E.C. 3.6.1.41.) (Guranowski et al., 1983) (Figure 1A). Finally, a phosphorolytic cleavage of Ap_4A to ATP and ADP was found recently in yeast, Saccharomyces cereviseae (Guranowski and Blanquet, 1985). The present paper describes phosphorolytic cleavage of Ap_4A in the protozoa Euglena gracilis and Acanthomoeba castellanii, and properties of partially purified Ap_4A phosphorylase from E. gracilis.

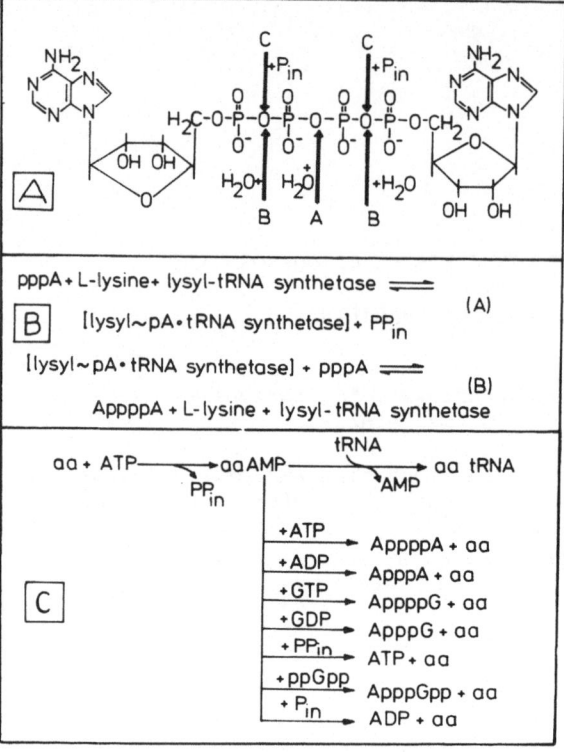

Fig. 1. Ⓐ Formula and sites of cleavage of Ap₄A.
Ⓑ Two-step sequence of Ap₄A synthesis.
Ⓒ General scheme of synthesis of adenylated nucleotides.
Ap₄A-cleaving enzymes in Ⓐ : A = dinucleoside tetraphosphate
(symmetrical) hydrolase; B = dinuculeoside tetraphosphate (asym-
metrical) hydrolase; C = dinucleoside tetraphosphate phosphorylase.

MATERIALS AND METHODS

Cultivations were performed as described by Wasternack (1975). The
enzyme assays, TLC methods and purification of the enzyme are described by
Guranowski et al. (1987).

RESULTS AND DISCUSSION

First information on a phosphorolysis of Ap₄A was drawn from a much
faster cleavage of the dinucleotide in the presence of P_i than in its
absence giving ATP and ADP in a homogenate of <u>Euglena</u> (both from the apoplastic mutant W₃BUL and from the photoorganotrophic cells) as well as of
<u>Acanthomoeba castellanii</u>. 675-Fold purification was performed from 42 g
(wet weight) frozen photoorganotrophic cells of E. gracilis by ammonium
sulphate fractionation and chromatography on DEAE Sephacel, Hydroxylapatite and Sephadex G-75 superfine (Table 1). The purified enzyme was
free of adenylate kinase and other enzymes catabolizing nucleotides and
dinucleotides. It could be stored in 25 mM HEPES/KOH buffer, pH 8.0,
containing 10 µM dithiothreitol and 50% glycerol stable for several months
at -20°C. Molecular weight of native Ap₄A phosphorylase was estimated as
about 30 kDal. pH-Optimum was 8.0 ± 0.2. The Ap₄A phosphorylase of E.
gracilis was absolutely dependent upon divalent metal ions reaching 94% of
maximum velocity at 5 mM MgCl₂. Among the following potential substrates
tested: Ap₂A, Ap₃A, Ap₄A, Ap₅A, Gp₄G, ATP, p₄A, AppCH₂ppA, ApppCH₂pA, and
ApCH₂ppCH₂pA, only Ap₃A, Ap₄A, Ap₅A, Gp₄G, and ApppCH₂pA behaved as

Table 1. Purification of Dinucleoside Oligophosphate
α,β-Phosphorylase from Euglena Gracilis

Fraction	Total Protein mg	Specific Activity[a] Units[b]/mg	Purification -fold
Crude extract	550	0.32	1
Dialyzed ammonium sulphate (50-70%) precipitate	180	1.98	6.2
DEAE-Sephacel	12	12	37.5
Hydroxylapatite	1.4	49	153
Sephadex G-75 superfine	0.04	200	625

a: Activity was assayed at 500 μM [^{3}H]Ap$_4$A as disappearance of Ap$_4$A because of ADP-degrading activities up to the fourth step of purification.
b: 1 Unit = amount of enzyme able to degrade 1 nmole of Ap$_4$A per min at 30°C.

substrates giving corresponding nucleoside diphosphates as one of the products. At 500 μM concentration, Ap$_4$A is cleaved 7-fold faster than Ap$_4$A. Phosphorolysis of Ap$_4$A and Ap$_3$A follows Michaelis-Menten kinetics in the range of 12.5-200 μM giving K$_m$ values of 27 ± 3 μM and 25 ± 3 μM, respectively. K$_m$ values for P$_i$ was 0.5 ± 0.04 mM at 100 μM [^{3}H]-Ap$_4$A.

Among several anions tested the following could substitute P$_i$ in the cleavage of Ap$_4$A or Gp$_4$G: arsenate, chromate, molybdate and vanadate. An anion-dependent dephosphorylation of NDPs formed during Ap$_4$A phosphorolysis was shown to be an inherent property of the yeast Ap$_4$A phosphorylase (Guranowski and Blanquet, 1986 a,b). We confirmed here that ADP and GDP incubated in the presence of Euglena Ap$_4$A phosphorylase and either arsenate, chromate, vanadate or molybdate are converted to AMP and GMP, respectively, and the conversion does not require divalent metal cations. Although similar in the pH optimum, cation requirement and anionolysis, the Ap$_4$A phosphorylases of Euglena and yeast differ with respect to substrate specifity and molecular weight (Table 2).

Table 2. Properties of Ap$_4$A Phosphorylase Purified from
Saccharomyces Cereviseae and Euglena Gracilis

	E. Gracilis	S. Cereviseae[a]
Substrates Natural:	Ap$_3$A, Ap$_4$A	Ap$_4$A
Artificial:	Ap$_5$A, Gp$_4$G, ApppCH$_2$pA	Ap$_5$A, Gp$_4$G
K$_m$ (Ap$_4$A)	27 μM	60 μM
K$_m$ (Ap$_3$A)	25 μM	-
K$_m$ (P$_i$)	0.5 mM	1 mM
Mol. Weight	30 kDal	40 kDal
pH Optimum	8.0	8.0
Cation Requirement	yes	yes
Anionolysis of NDPs	yes	yes

a: Data from Guranowski and Blanquet (1985).

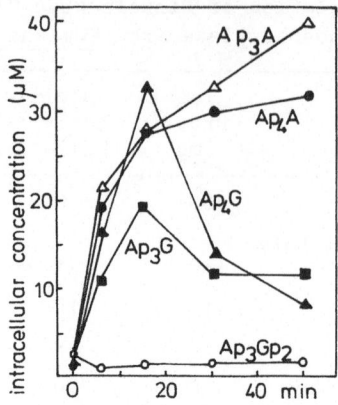

Fig. 2.   Changes in concentration of dinucleotides after heat shock
          (25 → 50°C) in Salmonella typhimurium (after Lee et al.,
          1983).

Studies on function of Ap₄A have to consider distinct Ap₄A-degrading enzymes: (asymmetrical) Ap₄A hydrolase, (symmetrical) Ap₄A hydrolase, or Ap₄A phosphorylase (as presented here). Until now, the role of Ap₄A is not known (cf. Wasternack, 1987). On the one hand, there are suggestions that Ap₄A acts as an alarmone at environmental stresses in prokaryotic (Lee et al., 1983) (Figure 2) and eukaryotic cells (Baker and Jacobson, 1986), although Ap₄A did not signal directly heat shock response (Guedon et al., 1985). On the other hand, Ap₄A can be a trigger of DNA replication both in vivo (Grummt, 1978) and in vitro (Zamecnik, 1983). This accords with the drastic changes of Ap₄A concentration during proliferation of various eukaryotic cells (Weinmann-Dorsch et al., 1984) (Figure 3).

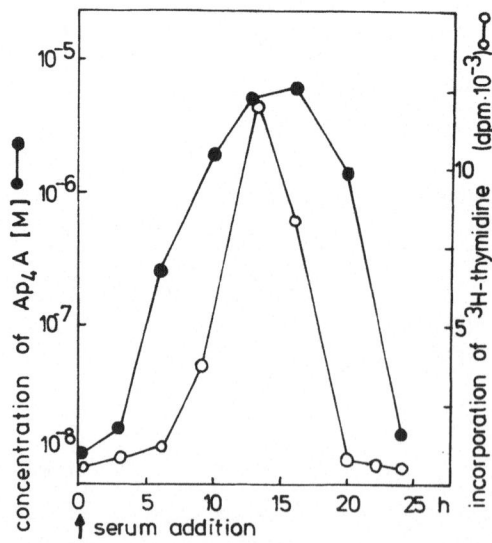

Fig. 3.   Cell cycle-dependent changes of Ap₄A concentration in BHK
          cells (after Weinmann-Dorsch et al., 1984).

Knowledge of Ap4A cleaving enzymes is a prerequisite for the understanding of the role of Ap4A.

REFERENCES

Baker, J. C. and Jacobson, M. K., 1986, Proc. Natl. Acad. Sci., USA., 83: 2350-2352.
Barnes, L. D., and Culver, C. A., 1982, Biochem., 24:6129-6133.
Brevet, A., Plateau, P., Cirakoglu, B., Pailliez, J. -P., and Blanquet, S., 1982, J. Biol. Chem., 257:14613-14615.
Garrison, P. N., and Barnes, L. D., 1984, Biochem. J., 217:805-811.
Goerlich, O., Foeckler, R., and Holler, E., 1982, Europ. J. Biochem., 126;135-142.
Grummt, F., 1978, Proc. Natl. Acad. Sci., USA., 75:371-375.
Guranowski, A., and Blanquet, S., 1985, J. Biol. Chem., 260:3542-3547.
Guranowski, A., and Blanquet, S., 1986a, Biochimie, 68:747-760.
Guranowski, A., and Blanquet, S., 1986b, J. Biol. Chem., 261:5943-5946.
Guranowski, A., Jacubowski, H., and Holler, E., 1983, J. Biol. Chem., 258:14784-14789.
Guranowski, A., Starzynska, E., and Wasternack, C., 1987, Int. J. Biochem., submitted.
Guedon, G., Sovia, D., Ebel, J.-P., Belfort, N., and Remy, P., 1985, EMBO J., 4:3743-3749.
Jacubowski, H., 1983, Acta Biochim. Polon., 30:51-69.
Lee, P. C., Bochner, B. R., and Ames, B. N., 1983, Proc. Natl. Acad. Sci., USA., 80:7496-7500.
Lobaton, C. D., Vallejo, C. G., Sillero, A., and Sillero, M. A. G., 1975, Europ. J. Biochem., 50:495-501.
Mayaux, J.-F., and Blanquet, S., 1981, Biochem., 20:4647-4654.
Plateau, P., Mayaux, J.-F., and Blanquet, S., 1981, Biochem., 20:4654-4662.
Rapaport, E., and Zamecnik, P. C., 1976, Proc. Natl. Acad. Sci., USA., 73:3984-3988.
Rauhut, R., Gabius, H.-J., Engelhardt, R., and Cramer, F., 1985, J. Biol. Chem., 260:182-187.
Wasternack, C., 1975, Plant Sci. Lett., 4:353-360.
Wasternack, C., 1987, Biol. Rdsch., 25 (5), in press.
Weinmann-Dorsch, C., Hedel, A., Grummt, I., Albert, W., Ferdinant, F.-J., Friis, R. R., Pierron, G., Moll, W., and Grummt, F., 1984, Europ. J. Biochem., 138:179-185.
Zamecnik, P., 1983, Anal. Biochem., 134:1-10.

EXPRESSION OF GENES CODING FOR NOVOBIOCIN BIOSYNTHESIS

AND NOVOBIOCIN RESISTANCE IN STREPTOMYCES NIVEUS

D. A. Ritchie, K. E. Cushing, P. G. Logan
and J. I. Mitchell

Department of Genetics, University of Liverpool
Liverpool L69 3BX, UK

Novobiocin is a clinically useful antibiotic produced by Streptomyces niveus and S. spheroides. It binds to the β subunit of DNA gyrase which prevents DNA unwinding with the consequent inhibition of DNA replication, recombination and transcription (Gellert et al., 1976). The novobiocin molecule consists of three ring components, ring A (a substituted benzoic acid), ring B (an amino coumarin), and ring C (a noviose sugar) as shown in Figure 1.

Rings A and B are derived from L-tyrosine and ring C is derived from uridine diphosphoglucose (Figure 2). From studies with radiolabeled precursors, analysis of the breakdown products of novobiocin, and the complementation of blocked mutants the pathway for novobiocin has been determined in some detail (Kominek and Sebek, 1973; Queener et al., 1978).

Fermentation studies have shown that carbon and nitrogen sources are used sequentially by S. niveus with citrate being used before glucose and ammonium sulphate before proline. Novobiocin production begins during the idiophase and is known to be regulated by citrate, which inhibits novobiocin production, and is repressed by nitrogen and phosphate (Kominek, 1982; Kominek and Sebek, 1973; Drew and Demain, 1977). It might be expected that control mechanisms operating on aromatic amino acid synthesis, particularly tyrosine, would also regulate novobiocin formation.

We are using the novobiocin biosynthetic pathway as a model system to analyze the flux between primary and secondary metabolism. In particular, we are interested in the regulatory mechanisms leading to the expression of the secondary pathway enzymes for novobiocin biosynthesis. The experimental approach is to clone the biosynthetic pathway genes and use them as DNA probes to follow their expression under different physiological conditions known to affect novobiocin production. In this paper we report our progress in three areas of this project.

NOVOBIOCIN PRODUCTION AND NOVOBIOCIN RESISTANCE

Novobiocin is produced in fermentation broth after about 90 hours incubation and reaches its maximum concentration of approximately 1 mg/ml by about 160 hours. The time of appearance of the antibiotic coincides with the exhaustion of the citrate in the medium and the utilization of

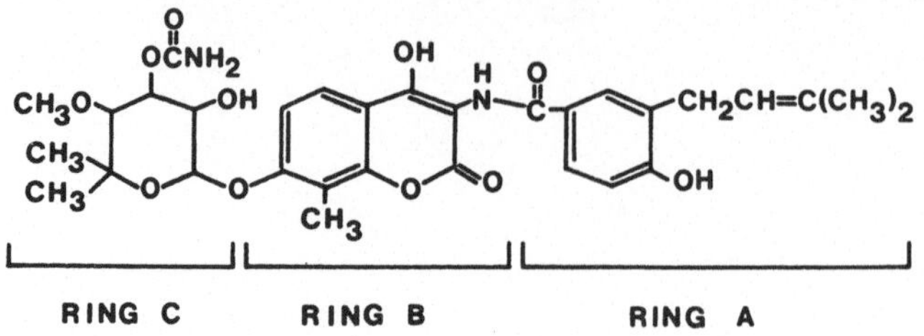

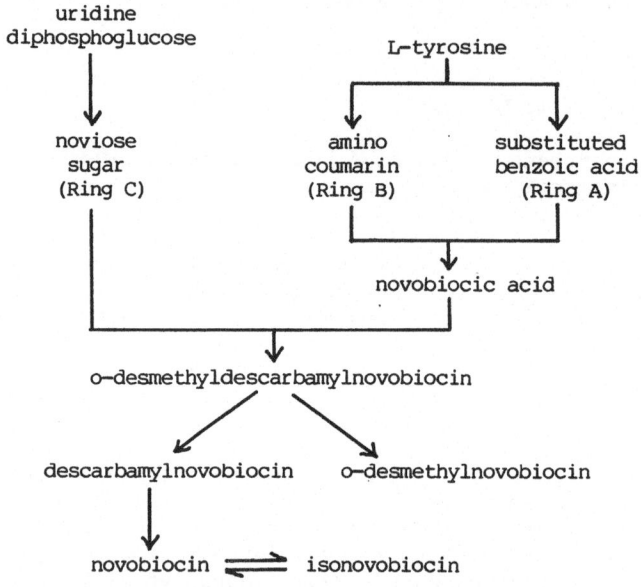

Fig. 1. Structure of novobiocin.

Fig. 2. Novobiocin biosynthetic pathway.

glucose as a carbon source. Our results confirm an observation of Kominek (1972) and demonstrate that novobiocin biosynthesis is repressed by citrate.

Spores of S. niveus wild type are resistant to novobiocin at a concentration of 25–50 µg/ml. During the antibiotic production phase the resistance of wild type mycelium increased from this basal level of 25–50 µg/ml to 500 µg/ml. Addition of citrate to the fermentation broth during the novobiocin production inhibited further production of the antibiotic and also inhibited the rise in novobiocin resistance. Furthermore, mutants blocked in novobiocin production show at all stages of growth the low basal level of resistance characteristic of germinating spores and young mycelium. These results suggest that novobiocin production and resistance may be functionally linked.

An increased level of resistance may be induced in germinating spores and young mycelium of S. niveus if they are grown in close proximity to novobiocin producing mycelium (Figure 3). Wild type spores take about 2 weeks at 30°C to produce a dense mycelium on plates containing 70 µg/ml novobiocin. When the spores were plated alongside wild type novobiocin producing mycelium the same amount of growth was observed within 3 days. Characteristically, the spores showed dense growth only on that side of the patched area next to the producing mycelium. This induction of novobiocin resistance by wild type mycelium was also observed for spores from mutants blocked in novobiocin production.

Proof that this induction effect is caused by the novobiocin-producing mycelium is strengthened by two further observations: (1) induction of resistance in wild type spores was not observed with mycelium from a mutant strain of S. niveus blocked in novobiocin production; and (2) induction of resistance occurred with the fermentation broth from which the novobiocin-producing mycelia had been removed, but unused broth had no effect.

Attempts to define the inducer molecules(s) have been unsuccessful so far, clearly it is not the novobiocin end product; however, it may be an intermediate in the biosynthetic pathway. With this in mind we have synthesized rings A and B and novobiocic acid but none will induce resistance in wild type spores. Similar tests with tyrosine and para-amino benzoic acid also gave negative results. Other intermediates will be tested as they become available but the possibility that the inducer is not part of the biosynthetic pathway must be considered.

CLONING NOVOBIOCIN RESISTANCE

In a number of instances where the genetic structure of an antibiotic pathway has been analyzed, it has been found that the genes coding for biosynthesis of the antibiotic and resistance to the antibiotic are closely linked. This is the case for methylenomycin (Chater and Bruton, 1983), oxytetracycline (Rhodes et al., 1984), streptomycin (Distler et al., 1985), puromycin (Vara et al., 1985), erythromycin (Stanzak et al., 1986) and actinorhodin (Hopwood et al., 1986). In the cases of actinorhodin and methylenomycin it is known that the resistance determinant is located within the biosynthetic cluster. This physical linkage has led to a general strategy of cloning antibiotic resistance determinants in order to select for cloned antibiotic biosynthetic gene sequences. We have adopted this method as our initial approach to cloning the novobiocin biosynthetic pathway genes.

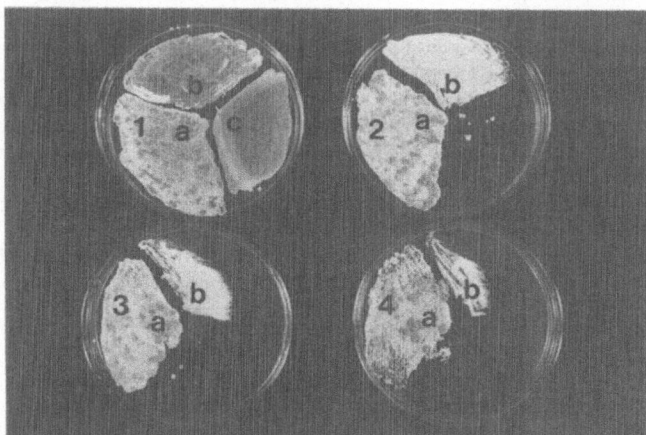

Fig. 3. Induction of novobiocin resistance by S. niveus mycelium. S. niveus wild type mycelium (a), S. niveus wild type spores (b), and S. lividans TK24 spores (c) were patched onto complete medium plates containing novobiocin at 0 μg/ml (1), 25 μg/ml (2), 50 μg/ml (3), and 70 μg/ml (4). The plates were incubated at 30°C for 3 days.

S. niveus wild type DNA was partially digested with the restriction enzyme BclI and the fragments were ligated in a shotgun reaction to the plasmid vector pIJ702 (Hopwood et al., 1985) cut with BglII. The ligated DNA was used to transform protoplasts of S. lividans TK24 which is sensitive to 10 µg/ml novobiocin and 50 µg/ml thiostrepton (the selectable antibiotic resistance marker on the vector). The transformed protoplasts were regenerated on medium containing thiostrepton and the resulting colonies were replicated to plates containing both thiostrepton and novobiocin. One transformant grew to form a colony in 3 days at 30°C on 50 µg/ml novobiocin. This thiostrepton and novobiocin resistant strain contained a plasmid (pGL101) of 10 kb in size suggesting an insert of about 4 kb into the vector.

Purified pGL101 retransformed S. lividans TK24 protoplasts to thiostrepton and novobiocin resistance with high efficiency (90% to 100%). Moreover, plasmid-cured derivatives obtained from regenerated protoplasts which had lost thiostrepton resistance had simultaneously lost resistance to novobiocin. On this basis we concluded that pGL101 contained novobiocin resistance determinant from S. niveus. A novobiocin resistance clone of similar size from S. spheroides was isolated by Murakami et al. (1983). A restriction map of pGL101 is shown in Figure 4.

Table 1 shows a comparison of novobiocin resistance levels for spores and mycelium of S. niveus wild type and S. lividans TK24 (pGL101). The patterns are generally similar and suggest that an increased level of novobiocin resistance also occurs with the clone; however, the maximum novobiocin concentration on which S. lividans TK24 (pGL101) grew was only 70 to 100 µg/ml and this required 7 days incubation.

TRANSFORMATION OF S. NIVEUS

An essential component of our cloning strategy is to transform mutants of S. niveus blocked in novobiocin production with cloned DNA sequences to look for those which restore antibiotic synthesis. This requires cloning vectors able to transform and be maintained in S. niveus. The phage vector Øc31 will not form plaques on this species and our attempts to transform with the plasmid vectors pIJ486, pIJ702 and pIJ940 using standard methods were unsuccessful. This indicates the presence of a strong restriction system in S. niveus. Attempts to reduce the level of restriction by using heat-treated protoplasts (Engel, 1987) or by adding carrier DNA to the transforming DNA were also unsuccessful. However, transformation of S. niveus wild type by pIJ702 has now been achieved by using a large excess of plasmid DNA. The initial transformation gave 27 colonies out of 5×10^6 regenerated protoplasts when replicated to plates containing thiostrepton. These transformants contained pIJ702 and were used to prepare DNA for a second transformation which gave a transformation frequency of about 10^3 cfu/µg DNA. While this transformation frequency is still not very high it

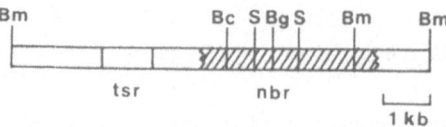

Fig. 4. Restriction map of pGL101. The plasmid is presented in a linear form with the plasmid vector pIJ702 unshaded and the S. niveus insert shaded. The locations of the thiostrepton resistance (tsr) and novobiocin resistance (nbr) determinants and cleavage sites for BamHI (Bm), BclI (Bc), BglII (Bg) and SalI (S) are indicated.

Table 1. Novobiocin resistance levels of spores and mycelium of S. niveus wild type and S. Lividans TK24 (pGL101), (++ = dense growth; + = sparse growth; - = no growth).

Novobiocin µg/ml	S. niveus		S. lividans (pGL101)	
	Spores	Mycel.	Spores	Mycel.
0	+ +	+ +	+ +	+ +
25	+ +	+ +	+ +	+ +
50	+ +	+ +	+ +	+ +
70	-	+ +	+	+ +
100	-	+ +	-	+ +
500	-	+	-	-

does offer the potential for further increases in efficiency and the prospect of detecting cloned novobiocin biosynthetic pathway genes.

Acknowledgements

This work was supported by grants from the Science and Engineering Research Council and the University of Liverpool.

REFERENCES

Chater, K. F., and Bruton, C. J., 1983, Gene, 26:67-78.
Distler, J., Mansouri, K., and Piepersberg, W., 1985, FEMS Microbiol. Lett., 30:151-154.
Drew, S. W., and Demain, A. L., 1977, Ann. Rev. Microbiol., 31:343-356.
Engel, P., 1987, Appl. Enviorn. Microbiol., 53:1-3.
Gellert, M., O'Dea, M. H., Itoh, T., and Tomizawa, J. I., 1976, Proc. Natl. Acad. Sci., USA, 73:4474-4478.
Hopwood, D. A. Bibb, M. J., Chater, K. C., Janssen, G. R., Malpartida, F., and Smith, C. P., 1986, in: "Regulation of Gene Expression - 25 Years On", I. R. Booth and C. F. Huggins, eds., Cambridge University Press, pp 251-276.
Hopwood, D. A., Bibb, M. J., Chater, K. C., Kieser, T., Bruton, C. J., Kieser, H. M., Lydiate, D. J., Smith, C. P., and Schrempf, H., 1985, "Genetic Manipulation of Streptomyces: A Laboratory Manual", The John Innes Foundation, Norwich.
Kominek, L. A., 1972, Antimicrob. Agents Chemother., 1:123-134.
Kominek, L. A., and Sebek, O. K., 1973, Devel. Ind. Microbiol., 15:60-69.
Murukami, T., Nojiri, C., Toyama, H., Hayashi, E., Katumata, K., Anzai, H., Matsuhashi, Y., Yamada, Y., and Nagaoka, K., 1983, J. Antibiot., 36:1305-1312.
Queener, S. W., Sebek, O. K., and Vezina, C., 1978, Ann. Rev. Microbiol., 32:593-636.
Rhodes, P. M., Hunter, I. S., Friend, E. J., and Warren, M., 1984, Biochem. Soc. Trans., 12:586-587.
Stanzak, R., Matsushima, P., Baltz, R. H., and Rao, R. N., 1986, Biotech., 4:129-232.
Vara, J., Malpartida, F., Hopwood, D. A., and Jimenez, A., 1985, Gene, 33:197-206.

CHROMOSOMAL REARRANGEMENTS AFFECTING THE REGULATION OF

URIDINE PHOSPHORYLASE GENE EXPRESSION IN ESCHERICHIA COLI

A. S. Mironov, S. T. Kulakauskas
and V. V. Sukhodolets

Institute of Genetics and Selection of
Industrial Microorganisms
Moscow 113545, USSR

The structural gene udp encoding uridine phosphorylase (EC 2.4.2.3) is located at 85 min on the E. coli chromosome (Bachmann, 1983). This enzyme catalyzes the phosphorolysis of uridine and also has some specificity for deoxyuridine and, to a lesser extent, for thymidine (Beachem and Pritchard, 1971). Furthermore, under conditions of constitutive synthesis in a thymine auxotroph deficient in thymidine phosphorylase (the deoA gene), uridine phosphorylase can catalyze the conversion of thymine to thymidine (Sukhodolets, Galeis and Smirnov, 1973).

Expression of the udp gene is negatively controlled by a repressor protein coded for by the cytR gene (Hammer-Jespersen and Munch-Petersen, 1975) and positively by the cyclic AMP (cAMP) - cAMP receptor protein (CRP) complex (Hammer-Jespersen and Nygaard, 1976).

A promoter region of the udp gene is adjacent to the neighboring metE gene and transcription of the udp occurs in the clockwise direction on the E. coli chromosome (Alkhimova, Mironov and Sukhodolets, 1981).

Transcription termination factor Rho is also involved in control of the udp gene: the rho15(ts) mutation in E. coli causes 8-10 fold increase in uridine phosphorylase activity which seems to be a consequence of readthrough transcription from the promoter of the adjacent metE gene (Mironov, 1982).

Selection for the efficient exogenous thymine utilization in thymineless (thy) strains deficient also in thymidine phosphorylase allows to isolate clones with an increased expression of the udp gene. By this way "promoter-up" mutations in the udp gene have been obtained (Mironov and Sukhodolets, 1979). The same approach has been used to select duplications of the udp gene with an increase in gene dosage. Two classes of duplications of the udp gene have been obtained: those involving possibly the udp gene alone and those involving in addition to this gene the adjacent metE gene (Alkhimova, Sukhodolets and Mironov, 1985).

Based on the data that the rho15(ts) mutation causes constitutive synthesis of uridine phosphorylase due to putative transcription from the metE gene, we have proposed that an enhancement of udp expression might be due to deletions which remove Rho-dependent termination site before the udp gene and fuse it to the metE gene. So we used the strain CM992 (HfrH thy deoC deoA1 metE: Tn10) carrying Tn10 inserted into distal part of the metE gene for selecting forms with increased level of udp expression. Mutants were scored on minimal glucose plates supplemented with thymine (20 μg/ml) on which the CM992 bacteria cannot grow due to the lack of thymidine phosphorylase and then tested on resistance to tetracycline.

We have isolated 14 independent mutants which exhibited sensitivity to tetracycline and retained methionine auxotrophy. These mutants presumably arising by deletions that fuse the udp gene to a new promoter were designed udpPf1, udpPf2, etc. However, we have found that 8 out of 14 mutants isolated cannot be transduced to metE$^+$ in P1 transduction crosses with different metE$^+$ donor strains suggesting that the metE gene is disrupted, probably as a consequence of chromosomal inversion.

Kleckner, Reichard and Bottstein (1979) reported that inversions constitute near 50% of the tetracycline-sensitive mutations isolated in strains carrying Tn10(Tet-r). One breakpoint of the inversions usually is the site of Tn10 while the other breakpoint is presumably chosen randomly.

The inversion that is induced by Tn10 located in the metE gene and simultaneously leading to high udp expression occurs in such a way that the inverted chromosomal segment is "before" the udp gene, providing for this gene a new promoter located within inverted segment. As a result of such an event the marker udp::Tn5 has become linked to the genes crp and rpsE, see Figure 1 and Table 1.

According to data given in Table 1 among 100 kanamycin-resistant (Km^R) transductants obtained in transductional crosses (P1)CM1052 (udpPf1 udp:: Tn5) x CM1131 (udpPf1 crp), 23 transductants inherited the donor marker crp$^+$. Likewise, of 106 Crp$^+$ transductants tested, 7 inherited the Tn5 marker Km^R of the donor strain. In a wild-type chromosome, there is no transductional linkage between the udp and crp genes: they are separated by 12 min of the genetic map. We infer from these results that the udpPf1 inversion extends over 12 min of the genetic map and includes crp gene and a proximal part of the metE gene. Accordingly, the metE:Tn5 marker representing a more proximal insertion into metE in relation to the metE::Tn10 (see Figure 1), has lost linkage with udp and zif9::Tn10 in strains carrying the udpPf1 inversion (data not shown).

Relatively low linkage between the udp and crp markers implies that the udpPf1 inversion may include also ribosomal protein cluster (73 min of the E. coli genetic map). Indeed, we have found 54% linkage between the markers Km^R(udp::Tn5) and Spc^R - spectinomycin resistance (the rpsE gene)

Table 1. Mapping of the udpPf1 Inversion Breakpoint adjoining to the udp Gene in Transductional (P1) Crosses. All strains carry the inversion udpPf1

Donor	Recipient	Selected Marker	Inheritance of Unselected Marker
CM1052 (udp::Tn5)	CM1131 (crp)	Km^R (Tn5)	Crp$^+$ - 23/100 (23%)
		Crp$^+$	Km^R - 7/106 (6.6%)
CM1052 (udp::Tn5)	CM1133 (rpsE)	Km^R (Tn5)	rpsE$^+$(Spc^S) - 64/119 (54%)

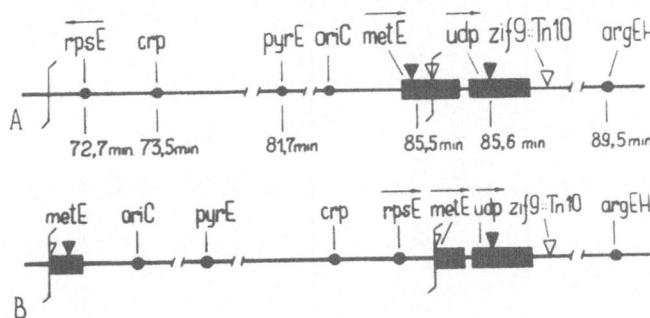

Fig. 1. Location of genetic markers on the E. coli chromosome: A – the wild-type chromosome, B – the inversion udpPf1. Arrows indicate the direction of transcription of corresponding genes; open triangles – positions of the Tn10 insertions; closed – Tn5 insertions. Broken lines mark the region of inversion.

Table 2. Levels of β-galactosidase in Strains carrying Fusion F(udp-lacZ) with and without the Inversion udpPf1. Activity of the Enzyme was determined according to Miller (1972)

Strain F(udp-lacZ)	Inducer (cytidine, 0.2 mM)	Activity of β-galactosidase (units per cell) in Bacteria grown on Minimal Medium with			
		Succinate	Glycerol	Glucose	Glucose + Casamine Acids (0.05%)
CM1054 (without inversion)	−	45	35	32	40
	+	778	375	122	160
CM1056 (udpPf1)	−	275	385	413	520
	+	n.d.	412	383	n.d.
CM1134 (udpPf1,crp)	−	170	n.d.	327	507

n.d. - not determined

in Pl transductional crosses (Table 1). From these data an assumption
was made that udp gene expression in strains carrying udpPfl might be under
the control of a ribosomal protein operon. In accordance to this, the
direction of transcription for ribosomal protein operons located at 73 min
on the E. coli chromosome is opposite to that of the udp gene, while in
the case of the udpPfl inversion the direction of transcription for these
genes should be the same.

In order to test this assumption the udp-lacZ fusion in which the
β-galactosidase (lacZ) gene is transcribed from the udpP promoter was
introduced into the genome of a strain carrying udpPfl (Kulakauskas,
Mironov and Sukhodolets, 1986). The activity of β-galactosidase in the
obtained strain CM1056 was then compared to the activity of the enzyme in
a control strain with the same fusion udp-lacZ but without inversion. It
may be seen from Table 2 that in the case of the control strain CM1054
cytidine serving as a specific inducer of the udp promoter causes 10 fold
increase in β-galactosidase activity on glycerol and about 20 fold increase
on succinate whereas on glucose cytidine causes only 4 fold increase in
enzyme activity that is due to sensitivity of the udp promoter to catab-
olite repression. In contrast, the strain CM1056 carrying udpPfl produces
essentially the same high level of β-galactosidase whether cytidine is
present or not. Furthermore, this level is not significantly changed after
introduction of the crp mutation (causing deficiency for Crp protein) into
the CM1056 genome.

Thus, expression of the udp-lacZ fusion in CM1056 is under the control
of a foreign promoter which seems to be provided by inversion.

In the control strain CM1054, the maximal level of enzyme induction
is found on succinate whereas the minimal level on glucose while in the
strain CM1056 carrying udpPfl enzyme activity exhibits an opposite dependence on the carbon sources used. This dependence is better seen in the
strain CM1134 because of the crp mutation that inhibits activity of the
udpP promoter: the level of enzyme progressively increases from succinate
(170 units) to glucose (327 units) and then to glucose with casamine acids
(527 units). Growth rate of bacteria increases in the same manner on these
carbon sources. To describe these growth-related effects the term "metabolic control" is sometimes used. This type regulation is characteristic
of rRNA and ribosomal protein operons.

The data obtained are compatible with the assumption that expression
of the udp gene in the inversion udpPfl mutant is under the control of a
foreign promoter provided probably by a ribosomal protein operon located
at 73 min on the E. coli chromosome.

REFERENCES

Alkhimova, R. A., Mironov, A. S., and Sukhodolets, V. V., 1981, Genetika,
     17:1730-1736.
Alkhimova, R. A., Sukhodolets, V. V., and Mironov, A. S., 1985, Genetika,
     21:756-762.
Bachmann, B. J., 1983, Microbiol. Rev., 47:180-230.
Beacham, I. R., and Pritchard, R. H., 1971, Mol. Gen. Genet., 110:289-298.
Hammer-Jespersen, K., and Munch-Petersen, A., 1975, Mol. Gen. Genet.,
     137:327-335.
Hammer-Jespersen, K., and Nygaard, P., 1976, Mol. Gen. Genet., 148:49-55.
Kleckner, N., Reichardt, K., and Botstein, D., 1979, J. Mol. Biol., 127:
     89-115.
Kulakauskas, S. T., Mironov, A. S., and Sukhodolets, V. V., 1986, Genetika,
     22:2649-2657.

Miller, J., 1972, Experiments in Molecular Genetics, Cold Spring Harbor, NY, Cold Spring Harbo Laboratory, pp 352–355.

Mironov, A. S., 1982, Genetika, 18:939–946.

Mironov, A. S., and Sukhodolets, V. V., 1979, J. Bacteriol., 137:802–810.

Sukhodolets, V. V., Galeis, V. P., and Smirnov, Yu. V., 1973, Genetika, 9(2):167–169.

REPEATED SEQUENCES IN CHROMOSOMAL DNA OF

STREPTOMYCES AUREOFACIENS 2201

A. Godány, P. Pristaš
J. Muchová and J. Zelinka

Institute of Molecular Biology, Slovak
Academy of Sciences, Dúbravská cesta 21
842 51 Bratislava, Czechoslovakia

Streptomyces aureofaciens 2201 produces an antibiotic substance
inhibiting the growth of Bacillus amyloliquefaciens and Micrococcus luteus.
After transformation of pSA 2201 DNA isolated from Streptomyces aureofaciens 2201 into Streptomyces lividans 66, transformants were obtained
which inhibited the growth of the mentioned testing microorganisms (Figures
1 and 2). The original host strain did not have this property.

Paper chromatography of butanolic extracts of disrupted cells of
Streptomyces aureofaciens 2201 and transformants of Streptomyces lividans
66/2201 indicated the identity of the antibiotic substance produced by
both microorganisms.

Chromosomal DNA of Streptomyces aureofaciens 2201 contains numerous
identical repeated DNA sequences of about 10 kb. On this DNA fragment,
many cleavage sites for several restriction endonucleases were found
(Figure 3).

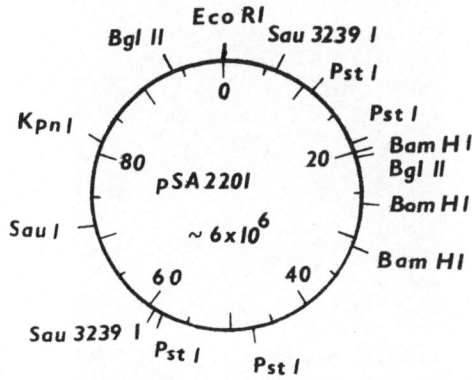

Fig. 1.   Restriction endonuclease cleavage
site map of pSA 2201 plasmid.

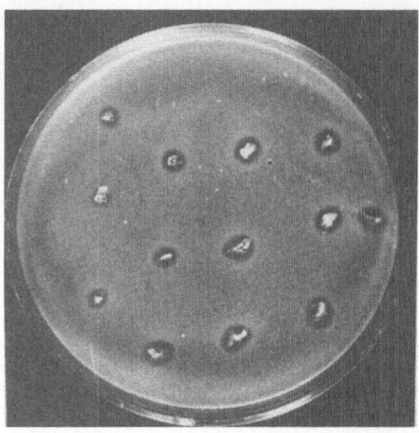

Fig. 2. Inhibition of Bacillus amyloliquefaciens
by clones of Streptomyces lividans 66
transformed with pSA 2201.

The fragments formed after cleavage of chromosomal DNA and of pSA 2201
are of nearly identical size. Cleavage of chromosomal DNA by restriction
endonucleases indicates that the repeated fragment is the integrated pSA
2201 plasmid present in many copies.

For the comparison of the biological properties of pSA 2201 and the
repeated sequence of the chromosomal DNA, parts of this molecule were
cloned into Escherichia coli vectors. From the chromosomal DNA by Bam HI

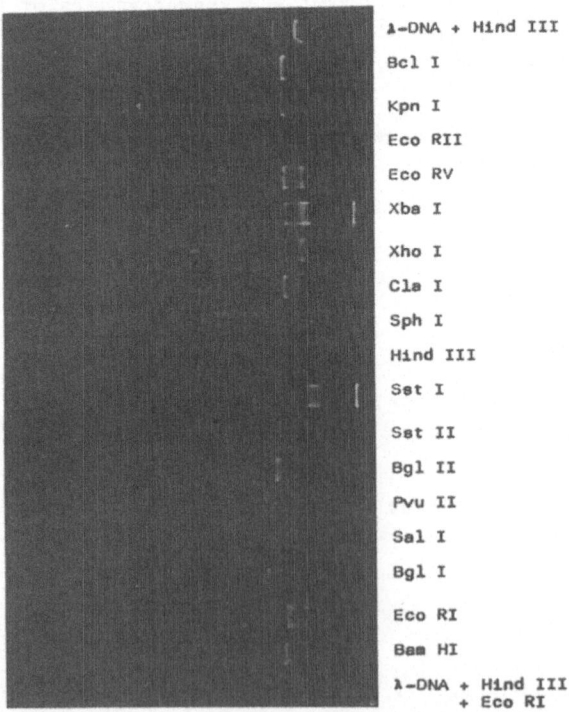

Fig. 3. Electrophoresis of chromosomal DNA of Streptomyces
aureofaciens 2201 after restriction endonuclease
cleavage.

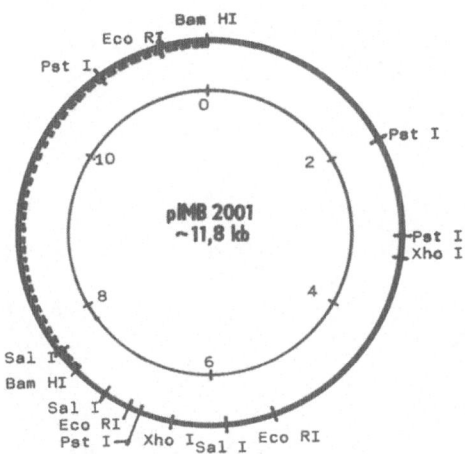

Fig. 4. Restriction endonuclease cleavage site map
of pIMB 2001.

a fragment about 9 kb was cleavaged out and cloned into the pBR 322 cleavage
site of Bam HI. Plasmids with both orientation of the insert were obtained
and designated as pIMB 2001 and pIMB 2007 (Figure 4) and were used for the
transformation of Streptomyces lividans 66. Transformants with both
orientation of the insert inhibited the testing microorganisms (Figure 5).

By these experiments we have proved the presence of the plasmid
replicon for streptomycetes as well as the production of the antibiotically
active compound coded by the repeated sequence of the chromosomal DNA of
Streptomyces aureofaciens 2201. For a more convenient characterization of
the repeated DNA sequence, plasmid pIMB 2050 was constructed consisting of
pUC 18, the gene of kanamycin Tn5 resistance and of the larger Xho I fragment of chromosomal DNA of Streptomyces aureofaciens 2201. This plasmid
is a shuttle vector for E. coli and Streptomyces lividans 66 (Figure 6).
Both hosts containing the pIMB 2050 plasmid produce the antibiotically
active substance.

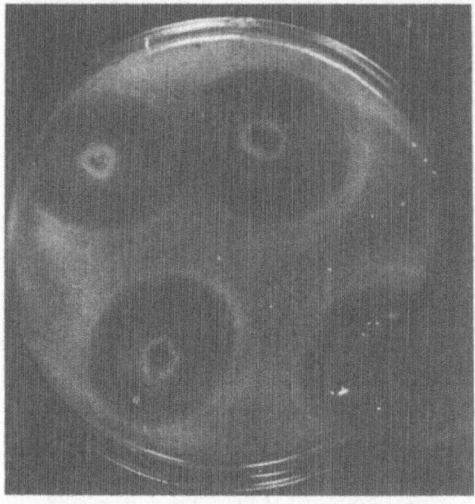

Fig. 5. Inhibition of Bacillus amyloliquefaciens by
clones of Streptomyces lividans 66 transformed
with pIMB 2001 or pIMB 2007.

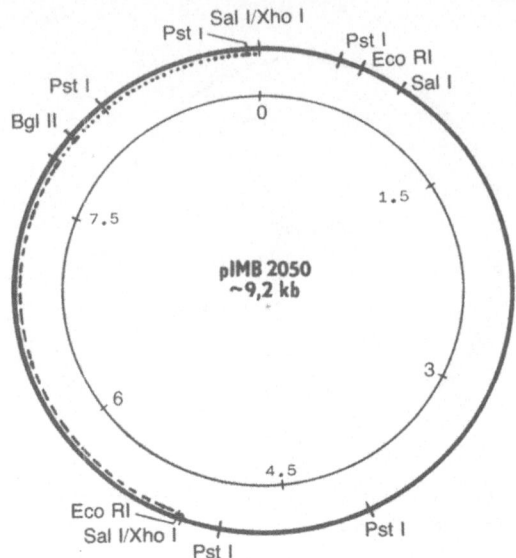

Fig. 6. Restriction endonucleavage site map
of pIMB 2050.

In E. coli the antibiotic substance is not released from the cells
and its production had to be proved by butanolic extraction of disrupted
cells; the extracts were then tested against the testing microorganisms
(Figure 7).

The results of our work could be summarized in the following way:

a) On the repeated segment of S. aureofaciens 2201 chromosomal DNA a
gene (or genes) responsible for the production of the antibiotic substance as well as the replicon for plasmid DNA in streptomycetes are
situated.

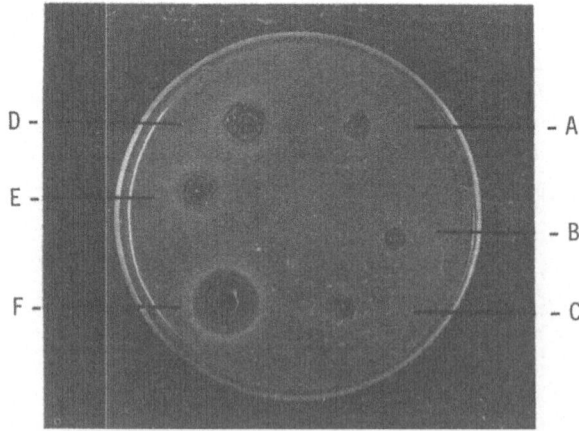

Fig. 7. Inhibition of Micrococcus luteus by
Escherichia coli extracts:
A – without plasmid DNA; B – pBR 322;
C – solvent; D – pIMB 2001; E – extract
from Streptomyces aureofaciens 2201;
F – pIMB 2050

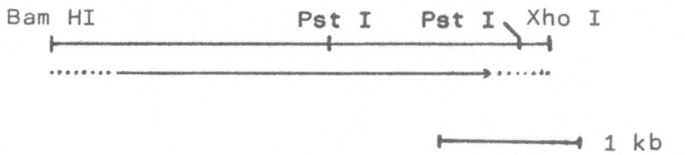

Fig. 8. Restriction endonuclease cleavage site map of
the DNA fragment coding for the production of
the antibiotic substance.

b)   It is evident that the repeated segment of about 10 kb in the chromosomal DNA of S. aureofaciens 2201 is the integrated pSA 2201 plasmid
in many copies.

c)   For the production of the antibiotic substance in E. coli and in
streptomycetes the fragment of DNA in Figure 8 is responsible.   In
E. coli the production of the antibiotic substance is directed by
promoters of the tetracycline and kanamycin resistance genes.   pIMB
2007 plasmid does not induce the production of the antibiotic substance.

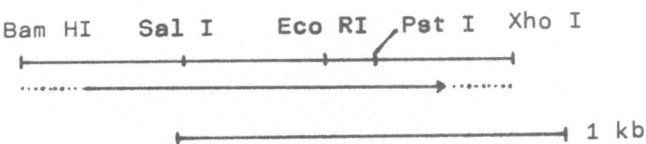

Fig. 9. Restriction endonuclease cleavage site of the
streptomycete replicon from S. aureofaciens
2201.

d)   For the autonomous replication of the plasmid DNAs in streptomycetes
the fragment of DNA on Figure 9 is responsible.

e)   pIMB 2001, 2007 and 2050 plasmid DNAs are stable in E. coli but are
rather unstable in S. lividans 66.   In the plasmids of this micro-
organism numerous deletions were observed.

STABLE GENE DUPLICATION BY "PUZZLE" SHOT GUN CLONING

IN INVERTED REPEATS OF A STREPTOCOCCAL PLASMID

G. S. Steinborn

Zentralinstitut für Genetik und Kulturpflanzenforschung
GDR Academy of Sciences
Gatersleben 4325, GDR

ABSTRACT

The amylase gene of Bacillus amyloliquefaciens was cloned by the usual shot gun method giving a highly expressed but unstable recombinant plasmid in Bacillus host cells. Stable recombinant plasmids were derived by recloning into the streptococcal plasmid pDB101 (18 kb, 5-7 copies per cell) containing long inverted repeats.

For recloning we established a new variant of the shot gun technique termed the "puzzle" method. With this method, a frequently cutting restrictase (e.g., Sau3A) was used for both random linearization of the large vector plasmid molecules as well as partial digestion of the donor plasmid DNA. Thereafter, ligation of donor and vector DNA resulted in a heterologous population of recombinant plasmids harboring different donor fragments in different sites of the vector plasmid. With respect to the level of amylase production and plasmid stability all expected combinations were found: (1) low level - unstable; (2) low level - stable; (3) high level - unstable; and (4) high level - stable. In this way, ligation and selection are comparable with a puzzle and, for application, highly expressed and simultaneously stable recombinant plasmids can be considered as genetically engineered constructs of high fitness. In the case of insertion into the inverted repeats of pDB101, stability of recombinant plasmids was attained by spontaneous, recE independent duplication of the donor fragment containing the amylase gene. Such gene duplication gave a remarkable gene dosage effect with 2-fold amylase production at high expression level.

DIFFERENT TYPES OF REPETITIVE ELEMENTS IN MAMMALIAN GENOME: CLONING AND CHARACTERIZATION OF cDNA COPIES OF TRANSCRIPTS CONTAINING REPETITIVE SEQUENCES

A. P. Ryskov, P. L. Ivanov, O. N. Tokarskaya,
A. V. Akulichev, E. T. Jumanova,
A. G. Jincharadze and L. V. Verbovaya

Institute of Molecular Biology
USSR Academy of Sciences, Vavilov str., 32
117984, Moscow B-334, USSR

INTRODUCTION

Mammalian genome contains different types of repetitive elements. Each of the repetitive families consists of many relative sequences. Primary structure of some of them has been determined and it was shown that homologous sequences could be found in different RNA species. Cloning of cDNA of these transcripts is a possible way of studying functional significance of repetitive elements.

MAJOR TRANSCRIPTS CONTAINING B1 AND B2 REPETITIVE SEQUENCES IN CYTOPLASMIC POLY(A)$^+$RNA FROM MOUSE TISSUES

From the cDNA library prepared on the basis of cytoplasmic poly(A)$^+$RNA of mouse liver, we selected clones hybridizing with the B1 or B2 sequence. The question arises as to the organization of these repetitive sequences in mature poly(A)$^+$RNAs. Figure 1 shows the nucleotide sequences of the two of such clones. They contain full-length copies of B1 (RTM35) and B2 (RTM54) which are located near the 3'-end of corresponding RNAs.

DNA inserts of the pRTM35 (containing B1) and the pRTM54 (containing B2) clones were further used for hybridization analysis. As probes, separated strands of B1 and B2 sequences were taken. Both strands of B1 and B2 hybridized fairly well to nuclear RNA of Ehrlich carcinoma cells (Figure 2). However, the cytoplasmic poly(A)$^+$RNA bound to the "-strand", i.e., it contained only the "+ strand" of B1 and B2 sequences. In cytoplasm of most cell types tested, both B2 and B1 "+ strand" were present in molecules of a messenger size and in the small molecular weight RNA fraction. Small poly(A)$^+$RNA containing B2 (more abundant) and B1 sequences are described elsewhere (Lekakh et al., 1984). At least the small B2 RNA is synthesized with the aid of RNA polymerase III. The content of small B2$^+$RNA varies strongly from one tissue to another being rather high in brain, testicles and negligibly low in lung and heart. Seven of eight tumors analyzed contained a high amount of small B2$^+$RNA. Its content in the cells of six tumors was higher than in such normal tissues as brain. Under the conditions used, small B1$^+$RNA was not recovered in normal tissues tested while in the tumors it was well detected. In general, the content

5'CCTTGTCTCAAAAACAAAAGACATGAGAGCAGGTATGAGGGAGGCATGGAAAGAAGTGGAGCTC

AGGTGCCGAGCAGTGGTGGCGCACGCCTTTAGTCCCAGCACTTGGGAGGCAGAGGCAGGTGGAT

TTCTGGGTTCCAGGCCAGCCTTGTCTACAAAGTGGGTTCTAGGACAGCTAGGGCTATACAGAGA

AACCCGGTCTCAAAACAAACAAACAAAAAAAAAAAAAAAAAAAAAAAAAA 3'   **a**

5'GCATTTTAATCTGAAGTCTCTGGTTGACCCTAAGGATATCGACCTCAGCCCTGTTACAATTGG

CTTTGGCAGTATCCCACGCGAATTTAAACTCTGTGTCATTCCTCGTTCATGAGACCTGCTGCC

CATCATTATCCCTCACAAAATGACTGTTTAAAAAAATGCCAAGCGGGCTGGTGAGATGGCTCA

GTGGGTAAGAGCACCCGACTGCTCTTCCGAAGGTCCGGAGTTCAAATCCCAGCAACCACATGG

TGGCTCACAACCATCTGTATCGAGATCTGACTCTGTCTTCTGGAGTGTCTGAAGACAGCTACA

GTGTACTTACATATAATAAATAAATAAGTCTTTA 3'   **b**

Fig. 1.   Nucleotide sequences of cDNA inserts of RTM35 (a) and RTM54 (b).
Repetitive elements are underlined.

of $B2^+$RNA was much higher than that of $B1^+$RNA. Messenger-size B1 and B2
containing RNAs are represented by few major bands and some smeared
material. Liver and kidney contain the 2 kb $B2^+$mRNA$_x$ described previously
(Ryskov et al., 1983). Other normal tissues tested and all tumors are
practically void of $B2^+$mRNA$_x$. Thus, the expression of corresponding gene
is tissue-specific. Interestingly, in regenerating liver the content of
$B2^+$mRNA$_x$ increases manyfold while in hepatoma cells it completely dis-
appears. Many tumors have a 1.6 kb B2-containing poly(A)$^+$RNA. Unlike
Brickell et al. (1983), we detected this RNA in several normal tissues
(testicles, kidney, lung), its content being practically equal for some
normal and tumor cells. Finally, in some tumor cells one could clearly see
1.4 kb and 0.7 kb B2-containing ply(A)$^+$RNAs. At the same time, only one
major band 1.4 kb long was observed among B1-containing poly(A)$^+$RNA (Figure
2A). The latter was found in tumor cells only. It was especially abundant
in colon carcinoma, plasmacytoma and hepatoma. This poly(a)$^+$RNA was absent
from cytoplasm of all normal tissues tested.

SEQUENCE ORGANIZATION OF SMALL POLYADENYLATED $B2^+$RNA

One may suggest that the formation of small $B1^+$ and $B2^+$RNAs is the
first step in the retroposition of the short repetitive sequences to the
new sites of the genome. Thus, the activation of the small B-type RNAs
transcribed by RNA polymerase III may induce the activation of retroposition. Therefore, the structure and heterogeneity of these small RNAs
was studied. From a cDNA library prepared from low-molecular-weight cytoplasmic poly(A)$^+$RNA of Ehrlich carcinoma cells, we have selected clones
hybridizing to B2 sequence. The DNA of ten clones was isolated and the
nucleotide sequence was determined (Kramerov et al., 1985). This found out
that all cloned sequences were highly homologous to the consensus genomic
B2 repeat. A sequence divergence ranging from 3 to 10% is revealed by the
comparison within any pair of sequences. The same differences also exist
among genomic copies of B2.

176

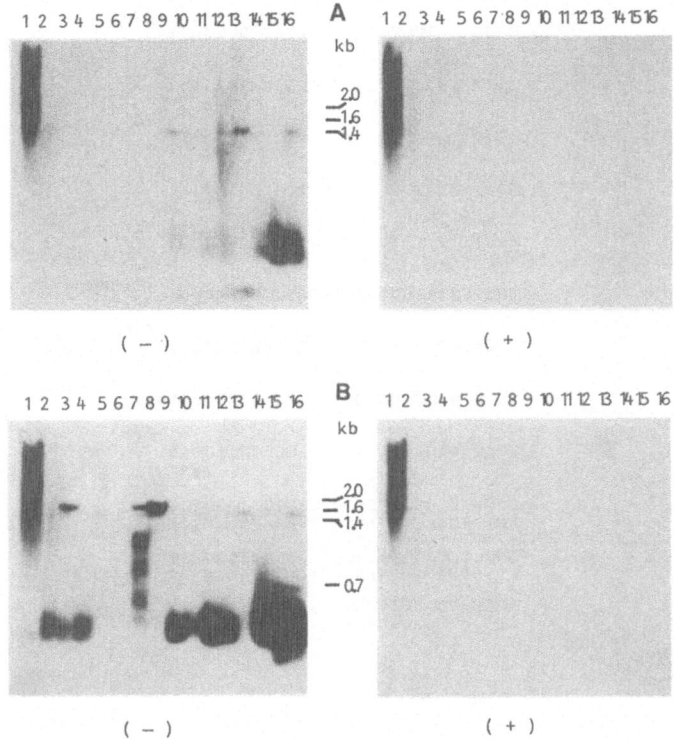

Fig. 2. Northern blot hybridization of mouse cell RNAs to separated DNA
strands (minus and plus) of the B1 (A) and B2 (B). 1: nuclear
poly(A)$^+$RNA from Ehrlich carcinoma cells; 2-6: cytoplasmic
poly(A)$^+$RNAs from testicles (2), kidney (3), brain (4), lung (5),
heart (6), liver (7), regenerating liver (8), hepatoma 22 (9),
mammary gland carcinoma (Ca 755), Lewis lung carcinoma (13),
lymphoma (14), Ehrlich ascites carcinoma (15), and MOPC
plasmacytoma (16).

One can conclude that the small B2$^+$RNA is transcribed from different
copies of the B2 family. All the clones contained a poly(A) tails ranging
in size from 10 to 50 nucleotides. Genomic B2 elements contain shorter
oligo(A). This suggests a post-transcriptional addition of poly(A) to the
small B2$^+$RNA. Primer extension mapping of the 5'-end of the small B2$^+$RNA
was done (Kramerov et al., 1985). As a result, it was shown that at least
a major part of the small B2$^+$RNA was initiated in vivo exactly at the
beginning of B2 element.

REPETITIVE SEQUENCES ACTIVELY TRANSCRIBED IN RAT
BRAIN CELLS: CLONING A cDNA FOR L1 SEQUENCE

In the rat brain cells Sutcliffe et al. (1984) found a set of mRNAs
containing the same element at the 3'-end which they designated as ID-
sequence. The latter has a 78% homology with the 35 bp sequence of 5'-end
of B2 element. The authors detected brain-specific small RNA hybridizing
to the ID-sequences and suggested that the ID$^+$RNA might play a role in
activating ID-containing brain-specific genes. However, further studies of

177

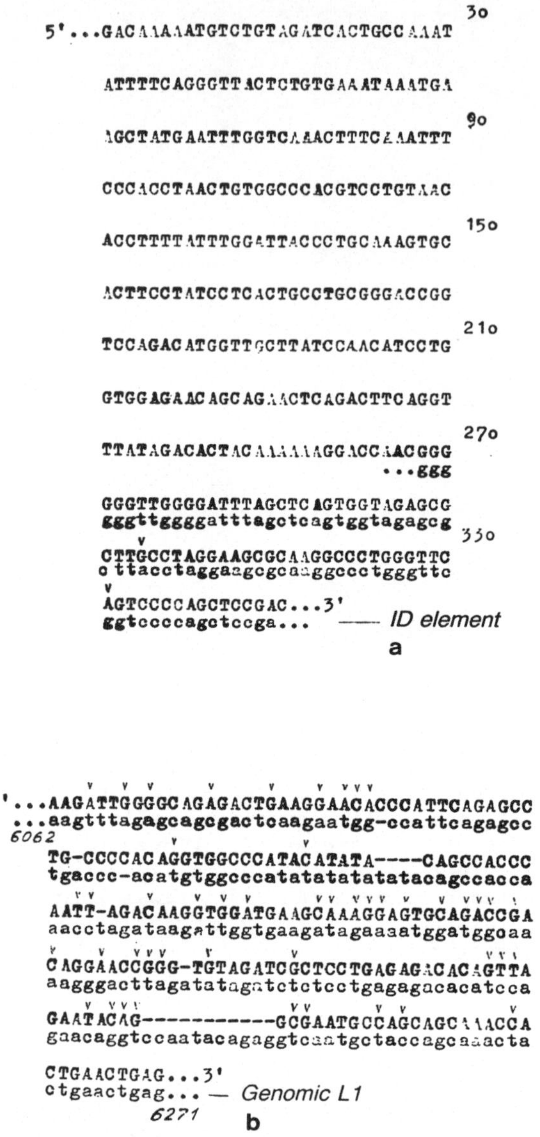

5'...GACAAAAATGTCTGTAGATCACTGCCAAAT
                                          30

ATTTTCAGGGTTACTCTGTGAAATAAATGA

AGCTATGAATTTGGTCAAACTTTCAAATTT
                                          90

CCCACCTAACTGTGGCCCACGTCCTGTAAC

ACCTTTTATTTGGATTACCCTGCAAAGTGC
                                          150

ACTTCCTATCCTCACTGCCTGCGGGACCGG

TCCAGACATGGTTGCTTATCCAACATCCTG
                                          210

GTGGAGAACAGCAGAACTCAGACTTCAGGT

TTATAGACACTACAAAAAAGGACCAACGGG
                                          270
                          ···ggg
GGGTTGGGGATTTAGCTCAGTGGTAGAGCG
gggttggggatttagctcagtggtagagcg
                                     330
CTTGCCTAGGAAGCGCAAGGCCCTGGGTTC
cttacctaggaagcgcaaggccctgggttc
                          v
AGTCCCCAGCTCCGAC...3'
ggtccccagctccga...   ──── ID element
                        a

       v v v   v   v   v vvv
'...AAGATTGGGGCAGAGACTGAAGGAACACCCATTCAGAGCC
...aagtttagagcagcgactcaagaatgg-ccattcagagcc
6062   v
    TG-CCCCACAGGTGGCCCATACATATA----CAGCCACCC
    tgaccc-acatgtggcccatatatatatatacagccacca
     v v   v   v vv vvv v   v vvv v
AATT-AGACAAGGTGGATGAAGCAAAGGAGTGCAGACCGA
aacctagataagattggtgaagatagaaaatggatggcaa
v   v vvv   v   v   v v   v   v v v
CAGGAACCGGG-TGTAGATCGCTCCTGAGAGACACAGTTA
aagggacttagatatagatctctcctgagagacacatcca
   v vvv   v v   vvv   v
GAATACAG----------GCGAATGCCAGCAGCAAACCA
gaacaggtccaatacagaggtcaatgctaccagcaaacta

CTGAACTGAG...3'
ctgaactgag...   ── Genomic L1
    6271   b

Fig. 3. Nucleotide sequences of cDNAs containing ID (a) and L1 (b).

the repetitive element transcription in brain cells is necessary. We have selected recombinants harboring repetitive elements from a cDNA library prepared from polysomal poly(A)$^+$RNA of rat brain. Hybridization and sequencing analysis revealed that all selected cDNAs might be subdivided into four types containing ID (0.5%), B2 (0.5%), L1 (0.1%) and the so-called simple sequences (0.3%). Figure 3 shows the examples of the cDNAs containing ID and L1. L1 is rate equivalent of human KpnI and mouse MIF I (long repetitive elements). In both cases, (+) strands of the repetitive elements are present in corresponding RNA transcripts. To our surprise, the contents of L1 transcripts in rat brain polysomal poly(A)$^+$RNA is rather high. Previously it had been shown for many cells studied that L1 transcripts were practically restricted to nuclei (see in Weiner et al., 1986). The cloned L1 RNA is corresponding to the 3'-end, the most

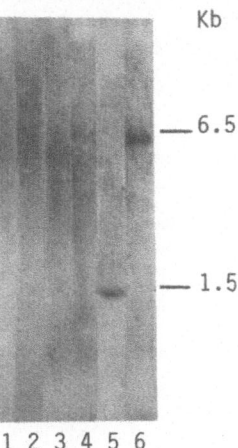

Kb

— 6.5

— 1.5

1 2 3 4 5 6

Fig. 4. Sputhern blot hybridization analysis of rat genomic DNA using L1
cDNA as a labeled probe. 1: BamHI; 2: EcoRI; 3: EcoRV; 4:
HindIII; 5: BglII; 6: BcnI.

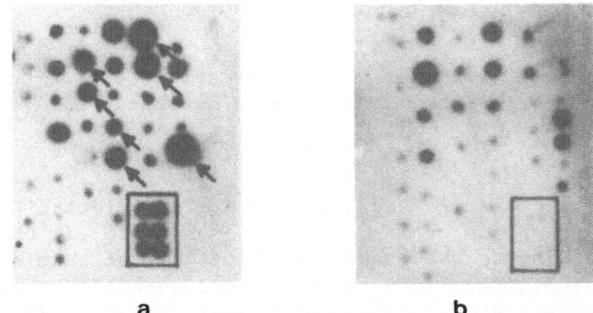

a   b

Fig. 5. Selection of mouse liver cDNA clones containing simple sequences.
Colony hybridization was done with ³²P-labeled DNA in the absence
(a) and in the presence (b) of salmon DNA. Control clones
containing simple sequences are boxed.

divergent zone of genomic L1 element. Indeed, in our preliminary experiments we found out that part of L1 was 20 times less frequent than the
central part of L1. Figure 4 shows a restriction analysis of rat genomic
L1 element. In two cases (BcnI and BglII restriction endonucleases), we
must have detected minor sub-families of genomic L1 having 1.5 bp and 6.5
bp in length.

(AC)$_n$·(GT)$_n$ IS A MAJOR ACTIVELY TRANSCRIBED
SIMPLE SEQUENCE OF MAMMALIAN GENOME

Simple sequences are widespread in a broad range of eukaryotic
genomes. Only recently they have excited interest because of their pro-
pensity to adopt the left-handed Z-DNA form. To identify cDNA clones
containing simple sequences, we have done colony hybridization with total
labeled DNA in the presence and in the absence of salmon DNA as a
competitor. All of the selected mouse liver cDNA clones can hybridize to
total DNA only in the absence of salmon DAN (Figure 5). Similar clones

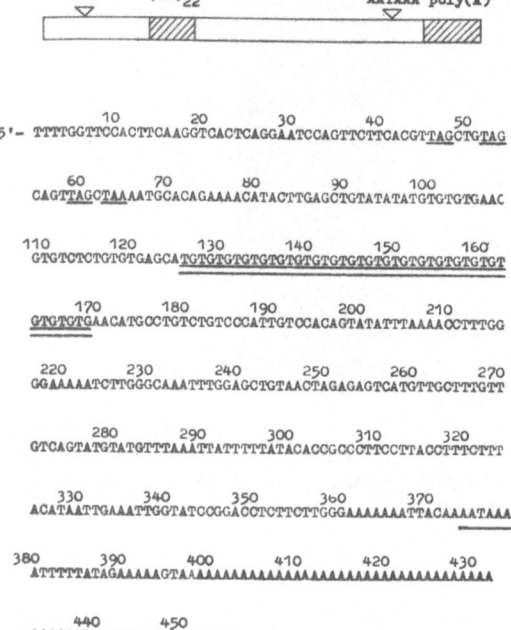

Fig. 6. Schematic representation of the organization and nucleotide sequence of cDNA containing simple sequence $(GT)_n$ and 3'-poly(A).

containing presumptive simple sequences were also selected from the cDNA libraries of rat and human brains. Their contents in all the cases were about 0.3%. According to the sequencing data, all of them contained $(CA)_n \cdot (GT)_n$. It means that $(CA)_n \cdot (GT)_n$ is the major type of the simple repetitive sequences transcribed in cytoplasmic poly(A)$^+$ RNA of the mouse liver and rat and human brains.

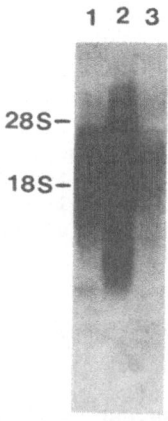

Fig. 7. Northern blot hybridization of mouse cell RNAs to $(TG)_n \cdot (CA)_n$. 1: cytoplasmic poly(A)$^+$ RNA of Ehrlich ascites carcinoma; 2: polysomal poly(A)$^+$ RNA of mouse liver; 3: cytoplasmic poly(A)$^-$ RNA of Ehrlich ascites carcinoma.

Figure 6 shows the organization and the nucleotide sequence of one of the clones. It contains long poly(A) corresponding to the 3'-end of mRNA. Polyadenylation signal AATAAA is located 15 nucleotides upstream the poly(A) stretch. $(GT)_n$ element is separated by about 200 nucleotides from 3'-poly(A) located in the same strand. That part of cDNA probably corresponds to the untranslated zone of mRNA. That is clear from the fact that the terminator codons can be found before $(GT)_n$. The nature of that mRNA remains unknown. Northern blot hybridization experiments were developed for RNAs of different mouse tissues (Figure 7). They revealed complex hybridization patterns consisting of heterodisperse and prominent bands of $(GT)_n^+$ and $(CA)_n^+$ RNAs specific for each tissue.

REFERENCES

Brickell, P. M., Latchman, D. S. Murphy, D., Willison, K., and Rigby, P. W. J., 1983, Nature, 306:756.

Kramerov, D. A., Tillib, S. V., Ryskov, A. P., and Georgiev, G. P., 1985, Nucl. Acids Res., 13:6423.

Lekakh, I. V., Kramerov, D. A., Markusheva, T. V., Ryskov, A. P., and Georgiev, G. P., 1984, Dokl. Acad. Nauk, USSR, 277:1006.

Ryskov, A. P., Ivanov, P. L., Kramerov, D. A., and Georgiev, G. P., 1983, Nucl. Acids Res., 11:6541.

Sutcliffe, J. C., Milner, R. G., Gottesfeld, J. M., and Lerner, R. A., 1984, Nature, 308:237.

Weiner, A. M., Deininger, P. L., and Efstratiadis, A., 1986, Annu. Rev. Biochem., 55:631.

NUCLEOTIDE SEQUENCE ANALYSIS OF THE 5'-END OF CHICKEN c-myb GENE

P. Urbánek, M. Dvořák, M. Trávníček, P. Bartůněk
and V. Pačes.

Institute of Molecular Genetics, Czechoslovak Academy
of Sciences, 16637 Prague, Czechoslovakia

INTRODUCTION

A normal cellular gene, c-myb, related to the v-myb oncogenes of acute
avian leukemia viruses (AMV and E26), has been found to be present in
genomes of vertebrates including man (Bergmann et al., 1981). A rather
distantly related gene has also been found in <u>Drosophila</u> (Katzen et al.,
1985).

The c-myb proto-oncogene is characterized by evolutionary stability and
preferential expression in immature cells of the myeloid and lymphoid
lineages (Gonda et al., 1982; Westin et al., 1982). This, together with
the fact that retroviruses carrying v-myb interfere with myeloid differentiation, suggests a role for c-myb in the control of normal hema-
topoiesis. However, rècent studies suggest two possible roles of c-myb in
cell growth and differentiation: a thymus specific function and a more
general function in cellular proliferation not restricted to hematopoietic
cells (Thompson et al., 1986).

Cloning and sequencing of the region of the chicken c-myb gene homologous to the v-myb sequences revealed that v-myb of AMV> has originated
by precise splicing of seven (EL - E7) c-myb exons (Klempnauer et al.,
1982).

However, comparison of the sizes of transcriptional and translational
products of v-myb (2.1 kb mRNA and 48 kDa protein) and c-myb (4 kb and 75
kDa) genes (Gonda et al., 1981; Gonda et al., 1982; Boyle et al., 1983;
Klempnauer et al., 1983) implies the presence of additional c-myb-specific
exons. The nucleotide sequences of cDNA derived from chicken c-myb mRNA
(Gerondakis and Bishop, 1986; Rosson and Reddy, 1986) clearly show that the
c-myb specific nucleotide sequences are located both upstream and down
stream from the region homologous to v-myb.

Here we characterize a chicken genomic DNA fragment (8.2 kb) containing nucleotide sequences transcribed into the c-myb mRNA and located in
the genome upstream from the region of homology with v-myb. By restriction
mapping of this fragment, hybridization of its subfragments to the c-myb
mRNA and, finally, by sequencing its selected parts we were able to characterize three additional c-myb-specific exons (designated below e_x, e_1, e_2)

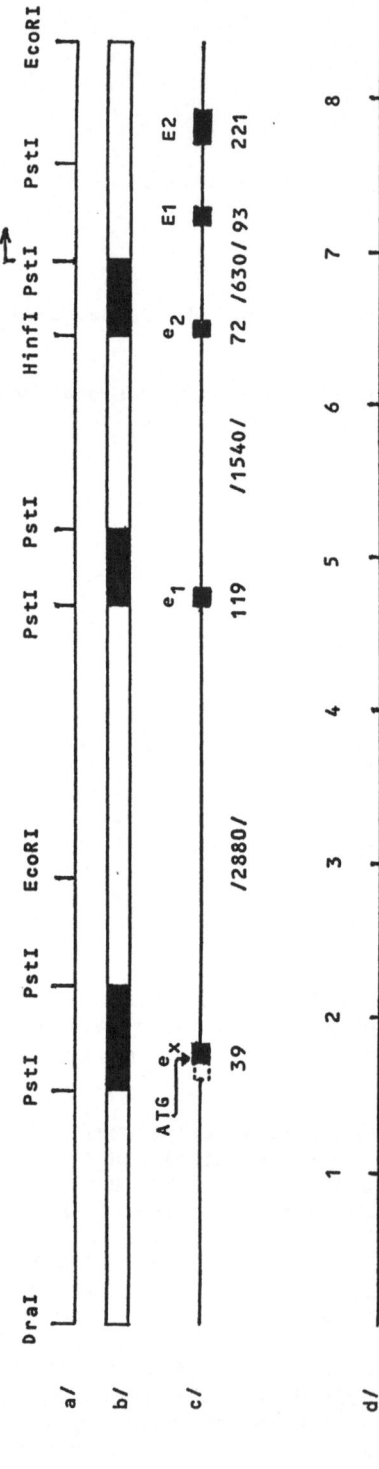

Fig. 1. The exon-intron structure of the chicken 8.2 kb c-myb DNA fragment. a) Complete EcoRI and pSTI and partial DraI and HinfI map of the fragment. The arrow indicates the c-myb region sequenced by Klempnauer et al. (1982). b) Hybridization of subfragments with the chicken thymus c-myb mRNA. Positive hybridizations are represented by black boxes (not shown for exons E1 and E2 – see paragraph c). c) The exon-intron structure of the fragment inferred from comparison of the nucleotide sequence presented in this paper with that of c-myb cDNA (Gerondakis and Bishop, 1986; Rosson and Reddy, 1986). The numbers indicate the length of exons in base pairs, except for exon e, where only the number of nucleotides presented in this paper (see Figure 2) is indicated. The putative initiation ATG codon (see text) is indicated. Numbers in brackets show the approximate length of introns. Exons E1 and E2 were detected and together with the surrounding introns were sequenced by Klempnauer et al. (1982). They are homologous to the v-myb oncogene. d) Scale in kb.

and determine the exon-intron structure of this part of the c-myb gene.
These results allow formulation of some evolutionary conclusions.

EXON-INTRON STRUCTURE

An 18-kb DNA fragment containing the 5'-end of the c-myb gene was
isolated from a chicken genomic DNA library (manuscript in preparation). A
partial restriction map of a portion (8.2 kb fragment) of this clone is
given in Figure 1a. Selected subfragments were used to probe chicken
thymus RNA Northern blots. Hybridization data confirmed the position of
two c-myb-specific exons (e_1, e_2) described in our previous paper (Dvorak
et al., 1987) and revealed in the presence of exon e_x located further
upstream (Figure 1b).

We determined the nucleotide sequence of this region. The precise
exon-intron structure was deduced from a comparison of this sequence with
the nucleotide sequences of c-myb cDNA (Gerondakis and Bishop, 1986; Rosson
and Reddy, 1986) (Figure 1c).

The nucleotide sequences of c-myb-specific exons e_1 and e_2 with the
flanking portions of introns and the 3'- end of exon e_x with the proximal
portion of the downstream intron and presented in Figure 2. When joined
together, the three exons encode 71 amino acids, which are identical with
the N-terminal portion of the c-myb protein predicted from the nucleotide
sequence of cDNA by Gerondakis and Bishop (1986). The nucleotide sequence
encolding these amino acids starts with the ATG codon in exon e_x and our
preliminary nucleotide sequences upstream from this ATG codon (data not
shown) do not reveal any other possible translation initiation site. Thus
we assume that this ATG codon is the initiation codon of the c-myb protein.
This is in agreement with the results of several other authors (Bender and
Kuehl, 1986; Gerondakis and Bishop, 1986; Majello et al., 1986). At present
no firm conclusion can be drawn about the length and exon-intron structure
of the leader portion of the c-myb mRNA upstream from the ATG codon.

The exon-intron junctions of e_x, e_1 and e_2 correspond to the consensus
sequences of donor and acceptor splice sites (Table 1).

EVOLUTIONARY CONSIDERATIONS

The amino acid sequences of c-myb proteins of man, mouse and chicken,
predicted from the corresponding cDNA sequences, are highly homologous (for
a comparison, see Majello et al., 1986). However, the actual degree of
evolutionary conservation of the exon-intron structures and of the intronic
nucleotide sequences of this gene remains to be elucidated. Comparison of
the nucleotide sequence of the exon e_x-intron junction in the chicken gene
with the corresponding sequence in the mouse gene (Bender and Kuehl, 1986)
shows a high degree of homology of the protein-coding exon regions and of
the donor splice sites (Figure 3). The exon e_x-exon e_1 junction is in the
same position in the c-myb mRNA sequences of chicken and mouse. The noncoding nucleotide sequences, both in exon e_x upstream from the ATG codon
and in the intron downstream from exon e_x differ between the two organisms
considerably (Figure 3). This indicates a high speed of divergence of the
noncoding regions of the c-myb gene during vertebrate evolution. A
detailed comparison would require more sequence data on the murine and/or
the human c-myb gene.

Evolutionary considerations can be extended further by comparing the
c-myb genes and amino acid sequences of vertebrates with those of the
Drosophila (Katzen et al., 1985; Gerondakis and Bishop, 1986). A region of

```
    ┌►┌──────┐
    │  │repeat 1│            exon e₁ ◄─┬─► exon e₂
  ┌─┴──└──────┘               ┌─┐      │
  │ L G K T R W T R E E│D E K │L K K L V E
  CTAGGGAAAACCAGGTGGACCCGTGAAGAGGATGAGAAACTGAAGAAACTTGTGGAAC
```

                                      exon e₂ ◄─┬─► exon El
  Q   N   G   T   E   D   W   K   V   I   A   S   F   L   P │ N   R   T   D   V
  AGAATGGCACAGAAGACTGGAAAGTCATTGCCAGTTTCCTTCCTAATCGGACAGATGT

```
                                            ┌►┌──────┐
                                            │  │repeat 2│
  Q   C   Q   H   R   W   Q   K   V   L   N   P   E│L   I   K   G   P   W
  TCAGTGCCAGCACCGGTGGCAGAAAGTATTAAACCCAGAACTTATCAAAGGTCCATGG
```

```
            exon El ◄─┬─► exon E2
                      │
  T   K   E   E   D   Q   R │ V   I   E   L   V   Q   K   Y   G   P   K   R
  ACTAAAGAGGAGGATCAAAGGGTAATAGAACTCGTGCAGAAATACGGTCCAAAGCGCT
```

  W   S   V   I   A   K   H   L   K   G   R   I   G   K   Q   C   R   E   R   W
  GGTCGGTCATTGCTAAGCATTTGAAGGGAAGGATTGGAAAACAGTGCAGGGAGAGGTG

```
                          ┌►┌──────┐
                          │  │repeat 3│
  H   N   H   L   N   P   E│V   K   K   T   S   W   T   E   E   E   D   R
  GCACAACCATCTGAATCCAGAAGTGAAGAAAACCTCCTGGACAGAAGAGGAAGATAGA
```

  I   I   Y   Q   A   H   K   R   L   G   N   R   W   A   E   I   A   K   L
  ATTATTTACCAGGCACACAAGAGACTGGGAAACAGATGGGCAGAAATTGCAAAGTTGC

```
  exon E2 ◄─┬─► exon E3
            │
  L   P   G   R │ T   D   N   A   I   K   N   H   W   N   S   T   M   R   R   K
  TGCCTGGACGGACTGATAACGCTATCAAGAACCACTGGAATTCCACCATGCGCCGGAAG
```

Fig. 2.   The nucleotide sequences of the c-myb-specific exons (in the
          5'-to-3' orientation of the mRNA-like strand). a) The 3'-end of
          exon eₓ with a portion of the downstream flanking intron. The
          presumable initiation ATG codon of the myb protein is underlined.
          b) Exon e₁ with flanking portions of introns. c) Exon e₂ with
          flanking portions of introns. For amino acid symbols see
          Figure 4.

Table 1. The exon-intron junctions of the c-myb-specific exons eₓ, e₁, e₂.
          The splice sites consensus sequences are from Mount (1982).

| | splice donor site | | splice acceptor site |
	exon	intron		exon
Consensus	$_C^A$AG	GT$_A^G$AGT............$(_T^C)_{15-30}$(N)$_T^C$AG	G
eₓ-intron-e₁	CAG	GTAACG......GCTGTGTTTCCTGCTGCAG		C
e₁-intron-e₂	GAG	GTAATT......TTTATTTTTAATGATTCAG		G
e₂-intron	CCT	GTAGGT......		

```
                                        exon e_x ◄─┬─► intron
                          M   A   R   R   P   R   H   S
ch    AGCGGCC-GCCGCGAGG[ATG]GCCCGGAGACCCCGGCACAGGTAACGGGGCG
mu    CCCGGCCCGCCTCG-CC[ATG]GCCCGGAGACCCCGACACAGGTAATGGGGAG
dif.  **    *  *  ***              *        *      *


ch    GGGGGTC-CGCGCTGTCAC---ACGGGGCGCGGGGCTGC
mu    GCTGAGAGGGCGGCCTCGGGCTAGGGCGCGCGGGGATCC
dif.  ** *****  ***  ***** *  *        *  *
```

Fig. 3. Comparison of the chicken DNA segment containing the exon e_x-intron junction with the corresponding nucleotide sequence of murine DNA (Bender and Kuehl, 1986). ch, chicken sequence; mu, murine sequence; dif., differences in the nucleotide sequences of the chicken and mouse (nucleotide replacements or deletions). The differences are indicated by asterisks. Insertion of dashes into the sequence signifies gaps, which has allowed the homology to be maximized. The presumable initiation ATG codon of the c-myb protein is boxed. The donor splice site is underlined. The amino acids encoded by the exon (identical in both organisms) are designated in the conventional single letter code (aligned with the second letter of each codon) (see Figure 4).

amino acid homology in c-myb proteins of these evolutionarily distant organisms is located in the N-terminal part of the vertebrate protein. It consists of three tandemly organized direct repeats of 51-52 amino acids each (Ralston and Bishop, 1983; Gonda et al., 1985; Gerondakis and Bishop, 1986). The amino acid sequence of this region is identical in man and mouse and only three amino acids (out of 155) are different in the chicken (Majello et al.,1986). Comparison with Drosophila shows 66.5% homology (Figure 4); when similar amino acids are considered homologous, 78% homology is obtained. Parallel alignment of the three repeats shows that there are 22 most conservative positions (out of 52) (Figure 4). The evolutionary conservation of this region indicates that it may be vital for c-myb protein function.

The exons e1 and e2 described above contain a part of this conserved c-myb domain. Since all exon junctions of this domain are now known, a conclusion about the origin of the c-myb domain. Since all exon junctions of this domain are now known, a conclusion about the origin of the c-myb introns in this region can be drawn. There are four exon junctions located within the repeats. Since they are in different positions in each repeat (Figure 5), it is likely that the ancestral intronless monomer was first triplicated and then introns were inserted into the repeats. This sequence of molecular events is supported by the finding that the homologous region of the Drosophila c-myb gene contains no introns (Katzen et al., 1985). Thus the whole evolutionary scenario might have been as follows: The monomer was triplicatd in the genome of a common ancestor of arthropods and chordates. Some 700 million years ago evolutionary divergence engendered these two phyla (Dobzhansky et al., 1977) and from this point the conserved domain began to evolve at least in two separate lineages. Only later the introns were incorporated into the c-myb domain of chordates, while this gene portion remains intronless in the Drosophila.

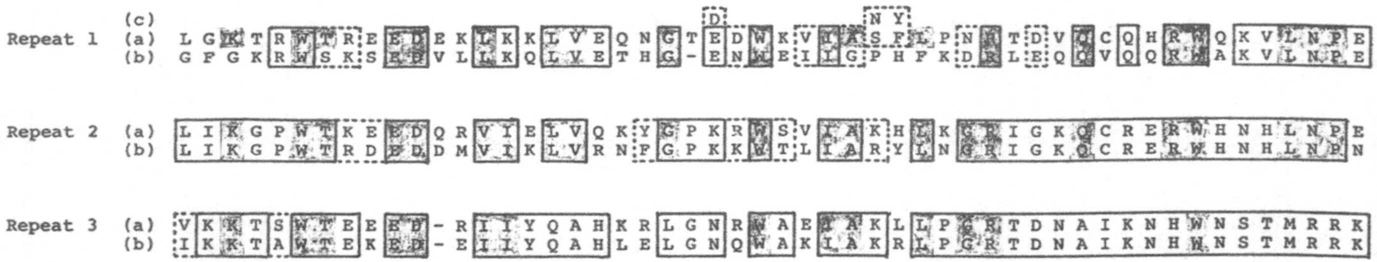

Repeat 1

(c) L G M TR TRE E KK KL VEQNGTEDWKV NY HASFLPN T:DVCQHRQK VLNPE

(b) G F G KR W.S.K.S E HVL LEQ EN HG-EN HPFKD LEQVQQA KVLNEE

Repeat 2

(a) L I K G P W K E E D Q R VIELVQ K:Y:AKH IGK CRERW HNHLNPE

(b) L I K G P W T R D D D M VIKLVRN:F:G P K K T:L:ARYLN RIGKQCRERW HNHLNPN

Repeat 3

(a) V K T S N T E E - R I:Y Q A H K R L G N R A E K L P G TDNAIKNHWWNS TMRRK

(b) I K T A E K E E - E I:Y Q A H L E G N Q W A K L A K R P G R TDNAIKNHWWNS TMRRK

Fig. 4. Amino acid homologies among the three tandem repeats of the chicken, mouse, man and Drosophila c-myb gene products. a) Chicken amino acid sequence (Gerondakis and Bishop, 1986; Rosson and Reddy, 1986). The human (Majello et al., 1986) and murine (Gonda et al., 1985; Bender and Kuehl, 1986) sequences are identical with that of the chicken except for three positions in the first repeat (line c). b) Amino acid sequence of Drosophila (Katzen et al., 1985).

The sequences are compared in two ways: (1) homology between vertebrate and Drosophila proteins (indentical amino acids, solid boxes; similar amino acids, dotted line boxes), (2) homology among the repeats (positions conserved in both vertebrate species and Drosophila in all three or only two repeats are shadowed). Similar comparisons have been presented by Katzen et al. (1985) and Gerondakis and Bishop (1986).

Amino acids encoded by the exons are designated in the conventional single-letter code: A, alanine; R, arginine; N, asparagine; D, aspartic acid; C, cysteine; E, glutamic acid; Q, glutamine; G, glycine; H, histidine; I, isoleucine; L, leucine; K, lysine; M, methionine; F, phenylalanine; P, proline; S, serine; T, threonine; W, tryptophan; Y, tyrosine; V, valine.

```
                          exon e  ←─┬─→  intron
               M   A   R   R   P   R   H   S
a)  AGCGGCCGCCGCGAGGATGGCCCGGAGACCCCGGCACAGGTAACGGGGCGGGGG

    GTCCGCGCTGTCACACGGGGCGCGGGGCTGC


                   intron  ←─┬─→  exon e₁
                            │   I   Y   S   S   D   D
b)  TTCCGCTCTGTGTCTCGCTGTGTTTCCTGCTGCAGCATATACAGCAGCGATGAC

     D   E   E   D   V   E   M   Y   D   H   D   Y   D   G   L   L   P   K
    GATGAAGAAGATGTTGAGATGTACGACCACGATTACGACGGCCTGCTTCCTAAG

     A   G   K   R   H   L   G   K   T   R   W   T   R   E   E
    GCTGGGAAACGTCACCTAGGGAAAACCAGGTGGACCCGTGAAGAGGTAATTGCC
                                    exon e₁  ←─┴─→  intron
    GGTTTGGTCATCTCCTGCATGGGGGA


                   intron  ←─┬─→  exon e₂
                            │  D   E   K   L   K   K   L   V   E   Q
c)  TGTTTATTTATTTTTAATGATTCAGGATGAGAAACTGAAGAAACTTGTGGAACA

     N   G   T   E   D   W   K   V   I   A   S   F   L   P
    GAATGGCACAGAAGACTGGAAAGTCATTGCCAGTTTCCTTCCTGTAGGTAGACA
                                    exon e₂  ←─┴─→  intron
    TTAGTTTCTGTTTCCAATTCTAGT
```

Fig. 5. Position of exon junctions within the nucleotide sequence of three
tandem repeats in the chicken c-myb cDNA (Gerondakis and Bishop,
1986, Rosson and Reddy, 1986). Exon junctions E1/E2 and E2/E3 are
from Klempnauer et al. (1982). For amino acid symbols, see
Figure 4.

Acknowledgements

     We thank Drs. Z. Hostomský, V. Pečenka and S. Hájek for helpful
comments.

REFERENCES

Bender, T.T., Kuehl, W.M. 1986, Proc. Natl. Acad. Sci. USA, 83; 3204-3208.
Bergmann, D.G., Souza, L.M., Baluda, M.A. 1981, J. Virol 40; 450-455.
Boyle, W.J., Lipsick, J.S., Reddy, E.P., Baluda, M.A. 1983, Proc. Natl.
     Acad. Sci. USA, 80; 2834-2838.
Dobzhansky, Th., Ayala, F.J., Stebbins, G.L., Valentine, J.W., 1977,
     Evolution, W.H. Freeman, San Francisco.
Dvorak, M., Travnicek, M., Sulova, A., Riman, J. 1987, Folia biologica
     (Prague) 33; 1-10.
Gerondakis, S., Bishop, J.M., 1986, Mol. Cell. Biol. 6; 3677-3684.
Gonda, T.J., Sheiness, D.K., Fanshier, L., Bishop, J.M., Moscovici, C.,
     Moscovici, M.G. 1981, Cell 23, 279-290.
Gonda, T.J., Sheiness, D.K., Bishop, J.M. 1982, Mol. Cell. Biol. 2;
     617-624.
Gonda, T.J., Gough, N.M., Dunn, A.R., de Blankier, J. 1985, Ebmo. J. 4,
     2003-2008.
Katzen, A.L., Kornberg, T.B., Bishop, J.M. 1985, Cell 41; 449-456.

Klempnauer, K.-H., Gonda, T.J., Bishop, J.M., 1982, <u>Cell</u> 31; 453-463.

Klepnauer, K. -H., Ramsay, G., Bishop, J.M., Moscovici, M.G., Moscovici, C., McGrath, J.P., Levinson, A.D. 1983, <u>Cell</u> 33; 345-355.

Majello, B., Kenyon, L.C., Dalla-Favera, R. 1986, <u>Proc. Natl. Acad. Sci. USA</u> 83; 9636-9640.

Mount, S.M. 1982, <u>Nucl. Acids Res.</u> 10; 459-472.

Ralston, R., Bishop, J.M. 1983, <u>Nature</u> 306, 803-806.

Rosson, D., Reddy, E.P. 1986, <u>Nature</u> 319; 604-606.

Thompson, C.B., Challoner, P.B., Nelman, P.E. Groudine, M. 1986, <u>Nature.</u> 319; 374-380.

Westin, E.H., Gallo, R.C., Arya, S.K., Eva, A., Souza, L.M., Baluda, M.A., Aaronson, S.A., Wong-Staal, F. 1982 <u>Proc. Natl. Acad. Sci.</u> USA 79; 2194-2198.

CLONING AND CHARACTERIZATION OF HUMAN GENES

EXPRESSED IN NEURAL CELLS

S. A. Limborska, P. A. Slominsky,
N. E. Maleeva and S. A. Korneev

Institute of Molecular Genetics
USSR Academy of Sciences
46 Kurchatov Sq., Moscow 123182, USSR

The study of genome activity in highly differentiated cells provides basic data on the selective use of genetic information in a concrete situation. It is particularly interesting to analyze the work of genes in the nerve tissue, which is the most complexly organized system of highly differentiated cells. Genetic engineering makes it possible to clone genes functioning in a particular kind of cells using appropriate mRNA or cDNA copies as starting material. It is important to analyze not only active unique genes but also the repetitive sequences working in differentiated cells which may be involved in the regulation of gene activity.

In the present study clones containing repetitive genome sequences of the Alu family have been obtained from a bank of cDNA from the human brain cells and skin fibroblasts. The primary structure of these clones is shown in Figure 1. In all cases the Alu sequence is located at the 3'-end of the RNA concerned. All the Alu segments proved to be typical representatives of this family of repeats, their deviation from the Alu-consensus amounting to 13% on an average (Deininger et al., 1981).

Computer sequence analysis revealed extended open translation frames in one direction. The short sequence GAGGCCGAGGT, which is similar to the T-antigen-binding region of SV-40 DNA, has one and the same point substitution T→C in all the clones that we examined (Rogers, 1985).

DNA insertions from the clones obtained were used for hybridization analysis of the Alu transcripts within cytoplasmic RNA from various human tissues. Figure 2 presents the results for cytoplasmic poly(A)-containing RNA from normal human tissues (spleen, stomach, brain) and tumors (adrenal cortex adenocarcinoma and neurinoma). The distribution of the hybridized material is heterogeneous in all samples. Note the pronounced hybridization zone in the 7S region, which is present in all cases except in the spleen RNA (the major hybridization band is located somewhat higher in the latter). Against the background of heterogeneous hybridization, the brain RNA preparation shows two hybridization zones, 7S and 12S. In a situation of high-efficiency hybridization (e.g., elevated radioactivity of labeled cDNA) additional Alu$^+$RNA hybridization zones can be detected, as shown in Figure 3. The figure presents the results of an experiment on the hybridization of cytoplasmic poly(A)$^+$RNA from different tissues with separated Alu-sequence chains. RNA preparations can be seen to bind only one chain

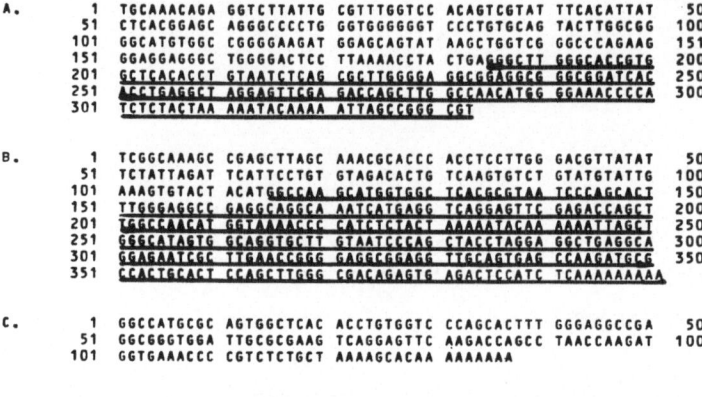

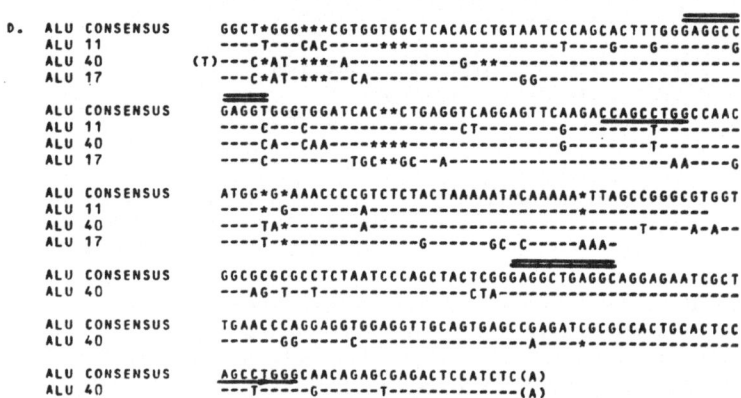

Fig. 1. Primary structure of the insertions in Alu-carrying clones.
A,B,C: insertions in clones pRHF11, pRHF40 and pBH17 respectively.
Alu sequences are underlined. D: Alu-consensus compared with the
Alu elements from clones pRHF11, pRHF40 and pRBH17. The double
lines indicate sections similar to the T-antigen-binding region of
SV40 DNA. Conserved regions typical of the Alu sequence are
underlined.

of the Alu sequence (the (-) chain), i.e., this type of RNA contains the
Alu element in one orientation which we refer to as canonical (Limborska et
al., 1987). To rule out the possibility that low-molecular Alu$^+$ RNA is not
polyadenylated but bound to poly(A)$^+$ RNA through noncovalent bonds, we
performed additional melting of the preparations followed by rechroma-
tography on poly(U) Sepharose. The results of this experiment are shown in
Figure 4. Rechromatographed material can be seen to contain small Alu$^+$ RNA
molecules, i.e., in all probability they are polyadenylated. Thus, we have
demonstrated the existence of a new class of small Alu$^+$ poly(A)$^+$ RNA in the
cytoplasm of human cells from various tissues.

Some of the clones from the human brain cDNA bank were analyzed by
colony hybridization with labeled cDNA of the brain, spleen, stomach and
adrenal cortex adenocarcinoma used as probes. The clones most actively
hybridizing with brain cDNA were selected. The insertion of one of these
clones (clone pRBH71) was sequenced; Figure 5 presents the 447-bp-long
sequence. A comparison with the Heidelberg sequence bank (performed by a
computer) failed to reveal any appreciable homology between clone pRBH71
and the decoded eukaryotic genes. Figure 6 shows the results of hybrid-

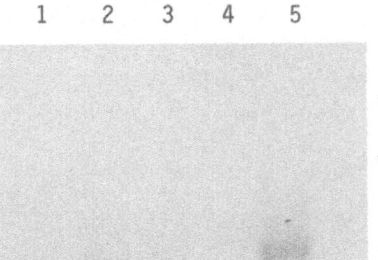

Fig. 2. Hybridization of the Alu-containing insertion of clone pRHF11 with poly(A)$^+$RNA from various human tissues. Cytoplasmic poly(A)$^+$RNA from the spleen (1), adrenal cortex adenocarcinoma (2), stomach (3), neurinoma (4), cerebral cortex (5).

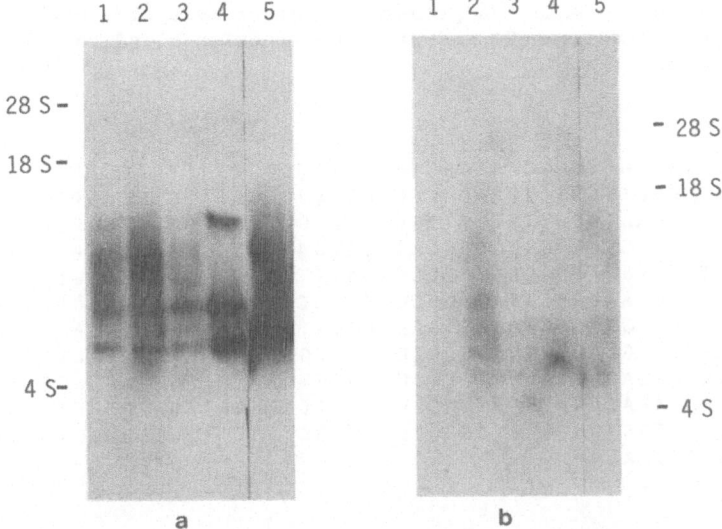

Fig. 3. Hybridization of separated chains of the insertion of clone pRHF11 with poly(A)$^+$RNA from various tissues: a: hybridization of RNA with the (−) chain of clone pRHF11; b: hybridization with the (+) chain. Poly(A)$^+$RNA from neuroblastoma (1), uterus (2), skin fibroblasts (3), stomach (4), cerebral cortex (5).

ization between this clone and cytoplasmic poly(A)$^+$RNA from brain and placenta cells. One can see that brain RNA carries this sequence in the 20S RNA region. No distinct zones of hybridization with this clone were found in placenta RNA.

We selected from the brain cDNA bank those clones that contained insertions but did not hybridize with the cDNA of the four tissues (brain,

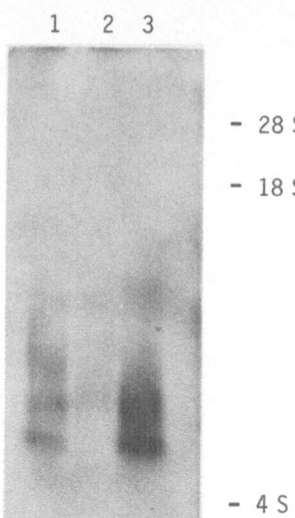

Fig. 4. Hybridization of the insertion of clone pRH11 with cytoplasmic poly(A)$^+$RNA from adrenal cortex adenocarcinoma before melting (1), and with the same RNA preparation after melting – both with the unbound fraction (2) and with the fraction bound to poly(U)-Sepharose.

```
RBH71

  1 GGCAAGGTTG CCAATGGGTA AGTAGCGCCT GTCACCAAGG  40
 41 ATACCGTGGT CGGAACGCCG GAGAGCCGAT TCCGAAGTAT  80
 81 CGCCGAGTGG TTGTGTTCGG AGGCGGACAC GCGATCGAGG 120
121 GCGAAACCGT CGCGGCGGAT GGTGATTCGG TTCGATTATA 160
161 TTCTGTGTTT GCCGGGTGCC GAGTTGTTGC ACCCGCCGTT 200
201 GGTTCGAACG CGAAGAGCCG GTTCCGAAGG TCCACAGGTA 240
241 GTTCGTGTTC GAAGGTTGAA CGCGTGGCCA AGGGCGAAGC 280
281 CGTCATGAAC GGCGGGCCCG GCCCGCAGGG CTGCCTACAC 320
321 CCACTGAGGT AACGTGCTCG GGGGTGAACA CGTGACCAGG 360
361 AACGAAACCA TCATGGACGG CGAACCCACA GGACATCCTA 400
401 CACCTACTGG GTGTCGGGCC GCGGCTACCT ACCGGTCGAG 440
441 GAATCAA
```

```
RBH3

  1 CAACACGGGG GGGCTTTTTG TTGTTTTTGT GGTTGCATTA  40
 41 GTCCTCACTG AAACCAAAAA AAAAAAAAAA A
```

Fig. 5. Nucleotide sequence of the insertions in clones pRBH71 and pRBH3.

spleen, stomach, adrenal cortex adenocarcinoma). It seems that these clones carry "minor" genes whose transcripts account for a fairly small part of the cell RNA. One such clone (pRBH3) was subjected to a more thorough analysis, and the primary structure of its 70-bp-long insertion is shown in Figure 5. Figure 7 illustrates the hybridization of this "minor" sequence with the RNA of various tissues. One observes hybridization with poly(A)$^+$RNA of the brain and of the tumor tissues (neurinoma and adrenal

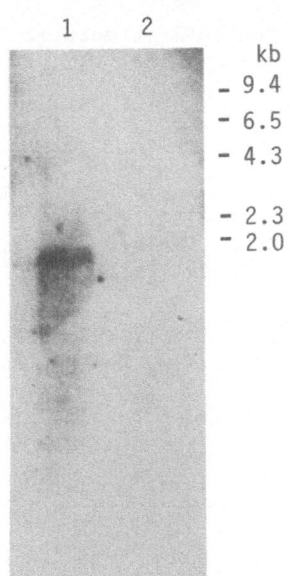

Fig. 6. Hybridization of clone pRBH71 with poly(A)$^+$RNA from the cerebral cortex (1) and placenta (2).

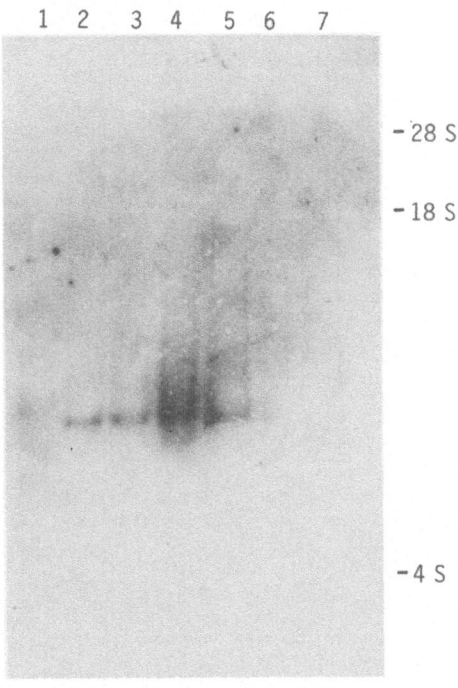

Fig. 7. Hybridization of clone pRBH3 with poly(A)$^+$RNA from the placenta (1), cerebral cortex (2), adrenal cortex adenocarcinoma (3), stomach (4), neurinoma (5), skin fibroblasts (6), spleen (7).

cortex adenocarcinoma). The hybridization zone is roughly in the 10S RNA area. The poly(A)$^+$RNA of the placenta, stomach, spleen and fibroblasts fail to hybridize with clone pRBH3, in all probability this gene does not get expressed in these tissues. Further studies involving the hybridization of these sequences with histological preparations will furnish more precise information about the cell structures in which these genes are expressed (Sutcliffe and Milner, 1984).

REFERENCES

Deininger, P. L., Jolly, D. J., Rubin, C. M., Friedman, T., and Schmid, C. W., 1981, J. Mol. Biol., 151:17-33.
Limborska, S. A. Korneev, S. A., Maleeva, N. E., Slominsky, P. A., Jincharadze, A. G., Ivanov, P. L., and Ryskov, A. P., 1987, FEBS Lett., 212:208-212.
Rogers, J. H., 1985, Int. Rev. Cytol., 93:188-279.
Sutcliffe, J. G., and Milner, R. J., 1984, TIBS, 30:95-99.

DETECTION OF A β-GLOBIN INTRON MUTATION IN

A β-THALASSEMIC PATIENT FROM AZERBAIJAN

S. A. Limborska, A. N. Fedorov,
V. L. Bukhman and M. I. Prosniak

Institute of Molecular Genetics
USSR Academy of Sciences, Moscow
Kurchatov sq. 46, 123182, USSR

Studies of β-thalassemia, a hereditary blood disease associated with deficient expression of the β-globin gene, have revealed a variety of mutations underlying this condition. So far 40 such mutations have been revealed by various molecular genetic techniques (Kan, 1986). They include point substitutions, deletions and insertions, which affect different stages of the functioning of the β-globin gene. β-thalassemia is believed to be caused by different sets of mutations in different ethnic groups. It is important to take this into account when developing diagnostic and prevention approaches for a particular region (Antonarakis et al., 1985).

In the Soviet Union β-thalassemia has been found to occur in Transcaucasia, in Central Asia and in the southern European part of the USSR, the incidence being especially high in Azerbaijan (Rustamov et al., 1981).

In an earlier study we examined a number of β-thalassemic cases from Azerbaijan, having undertaken an investigation of globin proteins, hybridization analysis of globin mRNA and restriction mapping of globin gene DNA. None of the cases showed any major deletions or rearrangements in the region of the β-globin genes. At the same time considerable heterogeneity was observed as to the degree of expression of the β-globin genes, manifesting itself in a substantial variation of the relative content of β-globin mRNA and in the varying repression of β-globin protein synthesis (Dergunova et al., 1984).

A comparative study of the β-globin sequences of the nuclear and cytoplasmic RNA in erythroid cells of the spleen was performed for one patient. In both cases the amount of β-globin sequences proved to be reduced by a factor of 5 as compared with α-globin sequences. It was supposed that the patient might have a β-globin gene transcription disorder or a disturbance of the earliest stages of the β-globin mRNA processing.

A DNA library was obtained on the basis of this patient's DNA, with λ EMBL4 bacteriophage DNA serving as vector. The library contained about 1 million recombinant clones, which number corresponds to about three human genomes. With this number of clones there is a good likelihood (90%) that any unique clone can be obtained from the library. The average size of human DNA insertions in the recombinant clones is 15-24 kb.

The search for the β-globin gene encompassed 150,000 recombinant phages (about half a genome). The method consisted in hybridizing phage plaques on nitrocellulose filters. The HhaI fragment of the Jw102 plasmid carrying human β-globin cDNA was used as the hybridization probe. The recombinant bacteriophage λA1β1 was obtained as a result of three cycles of hybridization.

Figure 1 shows a diagram (based on restriction data) of the human DNA region inserted in the λA1β1 phage. This region, which is 22 kb long, proved to contain a sizeable part of the human β-globin gene system including a portion of the δ-globin gene and the complete β-gene with adjacent regions, among them the Alu sequence located beyond the β-globin gene. DNA fragments of β-globin gene were subcloned within the pUC19 plasmid or phage M13mp10 and sequenced (Figure 1).

As a result we determined the primary structure of the entire β-globin gene and adjacent sequences, amounting to 2150 bp (Figure 2). Compared with the normal β-globin sequences, there is one point substitution in the minor intron; A instead of G at position 110. This part of the gene was thoroughly analyzed by the method of Maxim and Gilbert (1977) as well as by Sanger's technique (1977). Figure 3 presents the autoradiograph obtained in one of the sequencing experiments according to Sanger. The arrow indicates the A substituting the G.

One can see in Figure 4 that the mutation has given rise to an AG dinucleotide, the same as the one at the end of the intron and in very similar surroundings. The underlined segments are almost identical hepta-nucleotides with one substitution only. Thus, the mutation has created an additional splicing signal indicating the end of the intron. As a result the mechanisms responsible for removing the intron can do it in two alter-native ways: either by using the true signal or by operating on the

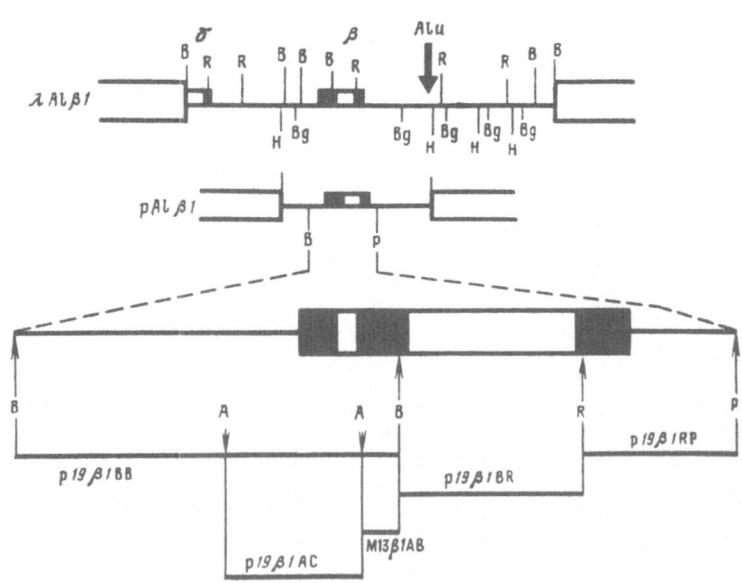

Fig. 1. Restriction map of the λA1β1 clone and its derivatives. Restriction sites: B - BamHI, R - EcoRI, H - HindIII, Bg - BglII, P - PstF, A - AccI. The arrow indicates the position of the Alu sequence.

```
                                        a gacaaaactc ttccactttt  -351
-350  agtgcatcaa tttcttattt gtgtaataag aaaattggga aaacgatctt  -301
-300· caatatgctt accaagctgt gattccaaat attacgtaaa tacacttgca  -251
-250  aaggaggatg tttttagtag caatttgtac tgatggtatg gggccaagag  -201
-200  atatatctta gagggagggc tgagggtttg aagtccaact cctaagccag  -151
-150  tgccagaaga gccaaggaca ggtacggctg tcatcactta gacctcaccc  -101
-100  tgtggagcca caccctaggg ttggccaatc tactcccagg agcagggagg   -51
-50   gcaggagcca gggctgggca tacaagtcag ggcagagcca tctattgctt    -1
 1    ACATTTGCTT CTGACACAAC TGTGTTCACT AGCAACCTCA AACAGACACC    50
 51   ATGGTGCACC TGACTCCTGA GGAGAAGTCT GCCGTTACTG CCCTGTGGGG   100
101   CAAGGTGAAC GTGGATGAAG TTGGTGGTGA GGGCCTGGGC AGgttggtat   150
151   caaggttaca agacaggttt aaggagacca atagaaactg ggcatgtgga   200
201   gacagagaag actcttgggt ttctgataag cactgactct ctctgcctat   250
251   tggtctattt tcccacccct agGCTGCTGG TGGTCTACCC TTGGACCCAG   300
301   AGGTTCTTTG AGTCCTTTGG GGATCTGTCC ACTCCTGATG CTGTTATGGG   350
351   CAACCCTAAG GTGAAGGCTC ATGGCAAGAA AGTGCTCGGT GCCTTTAGTG   400
401   ATGGCCTGGC TCACCTGGAC AACCTCAAGG GCACCTTTGC CACACTGAGT   450
451   GAGCTGCACT GTGACAAGCT GCACGTGGAT CCTGAGAACT TCAGGgtgag   500
501   tctatgggac ccttgatgtt ttctttcccc ttcttttcta tggttaagtt   550
551   catgtcatag gaaggggaga agtaacaggg tacagtttag aatgggaaac   600
601   agacgaatga ttgcatcagt gtggaagtct caggatcgtt ttagtttctt   650
651   ttatttgctg ttcataacaa ttgttttctt ttgtttaatt cttgctttct   700
701   ttttttttct tctccgcaat ttttactatt atacttaatg ccttaacatt   750
751   gtgtataaca aaaggaaata tctctgagat acattaagta acttaaaaaa   800
801   aaactttaca cagtctgcct agtacattac tatttggaat atatgtgtgc   850
851   ttatttgcat attcataatc tccctacttt attttctttt atttttaatt   900
901   gatacataat cattatacat atttatgggt taaagtgtaa tgtttttaata  950
951   tgtgtacaca tattgaccaa atcagggtaa ttttgcattt gtaattttaa  1000
1001  aaaatgcttt cttcttttaa tatacttttt tgtttatctt atttctaata  1050
1051  ctttccctaa tctctttctt tcagggcaat aatgataaca tgtatcatgc  1100
1101  ctctttgcac cattctaaag aataacagtg ataatttctg ggttaaggca  1150
1151  atagcaatat ttctgcatat aaatatttct gcatataaat tgtaactgat  1200
1201  gtaagaggt tcatattgct aatagcagct acaatccagc taccatttctg  1250
1251  ctttttatttt atggttggga taaggctgga ttattctgag tccaagctag  1300
1301  gcccttttgc taatcatgtt catacctctt atcttcctcc cacagCTCCT  1350
1351  GGGCAACGTG CTGGTCTGTG TGCTGGCCCA TCACTTTGGC AAAGAATTCA  1400
1401  CCCCACCAGT GCAGGCTGCC TATCAGAAAG TGGTGGCTGG TGTGGCTAAT  1450
1451  GCCCTGGCCC ACAAGTATCA CTAAGCTCGC TTTCTTGCTG TCCAATTTCT  1500
1501  ATTAAAGGTT CCTTTGTTCC CTAAGTCCAA CTACTAAACT GGGGGGATATT  1550
1651  ATGAAGGGCT TTGAGCATCT GGATTCTGCC TAATAAAAAA CATTTATTTT  1600
1601  CATTGCaatg atgtatttaa attatttctg aatatttttac taaaaaggga  1650
1651  atgtgggagg tcagtgcatt taaaacataa agaaatgaag agctagttca  1700
1701  aaccttggga aaatacacta tatcttaaac tccatgaaag aaggtgaggc  1750
1751  tgcaaacagc taatgcacat tggcaaca
```

Fig. 2. Nucleotide sequence of the β-globin gene of the β-thalassemic
         patient from Azerbaijan.
         Capital letters denote exon sequences, lower-case letters denote
         introns and adjacent sequences. The mutated nucleotide is framed.
         The initiating and terminating triplets are underlined. Arrows
         indicate the positions of normal nucleotide variations.

additional one (Forget, 1982). The additional signal, which comes first
in the RNA chain, proved to be far more active than the true signal.

    Thus, two types of molecules develop during the maturation of β-globin
mRNA under such conditions. In 10% of cases normal mRNA molecules are
formed, reflecting the activity of the true splicing signal in this
situation. The vast majority of mRNA (up to 90%), however, are the product
of the new signal and contain a small part (19 bp) of the minor intron.
Since it is not divisible by three, the additional 19 bp fragment considerably alters the translation properties of the next mRNA section,
causing the shift of the translation frame. A great many terminating
codons arise in the new reading frame, making this mRNA unable to ensure
the synthesis of normal β-globin proteins in the cell. In all probability,
it does not even get to conduct protein synthesis in the cell as, apparently,
it never gets into the cytoplasm but is destroyed in the nucleus soon after
it is synthesized. Therefore the patient has only normal β-globin mRNA in
the erythroid cell cytoplasm, but their amount is reduced ten times. As
a result ten times less β-globin protein is formed in the cells, which

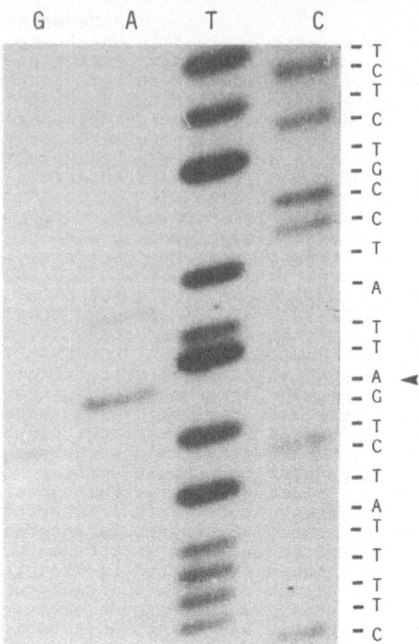

Fig. 3. Sequence analysis of the gene section containing the mutation.
The arrow indicates the mutated nucleotide.

causes a dramatic disturbance of the hemoglobin production as a whole and
eventually leads to the clinical manifestations of β-thalassemia.

Once the 2150 nucleotides of the Azerbaijan patient's β-globin gene
were sequenced, it could be compared with the previously decoded genes
from other populations. The β-globin genes analyzed by Orkin and co-workers
(1982) exhibited five points where nucleotide substitutions of the normal
DNA polymorphism type are possible. These points are indicated by arrows
in Figure 2.

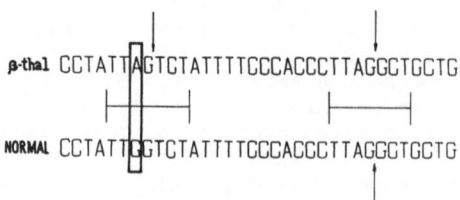

Fig. 4. Nucleotide sequence of the β-thalassemic patient's β-globin gene
in the vicinity of the mutation as compared with the normal
situation. The mutated nucleotide is framed; similar sequences
near the mutation and at the end of the intron are marked by
straight line segments. Arrows indicate the action sites of
splicing enzymes.

Three possible combinations of polymorphic points have been described for all the β-globin genes analyzed. These combinations are present to a varying extent in the Mediterranean population. The kind that occurs most frequently in this area is the β-globin gene of type 1: 53% of the normal cases and 66% of the β-thalassemic patients. The other two gene types, 2 and 3, occur in 28% and 19% of the normal cases, and in 20% and 14% of the β-thalassemic cases respectively.

The gene sequenced in the present study is of type 1, the most frequently occurring kind in the Mediterranean.

The G → A substitution at position 110 of the minor intron (the 252d nucleotide from the beginning of the gene) has been described in two other patients, both of them Cypriots, one of Greek and the other of Turkish origin (Fukumaki et al., 1982; Westaway and Williamson, 1981).

The decoding of this mutation and its discovery in Azerbaijan is of practical significance as well as of theoretical interest. On the other hand, it is the first step towards a molecular genetic study of our country's population, specifically the population of Azerbaijan. On the other hand, being aware of the occurrence of this mutation in Azerbaijan, one can develop rapid testing techniques for prenatal diagnosis, thus moving towards the prevention of the disease. Both the population analysis and the detection of the mutation for prenatal diagnosis require oligo-nucleotide probes carrying a radioactive or fluorescent label. Our further study will focus on obtaining such probes and using them for the analysis of this mutation.

REFERENCES

Antonarakis, S. E., Kazazian, H. H., and Orkin, S. H., 1985, Hum. Genet., 69:1-14.
Dergunova, L. V., Ryskov, A. P., Slominsky, P. A., Rustamov, R. Sh., and Limborska, S. A., 1984, Hematologia, 17:473-481.
Forget, B. G., 1982, Recent Progr. Hormone Res., 38:257-274.
Fukumaki, J., Ghosh, P. K., Benz, E. J., Reddy, V. B., Lebowitz, P., Forget, B. G., and Weissman, S. M., 1982, Cell, 28:585-593.
Kan, J. W., 1986, Amer. J. Hum. Genet., 38:4-12.
Maxam, A. M., and Gilbert, W., 1977, Proc. Nat. Acad. Sci., USA, 74:560-564.
Orkin, S. H., Kazazian, H. H., Antonarakis, S. E., Goff, S. C., Boehm, C. D., Sexton, J. P., Waber, P. G., and Giardina, P. V. J., 1982, Nature, 296:627-631.
Rustamov, R. Sh., Gaibov, N. T., Akhmedova, A. Yu., and Kulieva, N. M., 1981, Probl. Hematol. (in Russian), 9:12-16.
Sanger, F., Nicklen, S., and Coulson, A. R., 1977, Proc. Nat. Acad. Sci., USA, 74:5463-5467.
Westaway, D., and Williamson, R., 1981, Nucl. Acid Res., 9:1777-1788.

THE GENE OF THE YEAST RIBOSOMAL PROTEIN L32

M. D. Dabeva and J. R. Warner

Institute of Cell Biology and Morphology
Bulgarian Academy of Sciences, Sofia, Bulgaria and
Department of Cell Biology, Albert Einstein
College of Medicine, Bronx, NY, USA

The formation of ribosomes in eucaryotes requires simultaneous synthesis of four rRNAs and more than 70 ribosomal proteins (rP). In procaryotes the rP genes are clustered in large operons and their coordinate expression could be regulated at few control sites. In yeasts as in other eucaryotes these genes are scattered over the genome and their messenger RNAs are monocistronic. Most of the rP genes contain an intron, which are rare in yeasts. It seems that many of the rP genes are duplicated and both copies are expressed (see Warner, 1982; Fried and Warner, 1984).

These features point to a more complex regulation of their expression. In yeast S. cerevisiae the coordinate expression of rP genes is brought about by variety of mechanisms: equimolar transcription of mRNA (Kim and Warner, 1983), regulation of pre-mRNA processing, translational initiation and protein turnover (Warner et al., 1985; 1986; Planta et al., 1986).

In two cases it has been found that rP genes are physically linked (Molenar et al., 1984; Leer et al., 1985). Such sets of linked genes are especially suitable for analysis of their coordinate expression. We have found also a pair of rP genes, rP29* and rPL32 in a head-to-head array. The sequence of rP29 gene was reported previously (Mitra and Warner, 1984).

In this paper we report the sequence of rPL32 gene and predict the amino acid sequence of rPL32. We give the primary structure of the intergenic region and the results of our first experiments carried out to designate the sequences involved in rPL32 pre-mRNA transcription.

PRIMARY STRUCTURE OF rPL32 GENE

Figure 1A shows the arrangement of the two adjacent genes, rPL32 and rP29, their restriction map and the strategy used for sequencing of rPL32 gene by the dideoxy method of Sanger (Sanger et al., 1977). The entire sequence of the gene is given in Figure 2.

---

*rP29 is given according to the designation of Gorenstein and Warner (1976). It is not included in the nomenclature of Kruiswijk and Planta (1974).

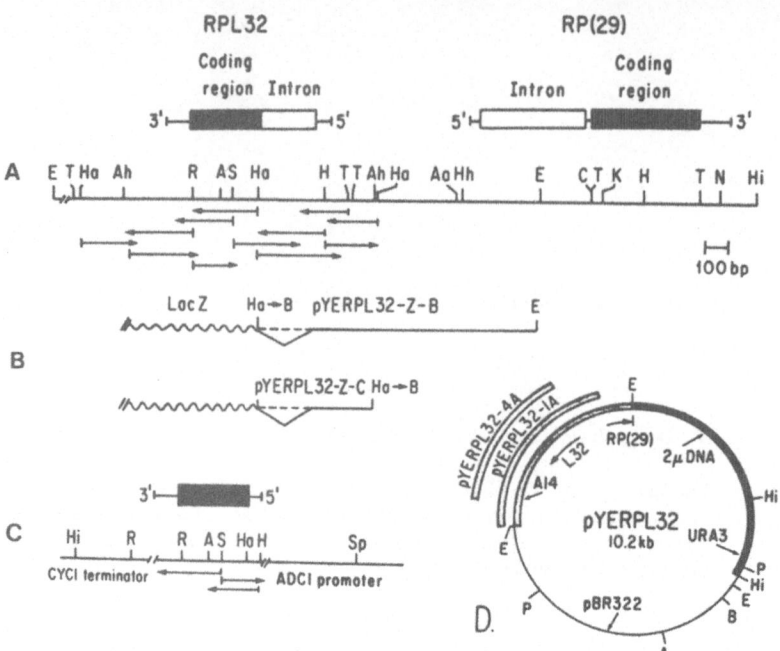

Fig. 1. Map of the two adjacent ribosomal protein genes rPL32 and rP29 and of the rPL32 clones used in this work.
A: Restriction map of rPL32 (left) and rP29 (right. Restriction enzymes; A, Accl; Ah, AhaIII; Aa, AhaII; C, ClaI; E, EcoRI; H, HpaII; Ha, HaeIII; Hi, HindIII; K, KpnI; N, NruI; R, EcoRV; S, Sau 3A; Sp, SphI; T, TaqI.
B: Map of the two rPL32-LacZ fusion genes used in this study. In pYErPL32-ZB the EcoRI-HaeIII (1 nucleotide downstream from the 3' end of the intron) fragment was inserted into EcoRI-BamHI sites of the LacZ fusion vector pSEY101 (Douglas et al., 1984) through BamHI linker. In pYErPL32-ZC the HaeIII fragment containing 278 nucleotides upstream from AUG codon was ligated through BamHI linker in BamHI site of pSEY101.
C: Map of the cDNA clones of rPL32 and the strategy of their sequencing. The interruption shown in the map indicates the oligo-C linker at the 5' end and the poly-A tail at the 3' end.
D: Diagram of the three plasmids containing the gene for rPL32. The shuttle vector YEp24' (Pearson et al., 1982) opened at EcoRI site was used for their construction pYErPL32 contains the 2.55 kb EcoRI-EcoRI fragment (see A); pYErPL32-1A contains an AhaII-EcoRI subfragment of the former; and pYErPL32-4A includes a TaqI subfragment bearing the entire rPL32 gene plus 76 nucleotides upstream of its single site of initiation.

THE INTRON

To determine the position of the intron within the gene we isolated two cDNA clones of rPL32 from a cDNA clone bank, kindly donated by McKnight (McKnight and McConaughy, 1983). Partial map of these plasmids and the strategy of their sequencing is given in Figure 2C. Figure 3 shows the splice junctions of the excised intron. The intron starts immediately after the first AUG codon and, as is the case with the other rP genes, is localized near the 5'-end of the gene. The 5' splice site is unusual in that the first 6 nucleotides are GTCAGT rather than the customary GTATGT. The 230-bp-long intron ends 39 nucleotides downstream from the consensus sequence TACTAAC.

```
-280            -270         -260         -250         -240         -230
GGCCGCATTT      AAATTTTTTT   TTTTTTCAAT   TTTTTTTCCT   ATTGGAGACG   GAATGGCAAA

-220            -210         -200         -190         -180         -170
GCTCACGCTG      GGCGATAGCA   CGTAACCACG   CTGGTGTATT   TTCATCCTAA   TCATTCTCTT

-160            -150         -140         -130         -120         -110
TCATCTATTT      CGATAAGCTA   CTTAATGTCG   AAGTTTTAAC   TTCCATTTGT   TGGAATGTTC

-100            -90          -80          -70          -60          -50
ATTACAGTTT      ATACTTACTC   CTTGCTCTTT   TATAATATAA   CCAAACAGAC   CGGAGTGTTT
                ‾‾‾‾‾‾‾‾‾                              ↑

-40             -30          -20          -10               10
AAGAACCTAC      AGCTTATTCA   ATTAATCAAT   ATACGCAGAG   ATG GTCAGTA
                                                      Met ‾‾‾‾‾‾‾


        20           30           40           50           60           70
TAACATGATT      TTATAACTAT   TTCGCTTTAT   TCTTTCATGG   GATTAGCAAG   AAGGCTTGGA

        80           90          100          110          120          130
ATACAAACAT      GAATTTTCAA   AGTTTATTCC   TGCTCTTGTG   TTTGAGAGAG   TCCGTTCAAA

       140          150          160          170          180          190
GGATGCTTCT      CCATGACTGT   GAAATAAAAG   TGAGATATGC   AAAAACTTTG   AATAATCTAC
                                                                         ‾‾‾
       200          210          220          230                      245
TAACAAGTTA      ATACGTTATT   TTTTATCGTT   TACATTTCAA   CAG   GCC CCA GTT AAA
‾‾‾‾‾‾                                                      Ala Pro Val Lys

                         260                   275                   290
TCC CAA GAA TCT ATC AAC CAA AAG TTG GCT TTG GTT ATC AAG TCT GGT AAG TAC
Ser Gln Glu Ser Ile Asn Gln Lys Leu Ala Leu Val Ile Lys Ser Gly Lys Tyr

     305                         320                   335                   350
ACC TTA GGT TAC AAG TCC ACT GTC AAG TCT TTG AGA CAA GGT AAG TCT AAG TTG
Thr Leu Gly Tyr Lys Ser Thr Val Lys Ser Leu Arg Gln Gly Lys Ser Lys Leu

                 365                   380                   395
ATC ATC ATT GCC GCT AAC ACT CCA GTT TTG AGA AAG TCC GAA TTG GAA TAT TAC
Ile Ile Ile Ala Ala Asn Thr Pro Val Leu Arg Lys Ser Glu Leu Glu Tyr Tyr

410                   425                         440                   455
GCT ATG TTG TCC AAG ACT AAG GTC TAC TAC TTC CAA GGT GGT AAC AAC GAA TTG
Ala Met Leu Ser Lys Thr Lys Val Tyr Tyr Phe Gln Gly Gly Asn Asn Glu Leu

           470                   485                   500                   515
GGT ACT GCT GTC GGT AAG TTA TTC AGA GTC GGT GTT GTC TCT ATT TTG GAA GCT
Gly Thr Ala Val Gly Lys Leu Phe Arg Val Gly Val Val Ser Ile Leu Glu Ala

                 530                   545                   560                   570
GGT GAC TCT GAT ATC TTG ACC ACC TTG GCT TAA AT  AAGGTAAGTT   CAAACGATTT
Gly Asp Ser Asp Ile Leu Thr Thr Leu Ala  *

     580          590          600          610          620          630
GTTGGAAGAC      AATTGGTTTG   ATGTGTATTT   TTCTATAATA   TATAAATCTT   AGACTATTCA

     640          650          660          670          680          690
TTAACCTTAT      TTTATTTACC   TTCTACCCGA   AAATGGGTTT   CTTTTGCTGA   AATAGTTCTT
        ↑                                                           ‾‾‾
     700          710          720          730          740          750
TCATGTAAAT      ATTGTTTAAG   CAGAAGGCAT   TCTTTGCCCT   CTTGCTCGTG   TGGGTTCCTT
‾‾‾‾‾‾‾‾          ‾‾‾‾‾

     760          770          780          790          800          810
TCCGCTAATA      ATCAATCATT   CTCATACTGT   TCAATATCAT   AATTATCTCT   GCGACGATTT

     820          830          838
TTCATCCGCC      TCCTCCGAAA   CTGGAAAA
```

Fig. 2. The primary structure of rPL32 gene and the predicted amino acid
sequence of its product. The 5' end and the 3' end of the gene as
revealed by S1 nuclease mapping are marked by arrows. The
asterisks mark the 5'- and the 3'-ends of the cDNS clones. The
consensus sequences upstream from the site of initiation of
transcription, in the intervenic sequences and in the 3' end
flanking region are underlined. Numbering begins with the first
initiating AUG codon.

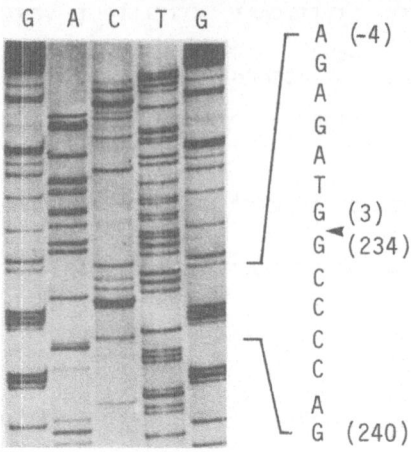

Fig. 3. Sequencing ladder of rPL32 cDNA. The Sau3A-SphI fragment (Figure 1C) was cloned and sequenced by the dideoxy method. The sequencing ladder is from the noncoding DNA strand. On the right is the sequence of the corresponding coding strand. The number of nucleotides is given in brackets. The arrow indicates the splice junction. The figure shows also the 5' end of the cDNA clone which begins with an A, 58 nucleotides upstream from the AUG codon.

THE 5'-END AND THE 3'-END OF THE GENE

The 5'-end of rPL32 gene was determined by S1 mapping and primer extension analysis. It was also deduced from the 5'-end sequences of the two cDNA clones.

The protected from S1 digestion 5'-end labeled DNA fragment gave a single band of about 486-490 nucleotides in length (Figure 4A). This result shows that the 5'-end of mRNA is rather homogeneous and is localized 54 to 58 nucleotides upstream from the AUG codon.

The 5'-end of the gene was mapped also using the primer extension analysis. The cDNA products obtained on the sequencing gel are 107 (for mRNA) and 206-210 (for pre-mRNA) nucleotides long (Figure 6B). The 5'-end of the two cDNA clones sequenced by us lies also 58 bp upstream from the initiation codon (Figure 3). Therefore we concluded that there is only one major site of initiation of transcription and it is mapped at position -58 (a) (see Figure 2). Interestingly, a single site of initiation of transcription is not common for most of the ribosomal protein genes sequences so far.

The 3'-end of rPL32 pre-mRNA was mapped using the S1 nuclease technique and by sequencing the cDNA clones of the gene.

Two major protected DNA fragments were found after S1 digestion: the first was doublet 218/219 nucleotides long, the second was more hetero-geneous - 204-211 nucleotides long (Figure 4B). Nucleotide 218 of the larger protected fragment corresponds to A at position 648 of the gene. The shorter heterogeneous fragment suggests a second putative termination site, 10 nucleotides upstream from the major one. It could be a result of S1 nibbling of the A-T rich end of the gene and should be accepted with caution (Hentschel et al., 1980). Although this region is AT rich, no AATAAA, the canonical poly(A) addition signal is present. It is noteworthy to mention that the consensus sequence TAG...TATGTA...TTT, proposed to play

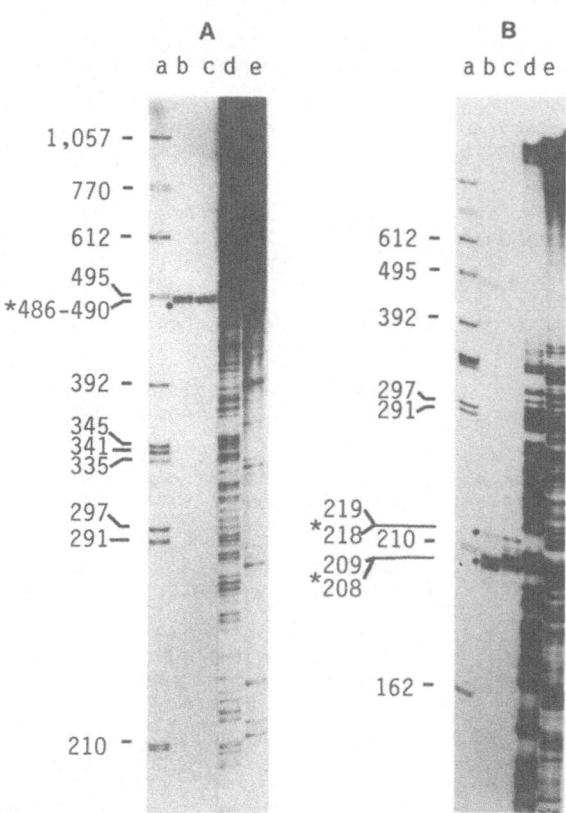

Fig. 4. S1 mapping of the 5'- and 3'-ends of the gene. A: Sequencing gel
showing the 5' end of rPL32 pre-mRNA. The plasmid pYErPL32
(Figure 1D) was digested with the restriction enzyme AccI, the
fragment containing the 5' end of the gene was labeled at its 5'
ends and cut with AhaII (see Figure 1A). The 702 bp DNA fragment
labeled at AccI site was hybridized to poly(A$^+$) RNA isolated from
a rna2^{ts} strain transformed with pYErPL32 after shifting the
culture to the restricted temperature. Under these conditions the
isolated RNA contains the rPL32 pre-mRNA and almost no mature
mRNA. Lane a - mol. w. marker; lane b - 30 min; lane c - 60 min
treatment with 100 u/ml of S1 nuclease at 37°C; lane d and e -
sequencing ladder. B. Sequencing gel showing the 3' end of rPL32
pre-mRNA. A 487 bp long AccI-AhaIII fragment containing the
distal part of the gene (Figure 1A) was 3' end labeled at the AccI
site and annealed to poly(A$^+$)RNA as described above. Lane a -
mol. w. marker; lane b - 60 min; lane c - 30 min treatment with
200 u/ml of S1 nuclease at 37°C; lane d and e - sequencing ladder.

role in termination of transcription in yeast (Zaret and Sherman, 1984)
begins 35 nucleotides downstream from the site of termination of transcription.

The 3'-end of mRNA was determined independently by sequencing the
3'-end of the two cDNA clones using synthetic oligonucleotides as primer.
In both cDNA clones the poly(A) tract starts at nucleotide 648 downstream
from the first codon, consistent with the result of S1 mapping.

```
        10          20          30          40          50          60
CATTTTATCA  CTCTCAAGTT  ATTGTCTTGG  TGTCTTCGCT  TAACTTGGCA  ATAGCAACAG
GTAAAATAGT  GAGAGTTCAA  TAACAGAACC  ACAGAAGCGA  ATTGAACCGT  TATCGTTGTC
    *                                            ←—rp29 —┘
        70          80          90         100         110         120
CGCAAACTGC  GTAAACGTAC  TAAAATACAC  ACCATCCGTT  GAATATATCA  TGCAGTACAT
GCGTTTGACG  CATTTGCATG  ATTTTATGTG  TGGTAGGCAA  CTTATATAGT  ACGTCATGTA

       130         140         150         160         170         180
TGACAGTATA  TCACTTCTGG  ATAGGACGCC  AACCGCCAGT  CTGTGTGGTA  TTTCCTCCCG
ACTGTCATAT  AGTGAAGACC  TATCCTGCGG  TTGGCGGTCA  GACACACCAT  AAAGGAGGGC
                               ③
       190         200         210         220         230         240
TCTCTGCGTG  CCAGACGGGA  TCAGCCCGTC  TATTCTCGTG  TCGTCCTGCC  TGGAGAGACT
AGAGACGCAC  GGTCTGCCCT  AGTCGGGCAG  ATAAGAGCAC  AGCAGGACGG  ACCTCTCTGA

       250         260         270         280         290         300
GAACCCCCAT  CTAGCACCGC  AGTAAGGATC  ATTACTTTGT  AGGCGGGAAA  TAAGAGGTTC
CTTGGGGGTA  GATCGTGGCG  TCATTCCTAG  TAATGAAACA  TCCGCCCTTT  ATTCTCCAAG

       310         320         330         340         350         360
AACAAGGACC  AAAAAAAAAA  CACTAAAATC  TGAGATCAAA  ATATGTGGTG  CACAGATGTA
TTGTTCCTGG  TTTTTTTTTT  GTGATTTTAG  ACTCTAGTTT  TATACACCAC  GTGTCTACAT
                                                                       ①
       370         380         390         400         410         420
ACGTTCCAAA  ATGTATGGAT  GGTAAGGTCT  TTTTCCAAGA  AACGTATCTT  TTTTTCAGCT
TGCAAGGTTT  TACATACCTA  CCATTCCAGA  AAAAGGTTCT  TTGCATAGAA  AAAAAGTCGA
                ──────────   ─────
       430   RPG   440         450         460         470         480
AGGGCCGCAT  TTAAATTTTT  TTTTTTTTCA  ATTTTTTTTC  CTATTGGAGA  CGGAATGGCA
TCCCGGCGTA  AATTTAAAAA  AAAAAAAAGT  TAAAAAAAAG  GATAACCTCT  GCCTTACCGT
              ①                                        ───       ──
       490         500         510         520         530         540
AAGCTCACGC  TGGGCGATAG  CACGTAACCA  CGCTGGTGTA  TTTTCATCCT  AATCATTCTC
TTCGAGTGCG  ACCCGCTATC  GTGCATTGGT  GCGACCACAT  AAAAGTAGGA  TTAGTAAGAG
                                          ─────   ────
           550   ②       560         570         580         590         600
TTTCATCTAT  TTCGATAAGC  TACTTAATGT  CGAAGTTTTA  ACTTCCATTT  GTTGGAATGT
AAAGTAGATA  AAGCTATTCG  ATGAATTACA  GCTTCAAAAT  TGAAGGTAAA  CAACCTTACA

       610         620        ③630          640         650  rp73 →   660
TCATTACAGT  TTATACTTAC  TCCTTGCTCT  TTTATAATAT  AACCAAACAG  ACCGGAGTGT
AGTAATGTCA  AATATGAATG  AGGAACGAGA  AAATATTATA  TTGGTTTGTC  TGGCCTCACA
                                                      ↓
       670         680         690         700     *   710         720
TTAAGAACCT  ACAGCTTATT  CAATTAATCA  ATATACGCAG  AGATG
AATTCTTGGA  TGTCGAATAA  GTTAATTAGT  TATATGCGTC  TCTAC
```

Fig. 5. Intergenic region between rPL32 and rP29. Upper – rPL32 coding
         strand and lower – rPL29 coding strand. The consensus sequences
         are designated 1 through 6 (see Teem et al., 1984) and presented
         in boxes. The consensus sequence RPG-box (Leer et al., 1985) on
         the coding strand of rP29 is marked with double line. Two common
         sequences on the coding strands of rPL32 and rP29 are underlined.

CODING REGION OF rPL32 GENE

     The single open reading frame found by us initiates one codon upstream
from the 5' splice junction of the intron. This is the first AUG codon
downstream from the site of initiation of transcription. The second exon
consists of 105 codons. The coding region of the gene conforms exactly
with that of the cDNA clones. The open reading frame stops at TAA 99
nucleotides upstream from the major site of transcription termination.

     The protein rPL32 is moderately basic compared with the other ribo-
somal protein genes which is in good correspondence with its mobility on a
two-dimensional polyacrylamide gel electrophoresis. The N-terminal part is
highly basic while the C-terminal part is nearly neutral. The amino acid
composition of the protein is not known so far, the predicted by us amino
acid sequence remains to be confirmed.

208

The intergenic region of 600 bp between the two rP genes is of special interest because it could reveal the regulatory sites controlling the expression of the two rP genes. The intergenic region contains a central part (nucleotides 330-390), surrounded by long runs of A and T. These runs are followed downstream in both directions by very rich in G and C sequences. Approximately 100 bp upstream of the sites of transcription initiation of the two genes begin A-T rich DNA segments.

Sequence comparison has permitted the identification of a number of common sequences upstream of most ribosomal protein genes (Teem et al., 1984; Leer et al., 1985). We did three kinds of comparisons: (1) search for these consensus sequences in the rPL32-rP29 intergenic region; (2) search for sequence homology between rPL32 and rP29 coding strand of the intergenic region; and (3) search for homologous sequences between the intergenic regions of rPL32-rP29 and rPL46-rPS24 (an other set of linked rP genes transcribed in opposite directions, Leer et al., 1985).

Some of the homologous sequences 5' to the initiation codon (designated 1 through 6, Teem et al., 1984) were found, others not (Figure 6), but none of them was perfect. HOMOLI: AACATC^{CG}/$_{TT}$ T^{A}/$_{G}$ CA^{TTT}/$_{CCA}$, repeated twice and RPG-box: ACCCTACAT^{TT}/$_{CA}$, overlapping one of the HOMOLI sequences were found oriented toward rP29 roughly in the middle of the intergenic region. These two consensus sequences were implicated in transcription (Rotenberg and Woolford, 1986; Woudt et al., 1986; Schwindinger and Warner, 1987).

Comparison of rPL32 and rP29 coding strands of the intergenic region revealed a number of common sequences from 6 to 9 nucleotides. Most of them were found to be repeated throughout the other part of the genes. The sequences that seem to be unique are underlined in Figure 6.

We compared the intergenic region of rPL32-rP29 with the sequence between rPL46 and rPS24 reported by Leer et al. (1985). Both regions are approximately 600 nucleotides in length and contain a single RPG-box, localized in the middle of the intergenic sequences. This box is on the coding strand of rPS24 and rP29 respectively. Other common sequences placed roughly at the same distance from the AUG codon were also found but their importance has not been shown.

PROMOTER REGION OF rPL32

It is tempting to speculate that a common control region can regulate the coordinate expression of the two rP genes rPL32 and rP29 (as is the case with GAL1-GAL10 promoter (West et al., 1984)). Although there is only one rPL32 gene and two rP29 genes in the genome, more extensive analysis of the intergenic region could reveal the promoter sequences and those involved in the coordinate transcription of mRNA for rPL32 and rP29.

The consensus sequences HOMOLI and RPG-box are localized on the coding strand of rP29. Deletion analyses have shown that they are active in both directions (Woudt et al., 1986). To answer the question whether these sequences are important for the expression of rPL32 gene we used two rPL32-LacZ fusion plasmids (Figure 1B). pYErPL32-ZB contained the whole intergenic region, pYRrPL32-ZC was a deletion, 40 kb downstream from HOMOLI and RPG-box. Yeast strain S150-2B ($\underline{a}_1$ trpl-289, his3-Δ1, ura3-52, leu2-3,11) was transformed with the two fusion plasmids and total RNA was extracted from cells cultured at 23°C and cells shifted for 15 min to 36°C. As a result of the temperature shock, rP mRNA synthesis is reversibly blocked for 10-25 min (Kim and Warner, 1983).

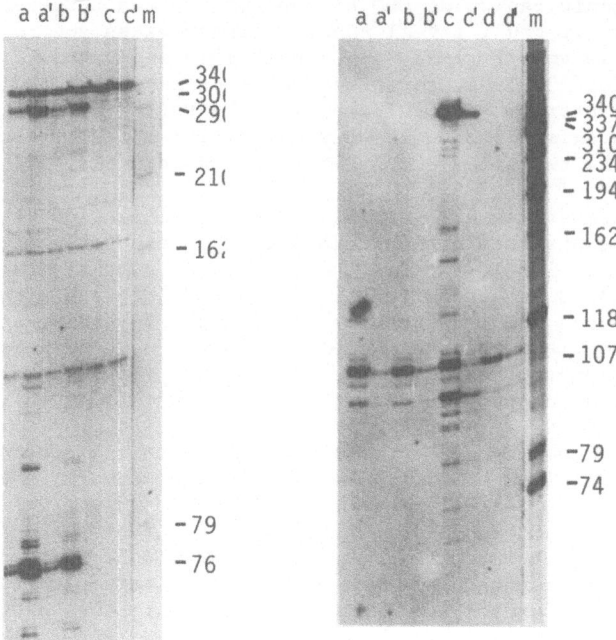

Fig. 6. Primer extension analysis of mRNA from cells transformed with
rPL32 plasmids.
cDNA synthesized by reverse transcriptase was analyzed on
sequencing gel. A, RNA, isolated from yeast cells transformed
with: a,a' – pYErPL32-ZB; b,b' – pYErPL32-ZC; c,c' – pSEY101
(fusion vector). a,'b',c', RNA isolated from cells grown up at
23°C, and a, b, c, after 15 min shift to 36°C. The primer was
synthetic oligonucleotides complementary to the fusion linker.
The lower band – cDNA of 76 nucleotides shows the position of
fusion mRNA, the upper band – cDNA of 306 nucleotides corresponds
to fusion pre-mRNA; m – mol, w. marker. B, RNA from cells
transformed with: a,a' – shuttle vector YEp24'; b,b' –
pYErPL32-4A; c,c' – pYErPL32-1A; d,d' – control cells; a, b, c,
RNA isolated from yeasts, cultured at 23°C, and a',b',c', after 15
min shift to 36°C. The primer was synthetic oligonucleotides
complementary to nucleotides 262-272 from exon 2. The lower band
– cDNA of 107 nucleotides corresponds to rPL32 mRNA, the upper
band – of 337 nucleotides – corresponds to unspliced rPL32
pre-mRNA; m – mol, w. marker.

     Fusion mRNA was analyzed by primer extension analysis using a syn-
thetic primer complementary to the fusion junction. The shortest cDNA, 76
nucleotides long corresponds to the mature fusion mRNA, the middle band,
230 nucleotides longer corresponds to the fusion precursor mRNA. These two
cDNA bands are absent in lanes c and c' where RNA from cells transformed
with the fusion vector pSEY101 was used. The longest cDNA resulting from
annealing of the primer to some other RNA was used as an internal quantitative marker. Comparing the amount of the cDNA products in lanes a, a', b
and b', it becomes evident that the rate of transcription of fusion m RNA
is much lower in cells transformed with pYErPL32-ZC, but in both strains

transcription is equally affected by the temperature shock. β-galactosidase assay showed that the enzyme activity is 60-80% lower in the strain bearing the plasmid with deletion in the upstream region. This result shows that RPG-box and/or HOMOLI play an important role in the transcription of the gene, but their deletion does not abolish the background transcription of the gene.

In the next experiment we deleted the upstream region of rPL32 till nucleotide-132 (74 nucleotides upstream from the site of transcription initiation), leaving intact the TATA-box in front of the gene (pYErPL32-4A, Figure 1D). To analyze the expression of the deleted gene we took advantage of the fact that if the gene for rPL32 is introduced on a multi-copy plasmid, unspliced rPL32 precursor mRNA accumulates in the cell. Normally, the cDNA corresponding to the precursor of rPL32 mRNA is a tiny band (Figure 6B, lane a and d). It increase 100 fold in strains where the gene is present on a multicopy plasmid (lane c). No accumulation of this band was observed with RNA from cells transformed with pYErPL32-4A (lane b). These data show that TATA-box is not sufficient for the expression of the gene and that at least 220 nucleotides in front of it are necessary for some transcription to occur.

REFERENCES

Fried, H. M., and Warner, R. J., 1984, in: "Recombinant DNA Approach to Studying the Control of Cell Proliferation", J. Stein and G. Stein, eds., Acad. Press, NY, pp 162-192.
Douglas, M. G., Geller, B. L., and Emr, S. D., 1984, Proc. Natl. Acad. Sci., USA, 81:3983-3987.
Gorenstein, C. G., and Warner, J. R., 1976, Proc. Natl. Acad. Sci., USA, 73:1547.
Hentschel, C., Irminger, J. C., Bucher, P., and Birnstiel, M. L., 1980, Nature, 285:147-151.
Kim, C. H., and Warner, J. R., 1983, J. Mol. Biol., 165:79-89.
Kim, C. H., and Warner, J. R., 1983, Mol. Cell. Biol., 3:457-465.
Kruiswijk, T., and Planta, R. J., 1974, Mol. Biol. Rep., 1:409-421.
Leer, R. J., Van Raamsdonk-Duin, M. M. C., Kraakman, P., Mager, W. H., and Planta, R. J., 1985, Curr. Genet., 9:272-277.
McKnight, G. L., and McConaughy, B. L., 1983, Proc. Natl. Acad. Sci., USA, 80:4412-4416.
Mitra, G., and Warner, J. R., 1984, J. Biol. Chem., 259:9218-9224.
Molenaar, C. M. T., Woudt, L. P., Jansen, A. E. M., Mager, W. H., Planta, R. J., Donovan, D., and Pearson, N. J., 1984, Nucl. Acids Res., 12:7345-7358.
Planta, R. J., Mager, W. H., Leer, R. J., Woudt, L. P., Raue, H. A., and El-Baradi, T. T. A. L., 1986, in: "Structure, Function and Genetics of Ribosomes", B. Hardesty and G. Kramer, eds., Springer-Verlag, pp 699-718.
Rotenberg, M. O., and Woolford, J. L., 1986, Mol. Cell. Biol., 6:674-687.
Sanger, F., Nicklen, S., and Caulson, A. R., 1977, Proc. Natl. Acad. Sci., USA, 74:5463-6467.
Schwindinger, W. F., and Warner, J. R., 1987, J. Biol. Chem., 262:5690-5695.
Teem, J. L., Abovich, N., Kaufer, N. F., Schwindinger, W. F., Warner, J. R., Levy, A., Woolford, J., Leer, R. J., Van Raamsdonk-Duin, M. M. C., Mager, W. H., Planta, R. J., Schultz, K., Friesen, J. D., and Roshbash, M., 1984, Nucl. Acids Res., 12:8295-8312.
Warner, J. R., Elion, E. A., Daveba, M. D., and Schwindinger, W. F., 1986, in: "Structure, Function and Genetics of Ribosomes", B. Hardesty and G. Kramer, eds., Springer-Verlag, pp 719-732.

Warner, J. R., Mitra, G., Schwindinger, W. F., Studeny, M., and Fried, H.
    M., 1985, Mol. Cell. Biol., 5:1512-1521.
West, R. W., Yocum, R. R., and Ptashne, M., 1984, Mol. Cell. Biol.,
    4*2467-2478.
Woudt, L. P., Smit, A. S., Mager, W. H., and Planta, R. J., 1986, EMBO J.,
    5:1037-1040.
Zaret, K. S., and Sherman, F., 1984, J. Mol. Biol., 176:107-135.

STRUCTURAL COMPARISON OF SOME MICROBIAL RIBONUCLEASES

C.P. Hill, G.G. Dodson, S.N. Borisova*, A.G. Pavlovsky*, K.M. Polyakov*, B.V. Strokopytov* and J. Ševčik**

Chemistry Department, University of York, York YO1 5DD, UK

*Institute of Crystallography, Academy of Science of the USSR, Moscow, 117333, USSR

**Institute of Molecular Biology, Slovak Academy of Sciences 842 51 Bratislava, Czechoslovakia

INTRODUCTION

Previous comparison of amino acid sequences and three-dimensional structures has shown that the microbial ribonucleases from an evolutionarily related family that is separate from that of the pancreatic ribonucleases (Hill et al., 1983). The microbial and pancreatic ribonucleases catalyse the same two step hydrolysis of RNA via a cyclic 2'3' intermediate. However, the specificities are different; the microbial enzymes prefer a purine base on the 3' side of the scissile phosphate, and in some cases are very guanine specific.

In recent years a number of microbial ribonuclease crystal structures have been refined to yield accurate and reliable coordinates. A collaboration has been undertaken to compare these refined coordinates and investigate in more detail the structural aspects of microbial ribonuclease evolution, catalysis and specificity. This work is in progress and we report here some preliminary observations.

STRUCTURAL COMPARISONS

The structures used here are listed in Table 1. These are not all of the microbial ribonucleaes whose structures have been solved, but they do provide representatives of the three Bacillus, Streptomyces and Eukaryotic subgroups. All the enzymes have the same topology with an alpha helix being stacked against a beta sheet.

The starting point for the comparisons has been to overlap just on the alpha carbon atoms and find the matrix for each pair of enzymes that puts the greatest number of corresponding alpha carbon atoms within 3A of each other. According to this the Bacillus enzymes, Ba and Bi, are homologous with each other over their entire length, as are the eukaryotic enzymes C2 and Pb1, except for a two residue deletion. The three subgroups are not so similar to each other, with essentially only the beta sheet and residues involved in substrate recognition being conserved.

Table 1. Coordinates used in the comparisons

		Enzyme source	Specificity	R(%)	dmin	
	Ba	Bacillus amyloliquefaciens	Non specific G>A>pyramidine	16.2	2.0	York
Prokaryote	Bi	Bacillus intermedius	Non specific G>A>pyramidine	21.0	2.0	Moscow
	Sa	Streptomyces aureofaciens	Guanine	17.5* 18.5	1.8	York/ Bratislava
	C2	Aspergillus clavatus	Guanine	20.0	1.3	Moscow
Eukaryote	Pb1	Penicillium brevicompactum	Guanine	16.7	1.7	Moscow

* The coordinates for both native (R=17.5%) and 3'GMP bound (R=18.5%) ribonuclease Sa were made available for this study. For all the other enzymes only the native structures were considered.

INVARIANT RESIDUES

Table 2 shows an alignment of the sequences on the basis of the structural similarities, in this alignment only five residues are invariant. Three of the invariant residues have been implicated in catalysis in several of the enzymes (references in Hill et al., 1983). How catalysis takes place has been suggested from the crystal complex of ribonuclease T1 with the inhibitor 2' GMP (Heinemann and Saenger, 1982). In this model the invariant glutamic acid abstracts a proton from the ribose 2' hydroxyl group, so activating the 2' oxygen to nucleophilic attack on the phosphorus. The phosphate group is bound by the invariant arginine, and the invariant histidine side chain is poised to donate a proton to the leaving group.

The crystal structure of the Sa – 3'GMP complex shows how Sa specifically recognizes the guanine base through four hydrogen bonds (Sevcik et al, 1987). All of these hydrogen bonds are to main chain amides, except for one to the invariant glutamic acid that has not been implicated in catalysis. This interaction has also been found in the complexes of ribonuclease T1 with guanine nucleotides (Sugio et al 1985a, Sugio et al 1985b). Inspection on the graphics suggests that none of the other RNA bases could make hydrogen bonds to this glutamic acid and still bind with the scissile phosphodiester bond in the same conformation at the active site. This invariant carboxylic acid is found on a small loop of the protein, the conformation of which is stabilized by hydrogen bonds between the main chain and the side chain of the invariant asparagine. Thus the two invariant residues that are not required for catalysis are conserved because of their importance in substrate recognition.

RELATIONSHIP BETWEEN THE SUBGROUPS

There are several observations that point to the Streptomyces enzyme Sa being more closely related to the Bacillus than to the eukaryotic subgroups. The first of these concerns the invariant asparagine, which has

Table 2. Alignment of the sequences on a structural basis

```
Pb1 A C A A T C G T V C Y  :T S S A I S S A Q A A G Y N L Y S T:
C2  D C D Y T C G S H C Y  :S A S A V S D A Q S A G Y Q L E S A:
Sa  D V S G T V C L S A L   :P P E A T D T L N L I A S D:
Bi      A V I N            :T F D G V A D Y L I R Y:    K R L P N D Y I T
Ba      A Q V I N          :T F D G V A D Y L Q T Y:    H K L P D N Y I T

                                                         +            +
Pb1                      N D D V      S N  |Y P H E Y H N Y E G F|
C2                       G Q S V G R S R   |Y P H Q Y R N Y E G F|
Sa                       G P F P Y S Q     |D G V V F Q N R E S V L P|
Bi  K S Q A S A L G W V A S K G D L A E V A P G K S I |G G D V F S N R E G R L P|
Ba  K S E A Q A L G W V A S K G N L A D V A P G K S I |G G D I F S N R E G K L P|

                *                                          *
Pb1 D F P V S |G T Y Y E F P I L K| S G K V Y T G S S |P G A N R V I F N D D|
C2  D F P V S |G N Y Y E W P I L S| S G S T Y N G G G |P G A D R V V F N D N|
Sa  T Q S Y G |Y Y H E Y T V I T| P G A |R T R G T R R I I C G E A|
Bi  S A G S |R T W R E A D I N| Y V S |G F R N A D R L V Y S S D|
Ba  G K S G |R T W R E A D I N| Y T S |G F R N S D R I L Y S S D|

                        *
Pb1 |D E L|  A |G V I T H| T G A S G N N |F V A C T|
C2  |D E L|  A |G L I T H| T G A S G D G |F V A C Y|
Sa  T |Q E D| |Y Y T G D H| |Y A T F S L I| D Q T C
Bi  |W L I| |Y K T T D H| |Y A T F T R I| R
Ba  |W L I| |Y K T T D H| |Y Q T F T K I| R
```

Residues are boxed when structurally equivalent between subgroups.
Structurally equivalence is assumed if alpha carbon atoms are within three
angstroms of each other after best global overlap.
Alpha helical residues are boxed with dashed lines.
* Denotes invariant residues that have been implicated in catalysis.
+ Denotes other invariant residues.

different conformations in eukaryotic and prokaryotic ribonucleases. The
result of the different side chain conformations is that the residues
hydrogen bonded by the asparagine side chain are displaced by on residue
between the prokaryotic and eukaryotic enzymes. A second demonstration
that Sa is closer to the Bacillus subgroup also involves the invariant
asparagine. In the prokaryotic enzymes this residue makes hydrogen bonding
interations through buried water molecules to the main chain close to the
invariant histidine. This interaction does not involve solvent molecules
in the eukaryotic enzymes but is mediated by direct Van der Waal's contact
between protein atoms.

     Another example of why Sa should be placed closer to the Bacillus than
to the eukaryotic enzymes is given by the substitution of a Bacillus
trypotphan (71 Ba numbering) of a tyrosine in both Sa and the eukaryotic
enzymes. Despite the amino acid identities the way in which this substitution is compensated for in neighbouring structure suggests that Sa is
closer to the Bacillus enzymes. The change in volume on going from
Bacillus tryptophan to Sa tyrosine is compensated for by a simple rotation
of a neighbouring tyrosine residue about the CA-CB bond. In the eukaryotic
enzymes however, this change in volume is compensated for by the insertion
of an extra residue. Another point is that the OH of the Sa tyrosine and
the NE of the Bacillus tryptophan both hydrogen bond the equivalent main
chain carbonyl oxygen. The eukaryotic tyrosine OH also hydrogen bonds a
main chain carbonyl group, but from a different residue.

CONCLUSION

     This more detailed analysis is confirming earlier results that the
microbial ribonucleases form a highly divergent family. Only five residues

215

are invariant; three of these are required for catalysis, the other two for substrate recognition. None of the residues in the hydrophobic interior are invariant. The family consists of at least three subgroups, formed by the Bacillus, Streptomyces and eukaryotic enzymes. Several factors place the prokaryotic Bacillus and Streptomyces subgroups closer to each other than to the eukaryotic subgroup.

REFERENCES

Heinemann, U. and Saenger, W. 1982, Nature 299; 27-31.
Hill, C. P.., Dodson, G.G., Heinemann, U., Saenger, W., Mitsui, Y., Nakamura, K., Borisov, S., Tischenko, G., Polyakov, K., Pavlovsky. 1983, Trends in Biochemical Sciences 8; 364-369.
Sevcik, J., Dodson, E.J., Dodson, G.G. and Zelinka, J. In this Volume.
Sugio, S., Amisaki, T., Ohishi, H., Tomita, K., Heinemann, U. and Saenger W. 1985, FEBS Lett 181, 129-132.
Sugio, S., Oka, K., Ohishi, H., Tomita, K., Heinemann, U. and Saenger, W. 1985, FEBS Lett 183; 115-118.

STRUCTURE BASES FOR NUCLEOTIDE RECOGNITION

BY GUANYL-SPECIFIC RIBONUCLEASES

A. G. Pavlovsky, S. N. Borisova, B. V. Strokopytov
R. G. Sanishvili, A. A. Vagin and N. K. Chepurnova

Institute of Crystallography, Academy of Sciences, Moscow
Institute of Molecular Biology and Biophysics
Georgian Academy of Sciences, Tbilisi
Institute of Molecular Biology, Academy of
Sciences, Moscow, USSR

Structural aspects of specificity and catalytic activity of enzymes are studied on various objects, in particular on ribonucleases. Remarkable progress has been achieved in understanding the pyrimidine specificity of RNAse A at molecular level. The tertiary structures of different microbial ribonucleases were solved at atomic resolution recently (Hill et al., 1983). These enzymes belong to the so-called guanyl-specific RNases such as the well-known RNAse T_1, which degrades single-stranded RNA to give oligonucleotides with terminal 3' guanosine monophosphate. In many of these, the cleavage occurs as a two-step mechanism via 2', 3' cyclic nucleotide intermediates. In order to reveal molecular bases for substrate specificity of RNAse T_1 the structures of the RNAse T_1- 2'- or 3'-GMP complex were studied at 2 Å resolution (Heinemann and Saenger, 1982; Sugio et al., 1985). Some important features of recognition patterns were obtained from these data. However, it was not possible to draw cleancut conclusions in respect to the molecular mechanism for substrate discrimination. For this reason we have undertaken a study of the crystal structure of RNAse Pb_1 (from Penicillium brevicompactum), a close analogue of RNAse T_1.

The crystals of RNAse Pb_1 were grown (Borisova et al., 1986) and the structure of this protein was solved by the molecular replacement method. As a reference structure we have used that of the closest analogue, namely RNAse C_2, kindly provided to us by Dr K. Poljakov. The structure of RNAse Pb_1 has been refined up to 1.4 Å (R = 0.184).

We also tried to search for growth conditions of complexes of RNAse Pb_1 with the nucleotides. The crystals of RNAse Pb_1 in complex with 3', 5'-guanosine diphosphate (pGp) and lately with 3'-GPM were obtained by co-crystallization technique in 1986. The crystals of complexes were of a better quality than the native crystal which permitted us to collect X-ray diffraction data up to 1.24 (1.48 Å for 3'-GMP complex). The strongest peaks in the difference electron-density map were those due to the base and 3'-phosphate moiety of the nucleotide. One can clearly see the bulges due to oxygen in position 6 and nitrogen in position 2 of the heterocycle (Figure 1). The density of the phosphate group had a typical tetrahedron shape with the phosphorus atom in the center and four oxygen atoms at its

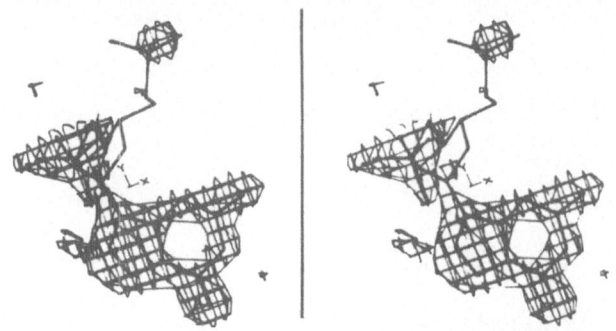

Fig. 1.   Stereodrawing of a part of the different electron density map in
          complex RNAse Pb_1 – pGp superimposed with the atomic model.

vertices. The density peaks due to the ribose ring and the phosphate group
in position 5' are less distinctive. The atomic model of the complex with
pGp was refined (R = 0.196).

The refined atomic model of the complex shows that the hydrogen bond
network system connects the base and 3'-phosphate group of the nucleotide
to the active site of the protein providing specific recognition. Unfortunately, even such a high resolution did not allow to localize hydrogen
atoms. Therefore a special computer program was written to locate hydrogen
atoms and analyze the configuration of hydrogen bonds. Table 1 lists the
geometrical parameters of hydrogen bonds formed in the complex. These are
hydrogen bonds which can "recognize" the guanine base (Figure 2). It is
clear that in the adenine base (an oxygen atom in position 6 is substituted
by the amino group and there is no amino group in position 2) only one
hydrogen bond (NH_{41} ... N7 ADE) is formed of five or six potential bonds
necessary for recognition of the base by the protein. The difference in
hydrogen bond network system in the enzyme-nucleotide complex formed by the
RNAse Pb_1 with adenine and guanine should be considered as a real molecular
base for substrate discrimination. The rate of hydrolysis of guanine and
adenine nucleotides by RNAses of this class differs by more than five
orders of magnitude.

The structure determination of the above complex was carried out
simultaneously with studying the crystal structure of RNAse Bi (binase) –
the bacterial protein (from Bacillus intermedius) having the same substrate
specificity. The binase structure was determined at 3.2 Å resolution by the
isomorphous replacement method in 1983 (Pavlovsky et al., 1983) and then
was refined up to 2.5 Å resolution. The crystalls were not of a quality
sufficient to collect experimental data with higher resolution. Neither did
we succeed in obtaining the binase-nucleotide complex. That is why we had
to search for other crystalline forms which were found recently. These
crystals permitted us to collect experimental data up to 2.0 Å resolution.
The structure of this form was solved by the molecular replacement method
and then refined by the lease-squares procedure. However, we were not able
to produce good crystals of the protein-nucleotide complex until now.
Therefore we tried to model the binase complex with a nucleotide using an
analogy with the well known RNAse Pb_1-pGp complex. To that end one should
find the closest structural features in the two RNAses. In accordance with
the method suggested in (Hill et al., 1983) we aligned the positions of Cα
–atoms of four β-strands in both enzymes. It is just in the β-sheet and
adjacent regions where the residues of the active site are located. But
since both proteins express similar substrate specificity it is natural to
suggest that such a feature has to be located in the recognizing part of

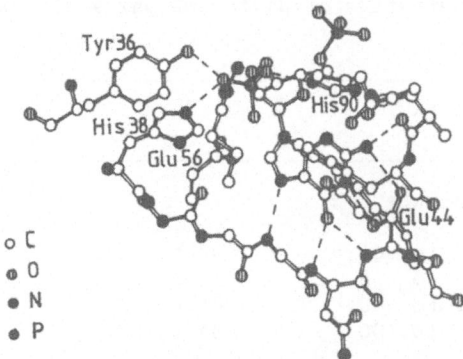

o C
⊙ O
● N
● P

Fig. 2. Active site of the RNAse Pb₁ in complex with pGp. Note the hydrogen bonds (broken lines).

Table 1. Geometry of the possible hydrogen bonds for the atoms in the active site of RNAse Pb₁ complexed with pGp (3'-GMP).

NUCLEOTIDE	ENZYME		DISTANCE Å	ANGLE grad.
ATOM	ATOM	RES.	H..O(N)	N...H...O
N1	OE2	GLU 44	1.8	154
N2	OE1	GLU 44	1.9	164
N2	O	ASN 96	2.0	148
O6	NH	ASN 42	2.2	110
O6	NH	TYR 43	1.8	169
N7	NH	HIS 41	2.0	174
O1P	OH	TYR 36	1.8	151
O2P	NE	HIS 90	1.8	170
O3P	ND	HIS 38	1.9	163

the active site. And at last we fitted two structures aligning the recognizing and catalytic parts of the active site. After the alignment we can accommodate the nucleotide in the active site of binase, in other words, to obtain the "theoretical" complex. These results were estimated from the difference between the coordinates of the phosphate group center of the nucleotide in the model complex and the phosphate binding site of binase localized from the electron density map. In the native crystal this site is

Table 2. Least square fits of different parts of the RNAse Pb_1 and Bi structures.

	Bi	Pb_1	$\langle \Delta r \rangle$ (Å)	$\Delta r\ PO_4$* (Å)
I. All $C\alpha$-atoms in β-sheet				
RESIDUES	51 to 61	36 to 46		
	68 to 76	52 to 60	1.4	2.1
	82 to 90	71 to 79		
	95 to 100	84 to 89		
II. "Recognizing" part of active site				
N, $C\alpha$, C, 0 of	55 to 60	40 to 45		
COO	GLU 59	GLU 44	0.7	1.4
III. "Recognizing" + catalytic part of active site				
N, $C\alpha$, C, 0 of	55 to 60	40 to 45		
COOH	GLU 59	GLU 44		
COOH	GLU 72	GLU 56	0.9	0.7
CNH_2NH_2	ARG 86	ARG 72		
	HIS 101	HIS 90		

* Difference between positions of 3'-phosphorus in modelled structure and center of the phosphate binding site of binase.

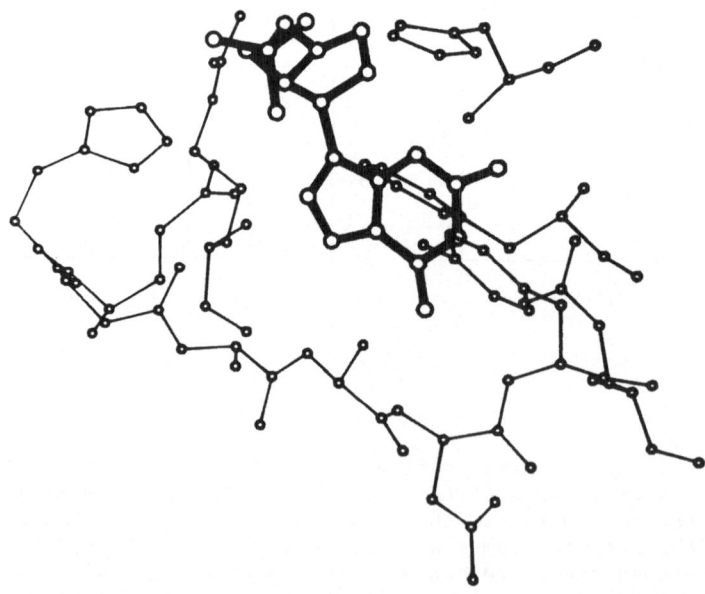

Fig. 3. A view of the active site residues in the RNAse Pb_1 3'-GMP complex.

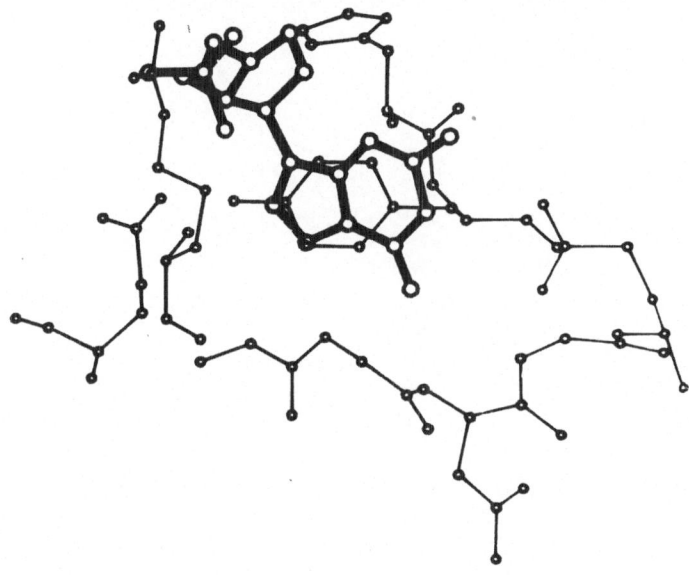

Fig. 4. A view of the active site residues in the modelled complex of
         RNAse Bi with 3'-GMP.

occupied by a sulphate ion. Table 2 shows the results of the three manners
of alignment. According to the criterion mentioned above the better predic-
tion is the last. Figures 3 and 4 show the parts of the active site in the
experimental complex of RNAse Pb_1 with 3'-GMP and in the modelled complex
of Bi with 3'-GMP. It is worth noting that the best fitting of atomic
positions is observed in the recognizing part of the two proteins.

Quite recently we managed to co-crystallize a complex of the second
form of binase with 3'-GMP and carry out an X-ray diffraction experiment.
It should be kept in mind that binase crystalls (in both forms) have two
protein molecules in the asymmetric part of the unit cell. We expected
nucleotides to be at the active site of both molecules, but actually the
electron density map showed only one molecule which is close to the predicted position. At present, we refine the coordinates of the binase complex atoms up to 2.0 Å resolution. Even at this initial stage of the
refinement it is obvious that the interaction between binase and nucleotide
atoms in the recognizing part of the active site is very similar to their
interaction in the complex RNAse Pb_1 -pGp (3'-GMP). At the same time the
interactions in the catalytic regions of the active site are essentially
different in the two proteins. The study of these differences is to be the
subject of our future investigation.

REFERENCES

Hill, C., Dodson, G., Heinemann, U., Saenger, W., et al. (1983) Tibs, 8,
    10, 364.
Heinemann, U., Saenger, W. (1982), Nature, 299, 5878, 27.
Sugio, S., Oka, K. -i., Ochishi, H., Tomita, K. -i., Saenger, W., (1985)
    FEBS Lett., 83, 1, 115.
Borisova, S. N., Vagin, A.A., Nekrasov, J.V., Pavlovsky, G.G. et al. (1986)
    Kristallografiya (USSR), 31, 474.
Pavlovsky, A.G., Vagin, A.A., Vainstein, B.K., Chepurnova, N. K.,
    Karpeisky, M. Ya. (1983) FEBS Lett., 162, 1, 167.

CRYSTALLIZATION AND PRELIMINARY X-RAY STRUCTURAL STUDIES

OF RNase Th$_1$ FROM <u>TRICHODERMA HARZIANUM</u>

K.M. Polyakov, B.V. Strokopytov, A.A. Vagin,
S.I. Bezborodova* and S.V. Shlyapnikov**

Institute of Crystallography, Academy of Sciences of
the USSR, Moscow

*VNII "GENETIKA", Moscow

**Institute of Molecular biology, Academy of Sciences
of the USSR, Moscow, USSR

RNase Th$_1$ is an alkaline guanyl-specific fungal ribonuclease. Primary structure of RNase Th$_1$ has been determined (Table 1). The homology of RNase Th$_1$ with other fungal RNases (T$_1$, C$_2$, Pb$_1$) is about 70%.

The crystals of RNase Th$_1$ were grown in dialysis tubes by equilibrium dialysis against 0.2 M K$_2$HPO$_4$ *3H$_2$O buffer (pH 7.1), containing 60% saturated ammonium sulphate and 4% v/v dioxane at 4°C. The crystals belong to the space group P3$_2$21 (or P3$_1$21). The unit cell parameters are: a=b=55.8 Å, c=80.0 Å.

Data collection was carried out with Syntex P2$_1$ diffractometer and KARD-4 diffractometer with position-sensitive detector (developed in the Institute of Crystallography). The intensities were measured up to 1.75 Å.

The main problem was to determine correctly the number of protein molecules per asymmetric unit. Two reasonable solutions were proposed with one or two molecules per asymmetric unit respectively. For the first case the ratio of the unit cell volume to molecular weight was equal to 3.4 Å^3/Da and for the second case it was equal to 1.7Å/Da. This ambiguity had been resolved by means of a special algorithm which gave us an opportunity to investigate different packing conditions in the unit cell concerned (Strokopytov, 1987). This investigation proved the theoretical impossibility for an asymmetric part of the unit cell to contain two molecules because of the existence of unreasonably short contacts (16 Å) between centres of molecules even for the best solution.

The structure of the RNase Th$_1$ was solved by molecular replacement method using RNase C$_2$ atomic coordinates (Polyakov et al., 1987). Rotation search was done using Crowther's fast rotation function (Crowther, 1971). Translation function for both possible space groups was computed by T3 program (Vagin, 1983). Joint analysis of packing and translation function clearly showed P3$_2$21 to be true space group. The initial set of coordinates for refinement was obtained from models of RNases C$_2$ and Pb$_1$ (Borisova et al., 1987). In this procedure we took into account the homology between these RNases. We used two programs Corels (Sussman, 1977) and Prolsq (Konnert and Hendrickson, 1980) for the crystallographic refinement. After preliminary refinement R-value was equal to 30.0% at 1.75 Å resolution.

Table 1.   Primary structure of RNase Th₁

Asp-Thr-Ala-Thr-Cys-Gly-Lys-Val-Phe-Tyr-
-Ser-Ala-Ser-Ala-Val-Ser-Ala-Ala-Ser-Asn-
-Ala-Ala-Cys-Asn-Tyr-Val-Arg-Ala-Gly-Ser-
-Thr-Ala-Gly-Gly-Ser-Thr-Tyr-Pro-His-Val-
-Tyr-Asn-Asn-Tyr-Glu-Gly-Phe-Arg-Phe-Lys-
-Gly-Leu-Ser-Lys-Pro-Phe-Tyr-Glu-Phe-Pro-
-Ile-Leu-Ser-Ser-Gly-Lys-Thr-Tyr-Thr-Gly-
-Gly-Ser-Pro-Gly-Ala-Asp-Arg-Val-Val-Ile-
-Asn-Gly-Gln-Cys-Ser-Ile-Ala-Gly-Ile-Ile-
-Thr-His-Thr-Gly-Ala-Ser-Gly-Asp-Ala-Phe-
-Val-Ala-Cys-Gly-Gly-Thr

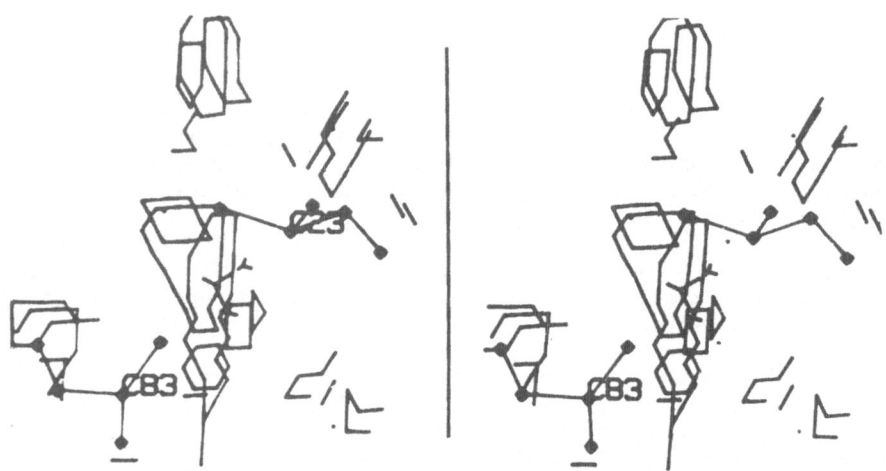

Fig. 1.   Stereoscopic drawing of the region of the difference electron density map ner Cys 23 and Cys 84.

The reliability of the final structure was proved by difference synthesis with ($|Fcal-Fobs|$) exp ($i\alpha$) coefficients computed for the region near residues Cys 23 and Cys 84. Sulphur atoms were clearly seen in the electron density maps (Figure 1) in spite of the fact that they were omitted in the atomic model of RNase Th₁.

The packing of RNase Th₁ molecules is shown on Figure 2. It is interesting to note that the active site region of molecules is fully accessible in this structure. That is why this crystalline form of RNase Th₁ seems to be preferable in the study of interactions of RNases with different substrate analogues.

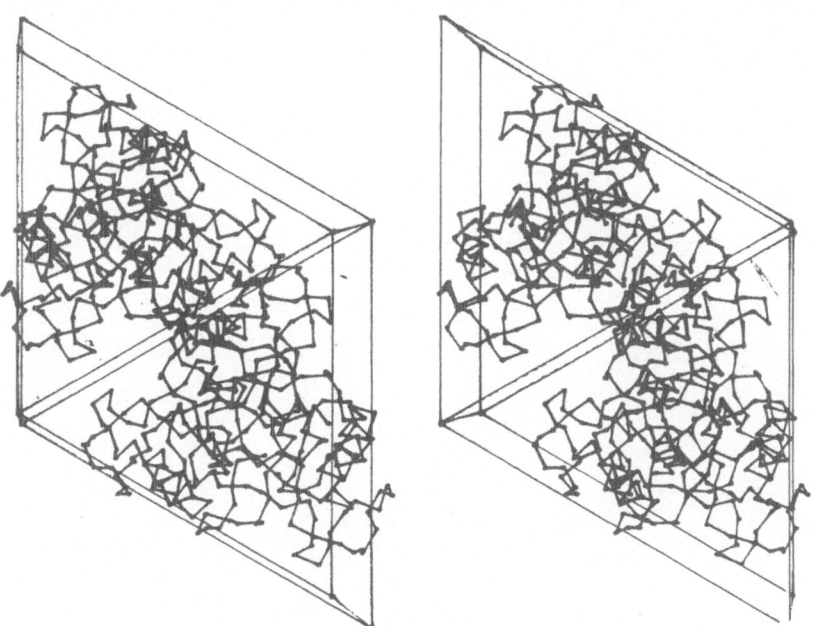

Fig. 2.   Stereoscopic drawing of the packing of RNase Th_1 molecules in the unit cell.

REFERENCES

Stokopytov, B.V. 1987, "Kristallografiya" (USSR), in press.
Sussman, J.L., Holbrook, S.R., Church, G.M., Kim, S.H. 1977, <u>Acta Cryst.</u>, A33; 80.
Konnert, J.H., Hendrickson, W.A. 1980, <u>Acta Cryst.</u>, A36; 344.
Growther, R.A. 1971, in: "Molecular Replacement Method," ed., Rossman, Gordon and Breach, N.Y.
Vagin, A.A. 1983, Ph.D. Thesis Institute of Crystallography.
Polyakov, K.M., Strokopytov, B.B., Vagin, A.A., Tishenko, G.N., Bezborodova, S.I., Vainshtein, B.K. 1987, <u>Kristallografiya</u>. USSR, 32:3
Borisova, S.N., Vagin, A.A., Nekrasov, J.V., Pavlovsky, A.G., et al. 1986, Kristallografiya USSR, 31; 474.

THREE-DIMENSIONAL STRUCTURE OF RNase C$_2$ FROM

ASPERGILLUS CLAVATUS AT 1.35 Å RESOLUTION

K.M. Polyakov, B.V. Strokopytov, A.A. Vagin,
S.I. Bezborodova* and L. Orna*

Institute of Crystallography, Academy of Sciences of the
USSR, Moscow

*VNII "Genetika", Moscow, USSR

Ribonuclease C$_2$ from Aspergillus clavatus is a guanyl-specific ribonuclease. The molecular weight of RNase C$_2$ is about 11000. Its primary structure was determined (Bezborodova et al., 1983).

The crystals of RNase C$_2$ were grown in dialysis tubes by equilibrium dialysis against 0.1 M phosphate buffer (pH 6.6), containing 60% saturated ammonium sulphate and 4% v/v dioxane at 4°C. The crystals of two different crystal forms were grown under the same conditions (Table 1). Usually the crystals of the first crystalline form were grown. But sometimes the crystals of the second crystalline form were obtained under the same conditions. There are two protein molecules per assymetric unit for the first crystalline form and one for the second crystalline form.

Data collection was carried out by Syntex P2$_1$ diffractometer and KARD-4 diffractometer with a position-sensitive detector (developed at the Institute of Crystallography). For the first form intensities were measured up to 1.75Å and for the second one up to 1.35 Å.

The structure of RNase C$_2$ for the first crystalline form was solved using the MIR method (Polyakov et al., 1983). Three heavy-atom derivatives were used for this purpose. Subsequently the electron density map was significantly improved by the averaging of the electron density of molecules related by non-crystallographic symmetry. The model was refined by the Jack-Levitt (Jack and Levitt, 1978) and Hendrickson-Konnert (Hendrickson and Konnert, 1980) methods. The final R-value for the first crystalline form was equal to 24.6% at 1.75 Å resolution.

The structure of RNase C$_2$ for the second crystalline form was determined using the molecular replacement method. Rotation search was done using Crowther's fast rotation function (Crowther, 1971). Translation function was calculated by the program T3 developed by Vagin (Vagin, 1983). Subsequently this structure was refined by the Hendrickson-Konnert method. The final R-value for the second crystalline form was equal to 19.8% at 1.35 Å resolution (Polyakov et al., 1987).

The RNase C$_2$ polypeptide chain folding is presented on Figure 1. The polypeptide chain folding of RNase C$_2$ highly resembles that of other fungal RNases (T$_1$, Pb$_1$). Schematic representation of hydrogen bonds between main

Table 1. Unit cell parameters for crystals of RNase C_2

	Sp. gr.	a, Å	b, Å	b, Å	$\gamma,^0$
I. form	$P2_1$	31.7	51.1	57.3	93.5
II. form	$P2_1$	30.6	31.9	49.7	116.0

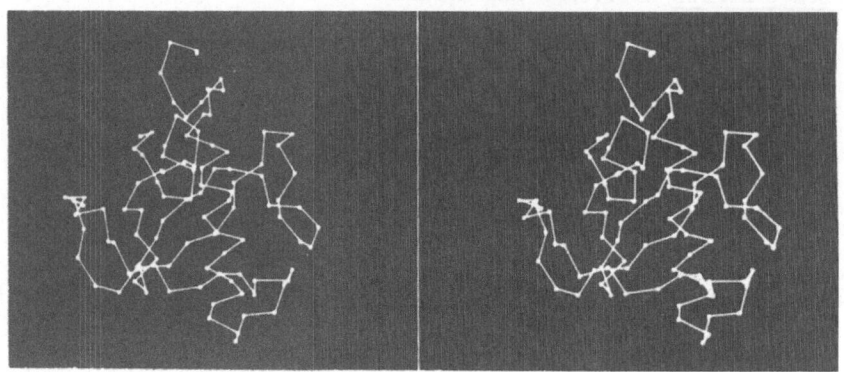

Fig. 1. Schematic drawing of the RNase C_2 main chain folding.

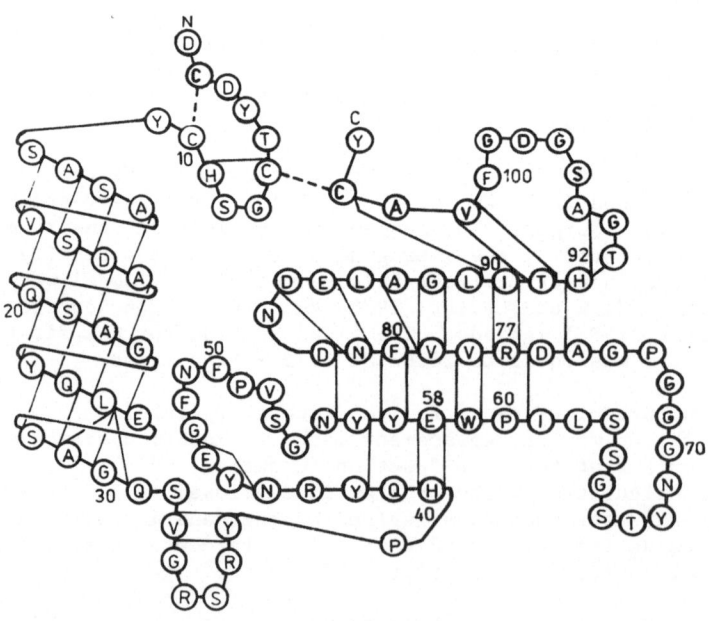

Fig. 2. Network of hydrogen bonds between main chain atoms of the RNase C_2 molecule.

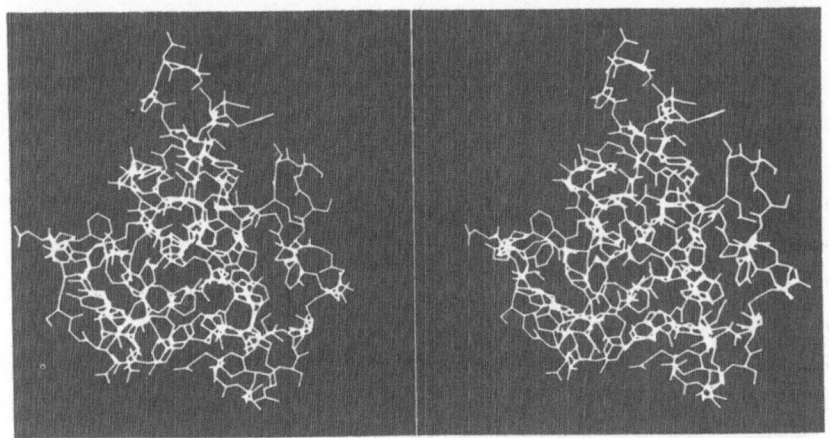

Fig. 3. Stereoscopic drawing of the RNase C$_2$ molecule.

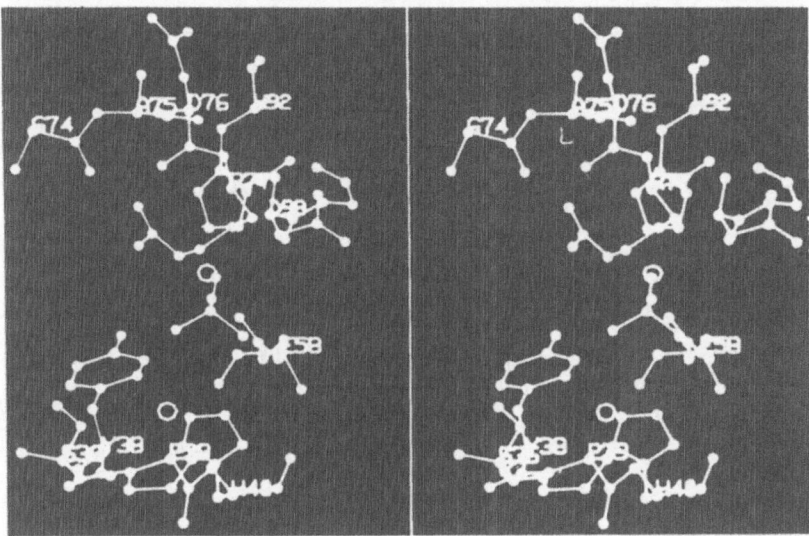

Fig. 4. Stereoscopic drawing of the active site of the RNase C$_2$ molecule.

chain atoms of the RNase C$_2$ is given on Figure 2. The RNase C$_2$ molecule contains one α–helix (residues 12-30) with regular network of hydrogen bonds and twisted β–sheet with four anti-parallel β–strands (residues: 40-42, 55-61, 75-82, 84-92). Hydrogen bond network in the β–sheet is quite regular too, except for the residue Val 79 forming a classical β–bulge with dipeptide Ala 87 – Gly 88. There are four turns with hydrogen bonds in the structure of RNase C$_2$ (residues: 6-9, 44-47, 81-84, 92-94).

The full atomic model of RNase C$_2$ is shown in Figure 3. The molecule of RNase C$_2$ contains 31 hydrophobic amino acids. These residues form four hydrophobic clusters in the structure.

The positions of about one hundred water molecules were detected.

The active site of RNase C$_2$ is formed by the following amino acids: His 40, His 92, Tyr 38, Glu 56 and Arg 77. These residues are placed on the β–sheet or near it.

Table 2. The possible hydrogen bonds for the atoms in the active site of RNase C_2

DONOR ATOM	RES.	ACCEPTOR ATOM	RES.	DISTANCE Å N..O	DONOR ATOM	RES.	ACCEPTOR ATOM	RES.	DISTANCE Å N..O
ND1	H 40	O	S 36	2.8	OH	Y 38	O2	PO4	2.6
ND1	H 92	O	D 98	3.1	NE2	H 92	O3	PO4	2.6
ND1	H 92	O	G 99	2.9	NE2	H 92	O4	PO4	2.8
NH1	R 77	O	G 74	2.9	OE2	E 56	O1	PO4	3.1
NH2	R 77	O	D 76	2.9	OH	WAT	O4	PO4	2.7
NE2	H 40	OE2	E 58	2.8	OH	WAT	O2	PO4	3.0
NE	R 77	OE1	E 58	3.1	OE1	E 56	O3	PO4	3.1

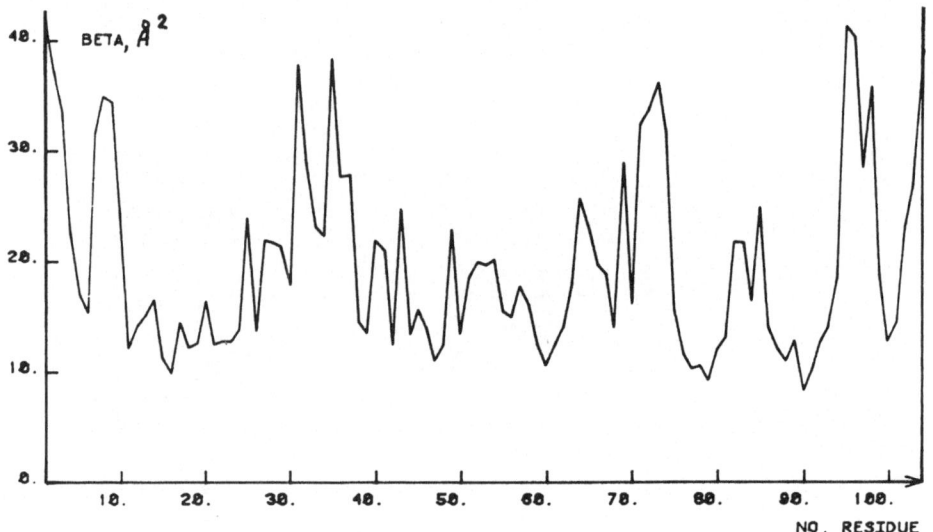

Fig. 5. The distribution of average atomic temperature factors for the amino acid residues of RNase C_2.

During the refinement process the position of the phosphate ion was located. It is bound in the active site of the RNase C_2 molecule by hydrogen bonds (Figure 4). The possible hydrogen bonds for the atoms in the active site of RNase C_2 are presented in Table 2. It is interesting to note that side chains of two histidine residues and the arginine residue may form hydrogen bonds with main chain atoms of neighbouring residues.

The distribution of atomic temperature factors is shown in Figure 5. The mean value of the temperature factors for residues belonging to the regular secondary structure is rather small. Temperature factors for amino acids forming the active site are also small enough.

The accessible surface area for water molecules was calculated. These values are strongly correlated with values of atomic temperature factors.

REFERENCES

Bezborodova, S.I., Hodova, O.M., Stepanov, V.M. 1983, FEBS Letters, 159:256.

Polyakov, K.M., Vagin, A.A., Tishenko, G.N., Bezorodova, S.I., Vainshtein, B.K., 1983; Doklady Akad., Nauk SSSR, 273:1383.

Jack, A., Levitt, M. 1978, Acta Cryst., A34:931-935.

Hendrickson, W.A., Konnert, J.H., 1980, Acta Cryst., A36:344.

Growther, R.A. 1971, in: "Molecular Replacement Method", ed. Rossman. Gordon and Breach, NY.

Vagin, A.A., 1983, Ph.D. Thesis Institute of Crystallography.

Polyakov, K.M., Strokopytov, B.V., Vagin, A.A., Tishenko, G.N., Bezborodova, S.I., Vainshtein, B.K. 1987, Kristallografiya (USSR), 32:3

AMINO ACID COMPOSITION AND INFLUENCE OF THE MODIFICATION OF SOME AMINO ACID RESIDUES ON THE CATALYTIC ACTIVITY OF THE SPLEEN EXONUCLEASE

A. T. Bakalova and L. B. Dolapchiev

Department of Enzymology, Institute of Molecular
Biology, Bulgarian Academy of Sciences
1113 Sofia, Bulgaria

Two main reasons make the investigation of the exonucleolytic enzymes of importance. Because of their specific mode of action, they are still an inevitable tool for some studies on the primary structure of number of natural and synthetic poly- and oligo-nucleotides. On the other hand, an open question for the structure-function relationship remains - which primary or higher structure of the enzyme active site defines such peculiar specificity - to attack only one, well determined end of the polymeric substrate molecule.

The nonspecific exonclease from mammalian spleens (spleen exonuclease or formerly phosphodiesterase II, EC 3.1.16.1.) is an orthophosphoric diester hydrolase, splitting the first phosphodiester bond of the poly ribo- and deoxyribo-nucleotides from the 5' end of the substrate molecule and releasing nucleoside 3'-monophosphates (Razzell and Khorana, 1961).

In the present contribution we describe the method for obtaining a homogenous preparation of the exonuclease from beef spleen, its physical and chemical properties and the influence of the modification of some amino acid residues on the catalytic activity of this enzyme.

PURIFICATION OF BEEF SPLEEN EXONUCLEASE

One of the main obstacles for the detailed study of the spleen exonuclease has been the lack of an easy and efficient method for its purification. Further, its noticeable instability hindered all attempts for investigation its molecular properties.

In our early studies we found a significant isoelectrofocusing heterogeneity (Bakalova and Dolapchiev, 1978), caused by the carbohydrate moieties, present in the exonuclease molecule (Silverman and Soll, 1977) (see below). This allowed us to propose an affinity chromatography step in the purification procedure, which not only reduced the purification time, but efficiently eliminated all possible contaminations with interfering enzymatic activities as endonuclease, non-specific 3'-nucleotidase etc. (Dolapchiev and Bakalova, 1982).

On Table I a summary of the purification procedure is given.

Table 1.

Step of purification	Protein A_{280}	Total activity	Specific activity	Yield %
Crude enzyme	540	108	0.2	100
ECTEOLA-Cellulose	67	67	1.0	65
Con A-Sepharose	8.62	52	6.0	48
Bio-Gel P-150	1.4	33.5	24.0	31

The preparation of the crude enzyme followed the method of Bernardi and Bernardi (1968). The second step – ion-exchange chromatography on ECTEOLA-Cellulose eliminates more that 90% of the contaminating proteins and lessens the activity of the nucleotidase more than 255 times. The next step – the affinity chromatography on Concanavaline A-Sepharose (Con-A) is very efficient for eliminating the endonucleolytic activity and the final molecular sieving step on Bio-Gel P-150 led to an enzyme homogenous by SDS disc electrophoresis.

MOLECULAR PROPERTIES OF THE BEEF SPLEEN EXONUCLEASE

The properties of the exonuclease are summarized on Table 2 and 3. Three independent methods were used for determining the molecular mass of the spleen exonuclease – molecular sieving, ultracentrifugation in density gradient and SDS disc electrophoresis. The enzyme consists in one single polypeptide chain which was proved by N-terminal amino acid analysis and with electrophoresis in denaturating conditions, with molecular mass of 98 kDa.

The isoelectrofocusing of the crude enzyme from beef spleen showed a significant heterogeneity of the activity profile towards different specific and non-specific substrates – thymidine 3'-monophospho p-nitrophenyl ester (p-NP.Tp), bis-p-nitrophenyl phosphate (bis-p-NPP) and RNA-core. This observations as well as the strong binding to Con A are caused by the significant amount of carbohydrates, covalently bound to the enzyme polypeptide chain. On the other hand the treatment with neuramini-dase led to one activity peak during isoelectrofocusing and in the same time did not affect the enzyme activity towards the substrates under study. It was also found that 10.15% w/w are carbohydrates – 6.5% are neutral sugars, 2.7% – amino sugars and 0.95% sialic acid (4 residues per mol enzyme).

The amino acid composition of the exonuclease isolated from beef spleen shows that the enzyme does not contain threonine, proline, meth-ionine and tyrosine. It is rich in glycine and alanine. There are 8 S-S bonds and 7 free -SH residues (Table 3). The amount of the triptophane residues was determined spectrophotometrically according to Balestrieri et al. (1978).

INFLUENCE OF THE MODIFICATION OF SOME AMINO ACID RESIDUES ON THE EXONUCLEOLYTIC ACTIVITY OF THE SPLEEN EXONUCLEASE

The treatment of a homogenous preparation of beef spleen exonucleas with 5,5'-dithio-bis (2-nitro-benzoic acid) (DTNB) showed a significant

Table 2.

Molecular mass	98 000 ± 3 000 Da
Protein	89.85%
-SH	7
-S-S-	9
Carbohydrates	10.15%
Neutral sugars	6.50%
Amino sugars	2.70%
Sialic acid	0.95%

Table 3.

Amino acid	μg amino acid per mg protein	No residues per mol protein
Asp	96	83
Thr	–	–
Ser	84	96
Glu	93	72
Pro	–	–
Gly	140	245
Ala	81	114
1/2Cys	26	25
Val	80	81
Met	–	–
Ile	32	28
Leu	80	71
Tyr	–	–
Phe	49	33
His	29	21
Trp	22	12
Lys	51	39
Arg	26	17

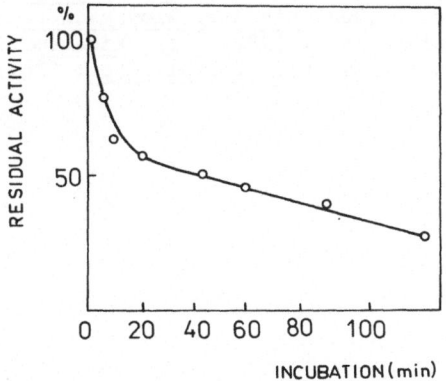

Fig. 1. Influence of the modification with DTNB on the exonuclease activity.

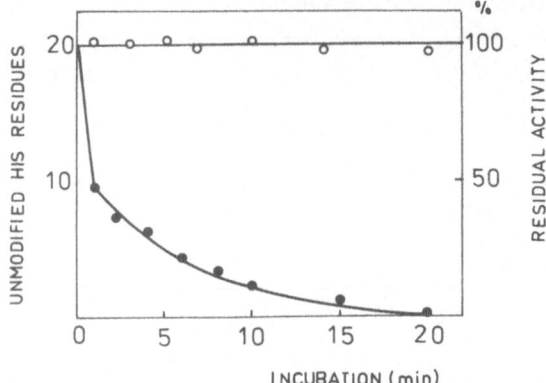

Fig. 2. Time course of modification of spleen exonuclease with DEP.
●────● number of unmodified hisitidines; ○ ── ○ enzyme activity.

loss of activity (Figure 1) – up to 40% inhibition after 20 min of modification and 70% after 100 min. The slow course of the inhibition of the
enzyme activity suggests that the –SH groups participate in the active site
of the enzyme. The further inhibition of the exonuclease may be due to
conformational changes of the whole molecule.

The modification of the histidine residues in the exonuclease (Figure
2) did not affect the activity of the enzyme towards p–NP.Tp. Under our
experimental conditions the reaction of the histidine residues with
diethylpyrocarbonate (DEP) went quite fast and all 21 histidine residues
determined in the amino acid composition, were modified without denaturation of the exonuclease. The determination of the kinetic parameters of
the hydrolysis of p–NP.Tp showed that with the course of the modification
the ratio K_{cat}/K_m did not change. Thus the histidine residues, although
easily reachable by DEP, either do not participate in the formation of the
active site directly or their modification does not change significantly
the active conformation of the spleen exonuclease.

REFERENCES

Balestrieri, C., Colonna, G., Giovane, A., Servillo, L. 1978, Eur. j.
     Biochem. 90:433–440.
Bakalova, A.T., Dolapchiev, L.B. 1978, C.R. Acad. Sci. Bulg. 31:1179–1182.
Bernardi, A., Bernardi G. 1968, Biochem. Biophys.Acta 155:360–370.
Dolapchiev, L.B., Bakalova, A.T. 1982, Prep. Biochem. 12:121–136.
Razzell, W.E., Khorana, H.G. 1961, J. Biol. Chem. 236:1144–1149.

BINDING OF THE METAL IONS IN THE ACTIVE SITE OF THE

EXONUCLEASE FROM CROTALUS ADAMANTEUS VENOM

R.A. Vassileva and L.B. Dolapchiev

Department of Enzymology
Institute of Molecular Biology
Bulgarian Academy of Sciences
1113 Sofia, Bulgaria

The very first studies on the exonuclease, isolated from different venoms, has shown that the chelation of the enzyme inhibits both phosphodiesterase and pyrophosphatase activities. Thus, while the prolonged dialysis against EDTA brought about its complete and irreversible inactivation (Bjork, 1967), the inhibitory effect of the chelator added to the incubation mixture could be overcome by an excess of Mg^{2+} (Razzell and Khorana, 1959). Further it has been shown that Mn^{2+}, Ca^{2+} and Zn^{2+} were equally efficient activators of the exonuclease.

In the course of studying the chemical properties of the exonuclease isolated from the venom of Crotalus adamanteus we showed that a homogeneous preparation of the enzyme contains 1 Zn^{2+}, 6 Mg^{2+} and 35 Ca^{2+} atoms per molecule (Dolapchiev at al. 1980).

In the present report the participation of the tyrosine and histidine residues in the active site of the exonuclease as well as their role in the coordination of the metal ions intrinsic for the enzyme is described.

EXPERIMENTAL

The exonuclease from Crotalus adamanteus venom was purified according to the previously described method (Dolapchiev et al., 1974) with modifications described later (Dolapchiev et al. 1980). The homogeneity of the enzyme thus obtained was controlled by SDS-disc electrophoresis, immunoelectrophoresis (Dolapchiev and Vassilev, 1981) and by N-terminal amino acid analysis (Dolapchiev et al. 1981).

The modification of the tyrosine residues with tetranitromethane (TNM) was carried out according to Sokolovski et al. (1966) and Riordan et al. (1967). The exonuclease (2 mg/ml) dissolved in 0.05 M Tris-HCl, pH=8.0 was treated with 8, 50 and 350 fold molar excess of TNM for 120 min at 20°C (Figure 1). The course of the nitration of the native enzyme and in denaturating conditions (8 M urea) was recorded spectrophotometrically and the amount of the nitroformiate ion formation was calculated using a molar extinction coefficient 14400 M^{-1} cm^{-1} from the increasing of the absorbancy at 350 nm (Figure 2) (Glover and Landsman, 1964).

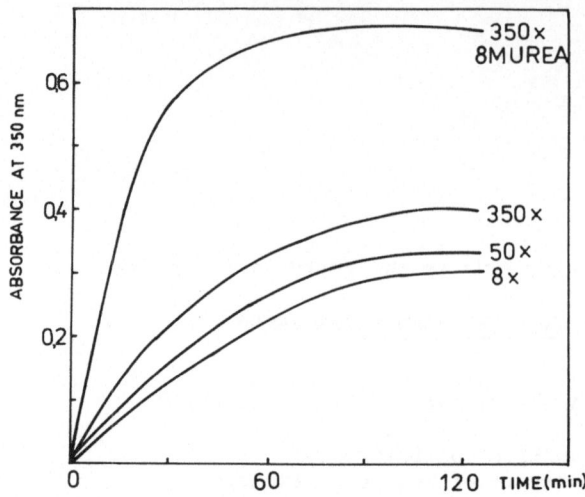

Fig. 1.   Spectrophotometrically recorded time course curves of nitration of
snake venom exonuclease with TNM.

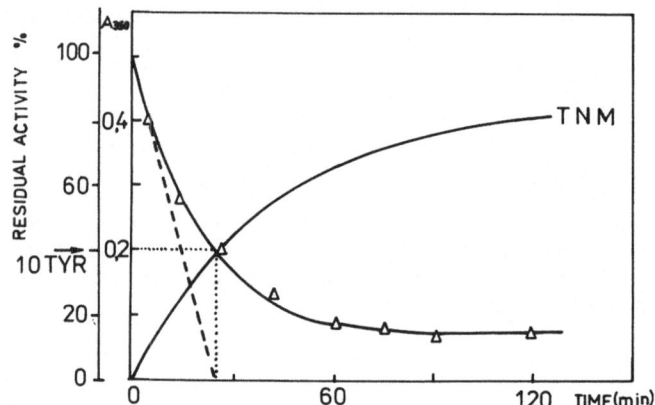

Fig. 2.   Courses of nitration and loss of activity of the exonuclease
nitration with TNM.

The modification of the histidine residues was performed with diethylpyrocarbonate (DPC) according to Ovadi et al. (1967) and Miles (1977).
The reaction was carried out in 0.1 M potassium phosphate buffer, pH=6.5 at
20°C and the carbethoxylation of the exonuclease (0.25 mg/ml) was evaluated
by difference spectroscopy measuring the absorbancy in the range 235–240 nm
(Figure 3), using 3200 M^{-1} cm^{-1} as a molar extinction coefficient for N-
carbethoxyhistidines (Ovadi et al. 1967).

Apo-enzyme, EDTA demetalyzed enzyme was prepared as reported earlier
(Dolapchiev et al. 1980).

The enzyme activity measurement as well as the kinetic investigations
were carried out titrigraphically and spectrophotometrically as described
before (Vassileva and Dolapchiev, 1982).

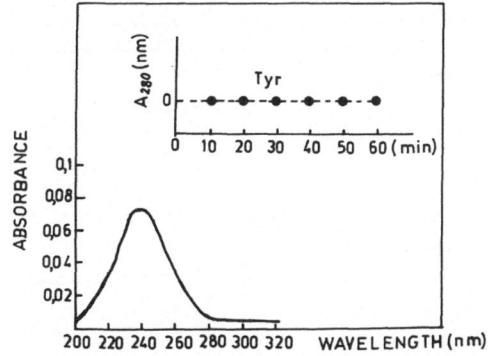

Fig. 3. Difference absorption spectrum of native and carbethoxylated exonuclease after 5 x DEP. The insert: absorbance change at 280 nm.

MODIFICATION OF THE TYROSINES OF NATIVE, CHELATED AND RECONSTITUTED EXONUCLEASE

Studying the amino acid composition we found that the snake venom exonuclease contains 34 tyrosine residues. The course of the nitration of the enzyme (Figure 1) shows that with 8 fold excess of TNM 15 residues are nitrated, with 50 - 17 residues and at 350 fold excess - 20 out of all present tyrosines are modified. Only under denaturating conditions total modification was achieved. The extrapolation of the residual activity curve (Figure 2) suggests that 10 to 12 of the tyrosines are involved in the pyrophosphatase activity of the exonclease.

Considering the wide specificity of the snake venom exonuclease different substrates were used - bis-p-nitrophenyl phosphate (bis-p-NPP), thymidine 5'-monophospho p-nitrophenyl ester (p-NP.pT) and poly U as substrates for the phosphodiesterase activity and ATP and NAD$^+$ - for the pyrophosphatase activity. Here (Figure 4), it is of importance to notice that at 8 fold excess of TNM the exonuclease looses its activity more towards the nucleotidic substrates with phosphodiester bond to be split (p-NP.pT and poly U) than towards the nucleotidic substrates with pyrophosphate bond (ATP and NAD$^+$). In the same time no change of the susceptibility of the non-nucleotidic substrate bis-p-NPP with the modified enzyme was observed. The further modification of the tyrosines (Figure 5) led to almost complete loss of both activities (80 to 97%) while again the non-nucleotide substrate bis-p-NPP was hydrolyzed with the nitrated enzyme as with the native one.

In regard to elucidate the possible coordination of the intrinsic metal ions by the tyrosines, the course of the nitration of native (K), chelated with EDTA (P) and subsequently reconstituted by the three different metal ions (P_{Mg}, P_{Ca}, P_{Zn}) was recorded spectrophotometrically (Figure 6). And taking into consideration the specific role of the Zn ion in the catalytic step the same time course curve of the nitration of a native and o-phenantroline (o-Ph) chelated exonuclease was recorded (Figure 7).

It is evident that the reaction goes identically for the native, chelated and reconstituted enzyme.

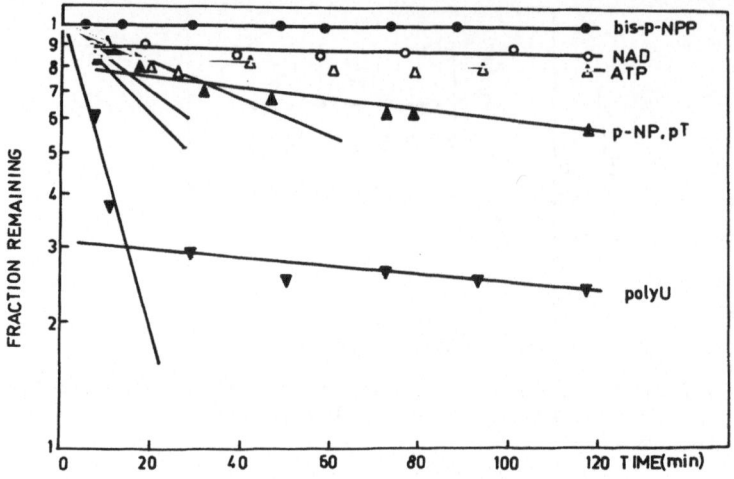

Fig. 4. Kinetics of activity loss during nitration with 8 x TNM.

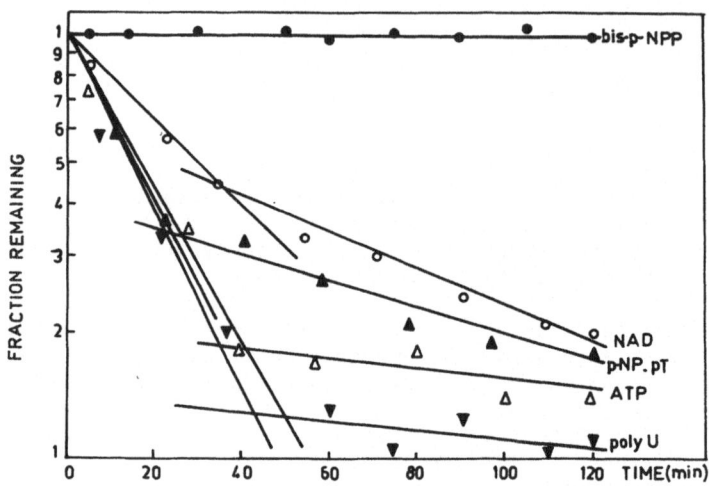

Fig. 5. Kinetics of activity loss during nitration with 350 x TNM.

From all these date we can conclude that the tyrosines do not coordinate directly the metals, present in the enzyme molecule. It is probable that some of them are involved in the binding of the nucleotidic substrates and/or together with carboxylic groups, located near-by, contribute to the activation of a water molecule for a nucleophilic attack of the phosphate ester bond to be split as known for other Zn-enzymes (Holms and Matthews, 1981).

MODIFICATION OF THE HISTIDINES OF NATIVE AND APO-ENZYME

The difference spectrum of modified venom exonuclease against native enzyme under the same conditions is shown on Figure 3. A single absorbance peak at 238-240 nm, characteristic for N-carbethoxyhistidine is observed. A possible modification of the amino acid residues is excluded, as no free -SH groups are present in the exonuclease and a possible modification of

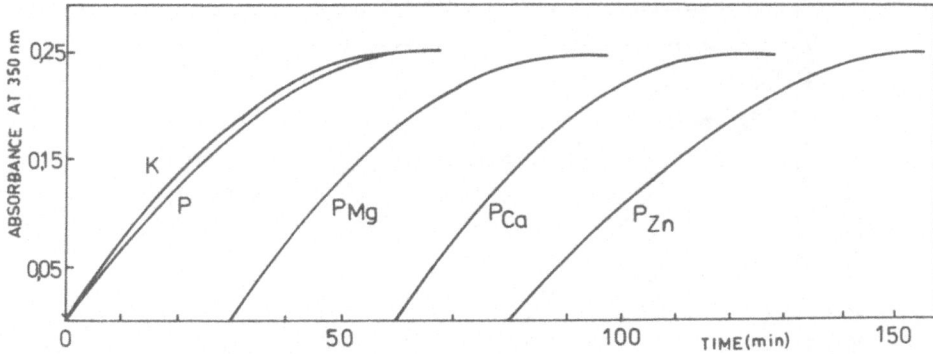

Fig. 6. Nitration of native (K), EDTA-treated (P) and reconstituted enzyme.

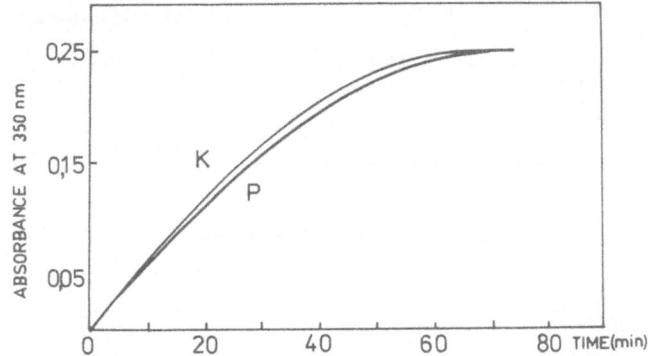

Fig. 7. Nitration of native (K) and o-Ph treated (P) enzyme.

tyrosines can not be expected as there is no change in the absorbancy at 280 nm (see the insert of Figure 3).

The influence of the modification of different numbers of histidine residues on the phosphodiesterase activity is shown in Fig. 8.

It is evident that in this case of modification the activity towards both non-nucleotidic and nucleotidic substrate with phosphodiester bond is inhibited equally down to ca. 50%. The study of the kinetic parameters showed that the loss of activity is mainly due to a change of K_{cat} but not of K_m.

The dependence of the number of N- carbethoxyhistidine residues in the exonuclease before and after chelation (native and apo-enzyme) on the different fold molar excess of DEP is shown on Figure 9.

One fold excess brought about modification of approximately 6 histidine residues, 2 fold - 10, 5 - 12 residues. The further increase of the modifier excess did not increase neither the number of N-carbethoxy-histidines for further inhibition of the enzyme activity. In the same time approximately 17 histidine residues were modified in the apo-enzyme.

Taking into consideration that 25 histidine residues were found by the amino acid analysis it can be concluded that there are two types of histidines - about 12 rapidly reacting or exposed to DEP and the remaining 50% do not react with the modifier. Some of the latter are either engaged in

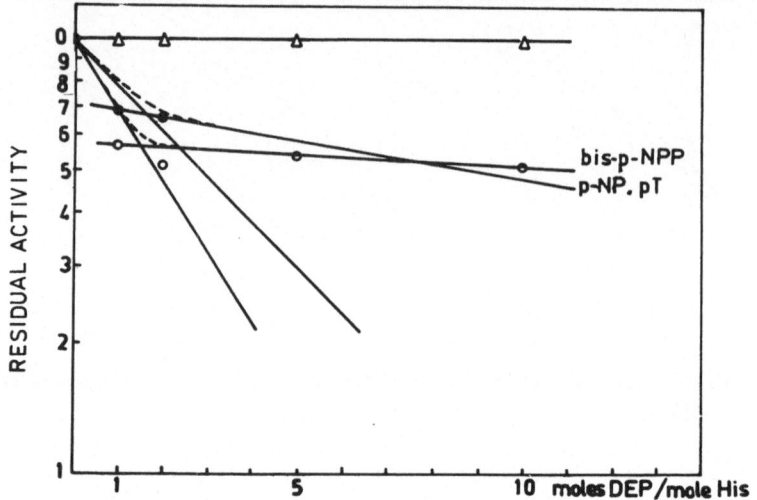

Fig. 8.   Inactivation of exonuclease by DEP.   Δ-Δ is the activity of a native enzyme.

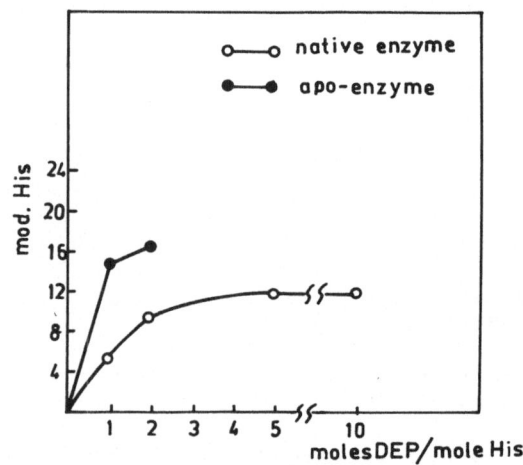

Fig. 9.   Histidine modification of native and apo-enzyme.

coordination of the intrinsic metal ions or are deeply buried and therefore unreachable by DEP. On the other hand from the difference of the modified histidine residues in the native and the apo-enzyme we can expect that 5 more residues modified are either metal-coordinated or disclosed by steric changes in the enzyme molecule after chelation. The remaining non-modified histidines are, most probably deeply buried in the exonuclease molecule.

REFERENCES

Bjork, W. 1967, Ark. Kem. 27:515-538.
Dolapchiev, L.B., Sulkowski, E., Laskowski, M., Sr. 1974, Biochem. Biophys. Res. Commun. 61:273-281.
Dolapchiev, L.B., Vassilev, L.T. 1981, Biochim. Biophys.Acta, 667:335-360.
Dolapchiev, L.B., Vassileva, R.A., Koumanov, K.S., 1980, Biochim Biophys Acta, 662:331-336.

Holmes, M.A., Matthews, B.W. 1981, <u>Biochemistry</u>. 20:6912-6920.

Glover, D.J., Landsman, S.G. 1964, <u>Analyt. Chem</u>. 36:1690-1695.

Miles, E.W. 1977, <u>Methods in Enzymology</u>, 47:431-442.

Ovadi, J., Libov, S., Elodi, P. 1967, <u>Acta Biochim. Biophys. Acad. Sci. Hung</u>. 2:455-458.

Razzell, W.E., Khorana, H.G. 1959, <u>J. Biol. Chem</u>. 234:2105-2113.

Riordan, J.F., Sokolovski, M., Vallee, B.L., 1967, <u>Biochemistry</u> 6:358-361.

Sokolovski, M., Riordan, J.F., Vallee, B.L. 1966, <u>Biochemistry</u> 5: 3582-3589.

Vassileva, R.A., Dolapchiev, L.B. 1982, <u>Proc. Forth. Int. Sympos. Metabol. Enzymol. of N.A. (Zelinka and Balan, Eds.)</u> 4:73-78.

RIBOSOMAL NUCLEASE WITH ACTIVITY TOWARDS DOUBLE-STRANDED RNA

M.A. Siwecka, M. Rytel and J.W. Szarkowski

Department of Comparative Biochemistry
Institute of Biochemistry and Biophysics
Polish Academy of Sciences, Warsaw, Poland

Ribonucleolytic activity has been suggested to be associated with plant ribosomes, but its role has not been elucidated (Hsiao 1968, Farkas 1982). According to Wilson (1982) nucleolytic ribosomal enzymes represent only a small fraction of the total nucleolytic activity of plant tissues. In our earlier investigations ribonucleolytic activity has been observed in nuclei, plastids, mitochondria, postribosomal supernatant and cytoplasmic ribosomes of rye embryos and 24-h seedlings. Specific ribonucleolytic activity was highest in the ribosome fraction from dry rye embryos (Siwecka et al., 1971). We have further demonstrated that deoxyribonucleolytic activity is also associated with rye embryo ribosomes. Both these activities are strongly associated with ribosomes and their subunits but a part of activity may be released by washing with 0.5 M ammonium chloride or 4 M urea (Siwecka et al., 1979). The NH_4Cl ribosomal wash seems to contain an enzyme hydrolyzing double-stranded RNA, namely, double-stranded viral RNA from Penicillium chrysogenum and the double-stranded poly(A)*poly(U) complex (Siwecka et al., 1982). So far there have been no reports on isolation from higher plants of enzymes hydrolyzing double-stranded RNA.

In view of this we have attempted to isolate enzymes degrading double-stranded RNA from the wash of cytoplasmic ribosomes of rye embryos. Optimal ionic conditions were established for investigation and isolation of these activities (Siwecka et al., 1983). The present paper describes the procedure for isolation and partial purification of enzymes hydrolyzing double stranded RNA structures.

MATERIAL AND METHODS

Plants. Commercial rye embryos (Secale cereale) were used as the source of ribosomes.

Isolation of Ribosomes. The embryos were homogenized in 0.01 M Tris-HCl buffer containing 0.25 M sucrose and 0.01 M $MgCl_2$, pH 8.5. The homogenate was filtered through three layers of nylon and then centrifuged at

This work was supported by the Polish Academy of Sciences within the CPBR 3.13 project.

12,000 g for 15 min. Triton X-100 was added to the supernatant to give a 0.5% concentration and the solution was then centrifuged at 40,000 g for 30 min. The pellet was discarded. The ribosomes were sedimented at 105,000 g for 2 h.

Release from Ribosomes of Proteins with Nucleolytic Activity. This was achieved by washing the ribosomes with 0.5 M NH$_4$Cl according to the procedure of Siwecka et al., (1977). The supernatants from two washes represented the starting material for purification of nucleases associated with the ribosomes.

Nucleolytic Activity Determination. This activity was determined by the method described by Bardon et al., (1976). One activity unit corresponded to the amount of enzyme causing under our experimental conditions an increase in VA$_{260}^{1cm}$ of 0.1. Specific activity was expressed in units/1 mg protein.

Protein Determination. Protein was determined by the method of Bradford (1976) with trypsin as standard.

Filtration through sephadex G-100. The ribosome wash containing about 40 A$_{280}$ units was applied onto a column packed with Sephadex G-100 (1.5 x 35 cm). Fractions containing proteins with activity towards poly(A)*poly-(U) complex were pooled and lyophilized.

Chromatography on CM-52 Cellulose. The fractions from Sephadex G-100 with the highest activity towards poly(A)*poly(U) were applied onto a column. Protein was eluted from the column with a NaCl gradient (0.0-0.3 M) in the above mentioned buffer.

Chromatofocusing. The fractions from the first peak of CM-52 cellulose were placed on a column (1 x 7 cm) packed with PBE gel. Separation was run in a pH gradient obtained by application of the following buffers: starting buffer A (0.025 M imidazole-HCl, pH 7.5) and Polybuffer 74-HCl (pH 4.0). Fractions exhibiting maximal activity towards poly(A)*poly(U) were collected and the mixed buffer was eliminated on a Sephadex G-75 column. The pooled fractions were applied once more onto a column with PBE 94. Separation was achieved by application of a gradient formed by starting buffer B (0.025 M ethanolamine-CH$_2$COOH) and Polybuffer 96-CH$_3$COOH (pH 6.0). Fractions with the highest activity towards poly(A)*poly(U) were purified on a Sephadex G-75 column, suspended in a small volume of starting buffer C (0.025 M triethylamine-HCl, pH 10.5) and applied onto a column (0.9 x 10.5 cm) packed with PBE 118 gel. Separation was achieved in a pH gradient of 10.5-8.0.

RESULTS AND DISCUSSION

For purification of proteins associated with the cytoplasmic ribosomes of rye embryos and exhibiting activity towards the double-stranded poly(A)-*poly(U) complex a five-step procedure was applied. The results of purification are shown in Table 1. The 0.5 M ammonium chloride wash is seen to exhibit a twice higher specific activity towards poly(A)*poly(U) and towards single-stranded RNA than the rye embryo ribosomes.

Proteins from this wash were separated by filtration on a Sephadex G-100 column, which yielded two protein peaks with activities towards poly(A)*poly(U), single-stranded RNA and denatured DNA. Fractions with the highest activity towards poly(A)*poly(U) were collected and chromatographed on a column packed with CM-52 cellulose. Three protein peaks were

Table 1. Purification of ribosomal nuclease with activity towards double-stranded RNA

Purification step	Protein (mg)	Activity towards poly(A)*poly(U)			Ribonucleolytic activity		
		Total act. (units)	Spec. act. (units/mg protein)	Purification factor	Total act. (units)	Spec. act. (units/mg protein)	Purification factor
Ribosomes	24.00	285.0	11.88	1.00	540.0	22.50	1.00
NH$_4$Cl wash I and II	18.00	400.0	22.22	1.87	720.0	40.00	1.78
Sephadex G-100	1.19	72.6	61.01	5.14	122.0	102.52	4.56
CM-52 cellulose	0.37	52.5	141.89	11.94	115.0	310.81	13.81
PBE 94 pH 7.5-4.0	0.19	39.0	205.26	17.28	110.0	578.94	25.73
PBE 94 pH 9.5-6.0	0.10	36.9	369.00	31.06	62.9	639.00	28.40
PBE 118 pH 10.5-8.0	0.08	24.0	400.00	33.67	39.0	650.00	28.89

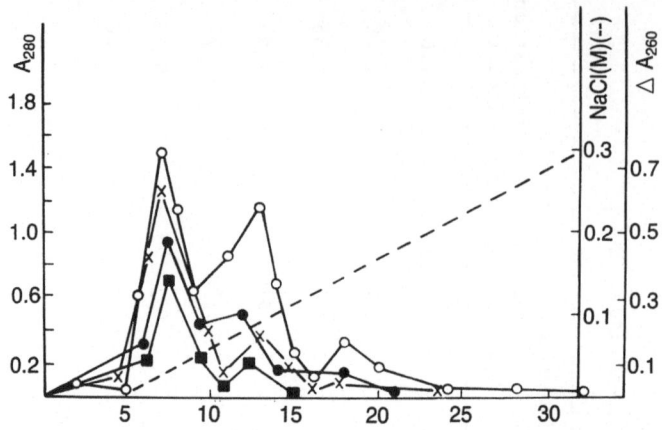

Fig. 1. CM-52 chromatography of material from a Sephadex G-100 column, in
0.0 - 0.3 M NaCl gradient. The activity towards the double-
stranded poly(A)*poly(U) complex was determined at pH 8.5, deoxyribonucleolytic activity - at pH 5.0 with the use of denatured
DNA from calf thymus, and the ribonucleolytic activity - at pH 7.8
with the use of highly polymerized yeast RNA. Protein was assayed
by absorption measurements at 280 nm. (o——o) protein, (x——x)
activity towards RNA, (●——●) activity towards DNA, (■——■)
activity towards poly(A)*poly(U). The NaCl gradient is shown by
the dashed line.

obtained (Figure 1), the first of which exhibited the highest activity
towards poly(A)*poly(U).

Separation of the latter peak on a column packed with PBE 94 gel in a
7.5-4.0 gradient is shown in Figure 2. The highest activity towards
poly(A)*poly(U) was found to be present in the fractions eluted at pH 7.4.
This indicates that the proteins active towards poly(A)*poly(U) are not
bound by the PBE 94 gel within the pH range tested. A large part, however,
of proteins without activity towards poly(A)*poly(U) remained on the
column. Fractions eluted at pH 7.4 were further purified on a column with
PBE 94 gel in a pH gradient from 9.5 to 6.0 (Figure 3). The highest
activity towards poly(A)*poly(U) was found in fractions eluted at pH 9.4.

These fractions were then placed on a column packed with PBE 118 gel.
Separations in a pH gradient of 10.5-8.0 yielded two fractions showing
activity towards poly(A)*poly(U). They were eluted at pH 9.8 and 8.6. The
protein eluted at pH 9.8 showed a higher activity towards the complex and
the data in Table 1 concern this fraction.

As a result of the above purification procedure we obtained from the
ribosomal wash a preparation 30 times more active than the rye embryo
ribosomes, irrespective whether the double-stranded synthetic complex, i.e.
poly(A)*poly(U) or single-stranded RNA were used as substrate. Copurifi-
cation of proteins with activities against the single- and double-stranded
polynucleotides suggests that both activities reside within the same
enzyme.

The present results indicate that two proteins hydrolyzing double-
stranded RNA structures are associated with rye embryo ribosomes. The

248

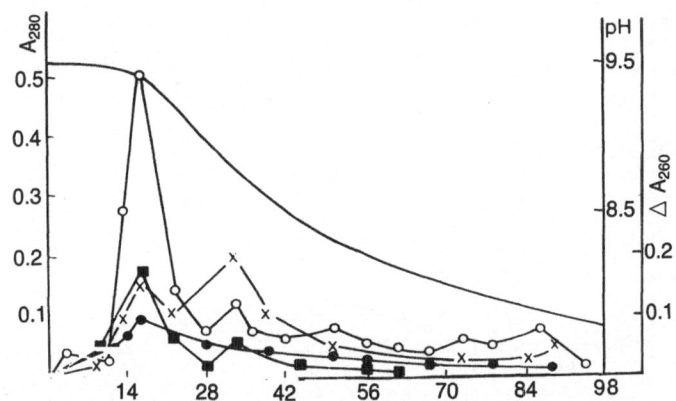

Fig. 2. Chromatofocusing, on a PBE 94 column, of the material obtained from
a CM-52 column. Elution conditions: starting buffer A: 0.025 M
imidiazole-HCl, pH 7.4; elution buffer: Polybuffer 74-HCl, pH 4.0.
(o——o) protein, (x——x) activity towards RNA, (●——●) activity
towards DNA, (■——■) activity towards poly(A)*poly(U).

results of chromatofocusing of the enzyme preparation provide evidence for
the strongly basic nature of these proteins.

The role of plant enzymes associated with ribosomes degrading double-
stranded RNA has not so far been elucidated. It is possible that these enzymes participate in regulation of the double-stranded RNA level and thus
in regulation of protein biosynthesis; the inhibitory effect of double-
stranded RNA on protein synthesis has so far been demonstrated only for
animal cells (Lengyel, 1982).

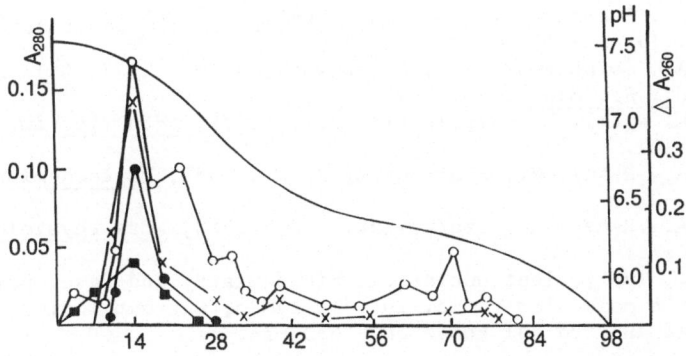

Fig. 3. Fractionation on a PBE 94 column in pH gradient of 9.5 - 6.0 of the
enzyme preparation obtained from a PBE 94 column (pH 7.5 - 4.0).
Elution conditions: starting buffer B: 0.025 M ethanolamine-
CH_3COOH, pH 6.0 (o——o) protein, (x——x) activity towards RNA,
(●——●) activity towards DNA, (■——■) activity towards poly(A)*
poly(U).

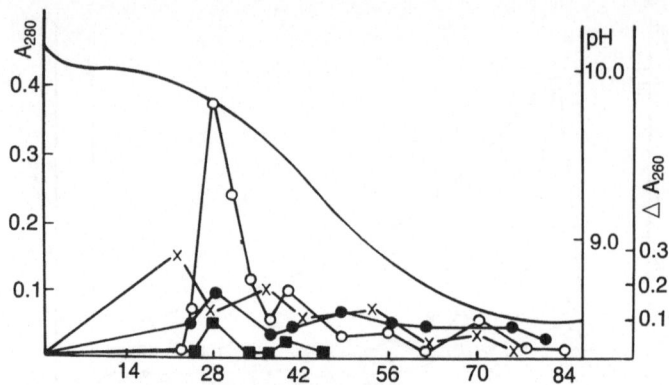

Fig. 4. Chromatography, on a PBE column in pH gradient 10.5 - 8.0, of the enzyme preparation obtained from a PBE 94 column. Elution conditions: starting buffer C: 0.025 M triethylamine-HCl, pH 10.5; elution buffer: Pharmalyte, pH 8.0 (o——o) protein, (x——x) activity towards RNA, (●——●) activity towards DNA, (■——■) activity towards poly(A)*poly(U).

Further purification and more complete characterization will be necessary in order to determine the role of these nucleases in RNA metabolism.

REFERENCES

Bardon, A. H., Sierakowska, H., Shugar, D., 1974, Biochim.Biophys.Acta 438:461-473.
Bredford, M. M., 1976, Anal.Biochem. 72:248-254.
Farkas, G. L., 1982, Ribonucleases and ribonucleic acid breakdown, in: "Encyclopedia of Plant Physiology," B. Parthier and D. Boulter eds., Vol. 148, Springer Verlag, Berlin - Heidelberg, p. 224-267.
Hsiao, T. C., 1968, Plant Physiol. 43:1355-1361.
Lengyel P., 1982, Biochemistry of interferons and their actions, in Ann. Rev.Biochem. Vol. 51:p. 251-282.
Siwecka, M. A., Szarkowski, J.W., 1971, Bull.Acad.Polon.Sci., Sér.Sci. Biol., 19:445-449.
Siwecka, M. A., Golaszewski, T., Szarkowski, J. W., 1977, Bull.Acad.Polon. Sci.Sér. Sci.Biol., 25:15-19.
Siwecka, M. A. Rytel, M., Szarkowski, J.W., 1979, Acta Biochim.Polon., 26:97-101.
Siwecka, M. A., Rytel, M., Szarkowski, J. W., 1982, Phytochemistry, 21:273-275.
Siwecka, M. A., Rytel, M., Szarkowski, J.W., 1983, Acta Physiologiae Plant, 5:105-111.
Wilson, C. M., 1982, Plant nucleases; biochemistry and development of multiple molecular forms, in: "Isozymes; Current Topics in Biological and Medical Research," 6:33-54.

INVESTIGATION OF IMMUNOGLOBULIN LIGHT AND HEAVY CHAIN GENES RESPONSIBLE FOR THE SYNTHESIS OF ANTIBODIES IN HYBRIDOMA PTF.02

S.M. Deyev, V.A. Ajalov, D.N. Urackov, A.G. Stepchenko,
F. Franĕk* and O.L. Polyanovsky

Institute of Molecular Biology
Academy of Sciences of the USSR, Moscow
*Institute of Molecular Genetics
Academy of Sciences of Czechoslovakia, Prague

The synthesis of immunoglobulin light and heavy chains is encoded by genes, each of which contains coding exons and sequences that carry out gene regulation in the cell. The immunoglobulin genes of some plasmacytoma cell lines have been studied in detail (Kabat et al., 1983). The structure of genes from mature lymphocytes that synthesize antibodies of a definite specificity are of great interest. The isolation of such genes and sequencing their structure allows one to establish the amino acid sequence of antibody polypeptide chains as well as the location and arrangement of regulatory sequences that exist in the promoter and enhancer regions of mature lymphocytes. This is a necessary step in studying the regulation of expression of genes coding for the light and heavy chains of antibodies. This study was made with habridoma PTF.02 characterized by a stable synthesis and a high yield of antibodies against pig transferrin (Bartek et al., 1982). This transferrin is known now to be an important cell growth factor and is used for the cultivation of myeloma and hybridoma cells in serum-free media.

The maturation of B-cells results in antibody synthesis and is followed by DNA structural rearrangement in the loci of genes for immunoglobulin heavy (H) and light (L) chains. These include V_L-J_L, D_H-J_H, V_H-J_H joints and DNA rearrangements in the locus of C_H-genes from class to class. Hybridomas possess a chromosomal set differing from the diploid set and the ploidy of chromosomes may vary.

The purpose of this work was to investigate the genes of heavy and light immunoglobulin chains responsible for the synthesis of antibodies in hybridoma PFT.02.

ANALYSIS OF THE LOCUS OF IMMUNOGLOBULIN J_H-C_H GENES

Since myeloma P3-X63-Ag8.653 cells were used for cell fusion with lymphocytes when obtaining hybridoma PTC.02, the analysis of the J_H-C_H locus began with that myeloma. Myeloma P3-X63-Ag8.653 was subcloned (Kearney et al., 1979) from the myeloma MOPC 21 P3-X63-Ag8 line (Kohler and Milstein, 1976). In order to measure the size of DNA fragments with $C\gamma_1$-genes, P3-X63-Ag8.653 DNA was digested with restriction endonucleases

EcoRI, BamHI, Hind III and hybridized with a labelled probe at the $C\gamma_1$-region, (Figure 1A). This probe allows one to describe sites neighboring the 5'-terminus of the $C\gamma_1$-gene for EcoRI and Hind III hydrolysates as well as sites at the 3'-end for BamHI hydrolysates. For comparison, DNA hydrolysates from mouse BALB/c liver were subjected to a similar analysis. The "non-stringent" hybridization conditions revealed fragments that carried all non-rearranged $C\gamma$-genes in the BALB/c DNA hydrolysates. Cross-hybridization of the $C\gamma_1$-probe with $C\gamma_{2a}$, $C\gamma_{2b}$ and $C\gamma_3$ genes may be attributed to their rather high homology (Yamawaki-Kataoka, 1979). The results of hybridization analysis are presented in Figure 1A. The structure and the restriction map of the non-rearranged $C\gamma_1$ locus are shown in Figure 2. Under the "stringent" and "non-stringent" conditions of hybridization, two fragments (23.0 and 6.6 kb similar in mobility with the embryonic fragment) were found in the EcoRI hydrolysate of P3-X63-Ag8.653 DNA. The fact that both $C\gamma_1$-loci are not rearranged at the 5' – adjacent to the $C\gamma_1$-gene region is demonstrated by analysis the HindIII hydrolysate of P3-X63-Ag8.653 DNA (Figure 1, HindIII, 1) that contained no 25 kb fragment corresponding to the embryonic state of this region. The differences in the intensities of bands for $C\gamma_1$, $C\gamma_{2a}$ and $C\gamma_{2b}$ genes can be explained by the allelic diversity of $C\gamma_1$ gene and its absence in the case of $C\gamma_{2a}$- and $C\gamma_{2b}$ genes. No fragments hybridising with the $C\mu$-probe were present in P3-X63-Ag8.653 DNA (Figure 1,B 2). It is likely that none of the allelic chromosomes in this myeloma contain the $C\mu$-locus.

The J_H-region was analysed using a 1.9 kb EcoRI-BamHI fragment with J_H3 and J_H4 exons in it (Figure 1C). Such a probe is universal for examining the structure of this region since the sequence neighboring the 3'-end of the J_H4-exon and containing a heavy-chain enhancer (Serfling et al., 1985) is retained in the genome as a rule from class to class. The P3-X63-Ag8.653 J_H-locus is characterized by a decrease in the J_H-probe hybridization zones MOPC 21, which can be seen from the comparison of BamHI hydrolysates (Figure 1C, BamHI. 2). The presence of only one 6.0 kb fragment allows one to put forward two suppositions about the structure of the J_H-locus in this myeloma: (i) the J_H-loci of both 12th chromosomes are structurally identical, which accounts for the similar hybridization pattern of their BamHI hydrolysates; (ii) the J_H-locus exists only in one chromosome whereas this region is deleted in the other one. Supposition (ii) is corroborated by comparing the HindIII hydrolysates of MOPC 21 and P3-X63-Ag.8.653 DNAs (Figure 1C, HindIII. 2,3) whose analysis is most effective for the structural studies of the J_H-loci. The HindIII recognition site is housed in the intron between the J_H3 and the J_H4 exons in the probe used (Figure 1C). This makes it possible to orient fragments revealed in the hybridization and to localize DNA fragments in the 5'- and 3'-direction from it. The absence of a 7.0 kb fragment from the HindIII hydrolysate of P3-X63-Ag8.653 in comparison with MOPC 21 correlates with the observation that Gl-chains are not synthesized in this myeloma. The 2.0 and 2.3 kb fragments common for the HindIII hydrolysates of MOPC 21 and P3-X63-Ag8.653 characterize the J_H-locus of the 12th chromosome (Zuniga et al., 1982) in which the J_H4-exon is present in the embryonal 2.3 kb fragment, and the J_H3-exon, in the rearranged but not expressed 2.0 kb fragment.

THE J_H-C_H LOCUS OF HYBRIDOMA PTF.02

Analysis of PTF.02 DNA $C\gamma$-genes by means of the $C\gamma_1$-probe showed a similarity of the hybridization maps with P3-X63-Ag8.653 DNA except an additional 14.0 kb fragment in the HindIII hydrolysate of PTF.02 DNA (Figure 1A, HindIII.3). The 25 kb fragment is associated with the non-rearranged region adjacent to the $C\gamma_1$-gene 5'-end. These observations allow one to consider two structural versions of the 12th chromosomes of

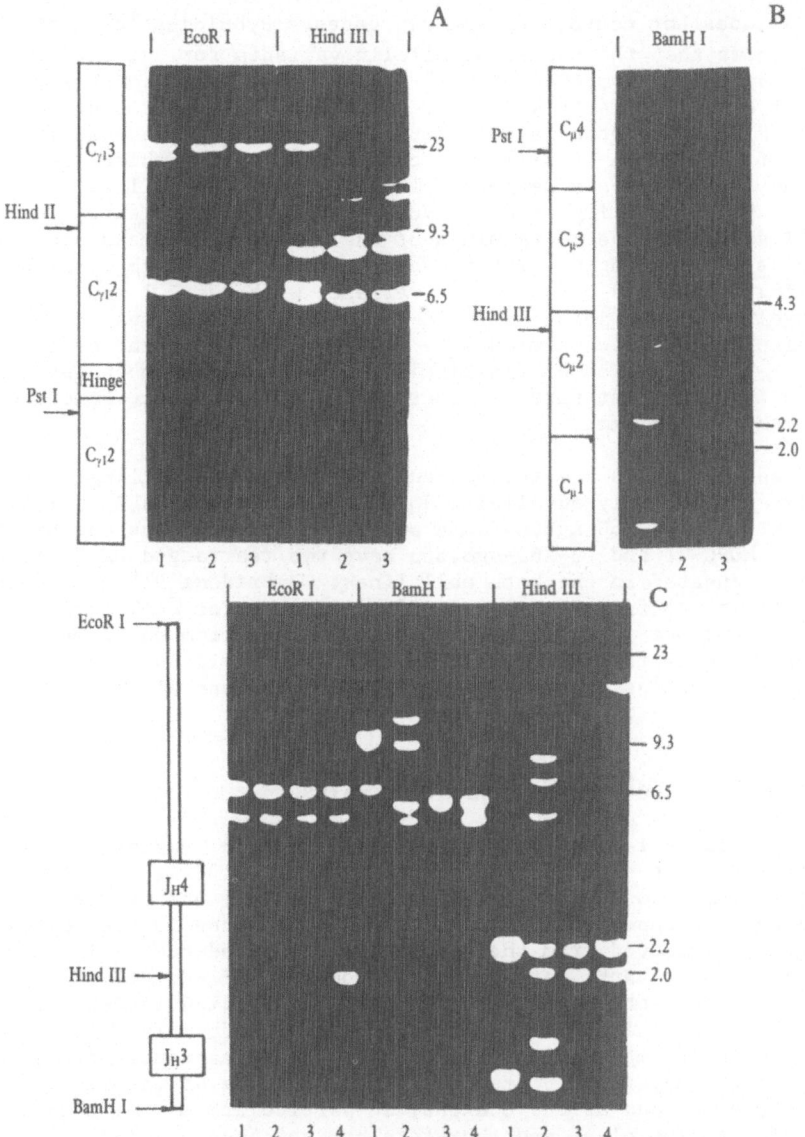

Fig. 1. Blot-hybridization of genomic DNAs with $C\gamma_1$, $C\mu$ and J_H probes. (The structure of each probe is given to the left of the corresponding autoradiogram). Southern blots are prepared from the hydrolysates of restriction endonucleases, DNA BALB/c (1), MOPC 21 (2), P3-X63-Ag8.653 (3), PTF.02 (4). The restriction endonucleases taken in the experiments are indicated above the respective units.

the lymphocyte involved in the hybridoma formation in PTF.02 cells: (i) both chromosomes are rearranged similarly and therefore an additional (in comparison with P3-X63-Ag8.653 DNA) 14.0 kb fragment is found in the HindIII hydrolysate of PTF.02 DNA; (ii) one of the chromosomes is rearranged whereas the other is deleted or does not contain the $C\gamma_1$-gene and is not revealed therefore in the hybridization. A priori, the first version is unlikely whereas the other is confirmed by the structural analysis of the J_H, $C\mu$ and $C\gamma_3$ regions. It has been established that

PTF.02 DNA does not contain $C\mu$ and $C\gamma_3$ genes. Hybridization with the J_H-probe shows that the number of allelic variants for the J_H-locus increases by one as compared with P3-X63-Ag8.653, which correlates with the ability of PTF.02 cells to synthesize G1-chains. The structure of the expressed PTF.02 J_H-locus is reconstructed, following the results of analysis of PTF.02 DNA HindIII and EcoRI hydrolysates, which revealed two new 15 and 2 kb fragments respectively (Figure 1C, HindIII.4, EcoRI.4). As mentioned above (Serfling et al., 1985), the recognition sites for HindIII and EcoRI adjoining the 3'-terminus of the J_H4-exon are retained when the J_H-C_H locus is rearranged. The presence of the EcoRI 2 kb fragment hybridizing with the J_H probe is at variance with the J_H1 and J_H2 exons being involved in the joint of V_H-D_H-J_H, since they are spaced by more than 2 kb from the EcoRI recognition site adjacent to the 3'-end of the J_H4-exon. The retention of the HindIII site located between the J_H3 and J_H4 exons is consistent with the fact that V_H-D_H exons are functionally linked with the J_H3 exon.

The above results are summarized in Figure 2 which illustrates the structures of the analysed allelic J_H-C_H loci of mouse BALB/c liver cells, myeloma MOPC 21, P3-X63-Ag8.653 and hydridoma PTF.02. One can see that the genomes of MOPC 21 and P3-X63-Ag8.653 have two rearranged C_H loci, one of them being repeated in the both cell lines. Hybridoma PTF.02 comprises three C_H loci: two from P3-X63-Ag8.653 cells used for cell fusion with lymphocytes and one (a functionally active locus) from the lymphocyte. The second J_H-C_H locus from the lympocyte is deleted. In the gene for immuno-globulin heavy chains expressed in PTF.02, the V_H and D_H exons are linked to the J_H3 exon and to the $C\gamma_1$ gene.

ANALYSIS OF GENES FOR IMMUNOGLOBULIN κ-CHAINS

The hybridization analysis of hybridoma PTF.02 κ-genes was reported elsewhere (Deyev et al., 1985; Stepchenko et al., 1986). The hybridoma was found to contain two rearranged genes for κ-chains; one of the genes was unfunctional and appeared in hybridoma cells from the initial myeloma. The second gene originated from the lymphocyte that produced antibodies against transferrin. Gene banks were obtained from MOPC 21 and PTF.02 genomic DNA hydrolysates, and both genes were isolated and studied in detail (Stepchenko et al., 1986). In this work the gene from lymphocytes has been analysed, with an emphasis laid on the nucleotide sequence of the variable gene and on the adjoining 5'-region. The structure of the intron between the J and C exons and of the C exone, in particular, is the same for all animals of the same class and therefore has not been studied.

Experiments were made with plasmid RTκ75 obtained earlier (Stepchenko et al., 1986). A 7.5 kb BamHI fragment with a gene responsible for the synthesis of antibody κ-chains was digested with restriction endonucleases EcoRI, HindIII, PstI and subcloned in plasmid pUC 9. Plasmids that contained different parts of the BamHI fragment were digested at pUC 9 linker sites, the fragments were electroeluted from the gel after prepative electrophoresis of the hydrolysates, and the nucleotide sequences of the variable gene and of the adjacent 5'-region accommodating regulatory trancription sequences were determined. In order to achieve the "over-lapping" and get a better variability of the analysis, both complementary chains were sequenced. Figure 3 depicts the sequence of 2075 nucleotides determined in this research.

The localization of the VJ exon and the corresponding amino acid sequence of this gene part were established by comparison with the known structures of the V and J segments (Kabat et al., 1983). The variable part of the κ-chain under study as well as other V genes that belong to the

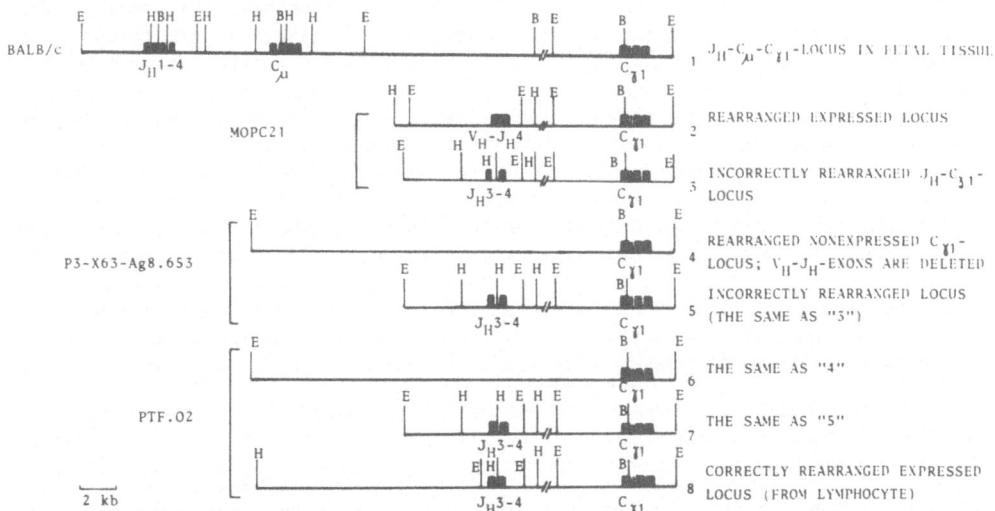

Fig. 2. Allelic variants of the J_H-C_H locus in the cells of BALB/c, MOPC 21, P3-X63-Ag8.653 and PTF.02 (restriction endonucleases: E-EcoRI, H-HindIII, B-BamHI)

same family consist of four relatively conservative zones (FR1-FR4) and of three hypervariable sites (CDR1-CDR3). The structure of FR4 corresponds to the lining J_5 segment. However, a difference was observed at position 475 from the embryo cell genome (Sakano, 1979), i.e. a C→G transversion, which may stem from a somatic mutation that predetermines a Gly substitution for Ala in the polypeptide chain. Similar mutations have been reported for variable and linking parts of some other immunoglobulin genes (Bothwell et al., 1981; Crews et al., 1981). These mutations appear to be one of the mechanisms that govern the diversity of antibodies. other DR sites of the V_K-gene are highly homologous (from 50 to 100%) to the corresponding FR sites of some mouse plasmacytoma κ-genes under study. The nucleotide sequence of the gene leader part has a significant homology (73%) with plasmacytoma TI and L6 L exons (Kabat et al., 1983). This homology should be assumed to be rather high since the leader peptides and gene fragments coding for them have a considerable variability and a low homology within the family of κ-gene L-sequences.

The promoter region was shown to house sequences involved in the regulation of κ-gene transcription: A TATA box (position-50), a CAT box (-38) as well as dc (-93) and pd (-125) oligonucleotide blocks (see Figure 3) similar to the conservative sequences found in plasmacytoma immuno-globuline κ-genes (Falkner and Zachau, 1984). Moreover a "shadow" promoter was discovered in the intron between the gene L and V fragments. In addition to the TATA box, it contains a sequence which is characteristic of a high-level homology (9 out of 10) with a dc consensus (TNATTTGCAT). The pd and dc motifs occur frequently in other parts of the gene as well, which seems to make it easier for specific protein factors to spot the promoter (Singh et al., 1986).

The SpI factor (Dynan and Tjian, 1985) which is responsible for the effective expression of many eukaryotic genes does not seem to be involved in the expression of the κ-gene. To have this factor activated, an ex-tended G/C-rich nucleotide sequence with a GC motif (CCGCCC or GGGCGG) is required.

```
                                  CTATCCTTACTTACCTTCTCATGAACA-1302
AACCCATATAATCTCTCATCACACCTCTACTAACATGCATGAACACACTCTCATCCTTACTCTCTATCCTATCCC-1275
CTTACAGTGTTTCTATATTCCCTTCCTAATACATTTCTCTTTTCAACAACCTAATTCTAATTGCCAAAAATCAGC-1200
CATCCAGAACTATGTTAGTTTTGATTGGGGAGAGTTTGTTGCTGGAAATTTATGAAGCTTTAGGAAACATAGAGG-1125
ATTTGAACGGGGGAATGGGCCATATTTAGAGAGACAGATTTCTAAATCTCTACTACCTAGAAAAAATGGTTTATG-1050
TAAAGAAACAGAAAGAAAGAACAATAATTGAGGCATAGCATGTATATTAGGCAGTATGTGTATAACATATATATT -975
ATATTCTCTATATGATAGAGTGATAATATGGCACAAATGGAAAAGAAAAAAATCAGGACTGTAGAGTCAGTCCAC -900
TTAGCTGTTCAGAATACTTGCTATTCTTGCAAACGATCTGGGTTAGGTTCCCAGCAAACACGTGAATATCTCCAT -825
CATTTGAACTTTAAAATCCTTTTCTGAATTCCCCTAGCAATAAATACATAGACAATGGGCATGCATACATACATC -750
AAGAGGTAAGCACTCATGCACATAAAATTACAGTTTTTGTTGTTGTTTTCATGTCATATTTTTTGAGAAGAGAAT -675
CAAACAGAAGTTCCTCTTTGGTAGTCAAAAGAATCATAAATAGATTAAGATGAATTGGACAGTGCTGGGAATGAA -600
TGGAGTGGTGATGTACTATCTTGTCATTCCTTAGAAGATGTGCCTCCCAATATTTTTCAATTTGAATTTTGATAA -525
ATACAGAAAGAGCTAGGTATAATCATAAACAAGAATGAAATTTTTTATGACCAGTGATATCTACAGATAACTATT -450
AATGAACATTTTGAAGTTTCTGCGAACATGCAAATCCTGAAACTCCCTAGCATATACAACATATAGGAATAGAAT -375
                    ━━━━c━━━━━
GATGTACTGAACAAGGGGGAAATTTTTAAAAAGCATTAAATAAATGCTTTTTAAGACTATTTTTTGTGCCTCCAA -300
AGATGAGCACCTTCATTGGAGAAAAATCTGTCTGTGCTGGTCTGATGTAACAAATCAGAGGTTGTTACGCTTTGG -225
TTGAGTTACTAAAATAAGCCTATTCTGCAGCTGTGCCCAGCACTGTTCAGATGAAGCTGATTTGCATGTGCTGAG -150
                           ━━━━━━pd━━━━━━                          ━━━dc━━━
ATCATATTCTACTGCCCCAGAGATTTAATAATCTGATCATACACACTTCATTCTTCCTCAGGAGACGTTGTAGAA -75
┌MetArgProSerIleGlnPheLeuGlySerCysCysSerGlyPheMetVal┐
│ATGAGACCGTCTATTCAGTTCCTGGGCTCTTGTTGTTCTGGCTTCATGGTAAGGAGTTTAACATTGAATATGCTA  75
└━━━━━━━━━━━━━━━━━━L━━━━━━━━━━━━━━━━━━┘
AAAAGAGTATGTGATCAGGAATTTCTGGTCCTTCAGAAAAATCTTCTTTGCATATAATTAATTTCATGGGGATTT 150
                                ┌AspIleGlnMetThrGlnSerProSerSerLeuSerAlaSerLeu┐
GTGTTCTTTTTAATTATAGGTCCTCAGTGT│GACATCCAGATGACACAGTCTCCATCCTCACTGTCTGCATCTCTG 225
                              └━━━━━━━━FR-1━━━━━━━┘
┌GlyGlyGlyLysValThrIleThrCysLysAlaValGlnAspIleLeuLysLysTyrIleProTrpPheGlnArgLys┐
│GGAGGGCAAAGTCACCATCACTTGCAAGGCAGTCCAAGACATTTTGAAGAAATATATACCTTGGTTCCAACGCAAG 300
│           └━━━━━━━━━━CDR-1━━━━━━━━┘              └━━━━━━━FR-2━━━━
┌ProArgArgGlyProArgLeuLeuIleHisTyrThrSerThrLeuGlnProGlyIleProSerArgPheGlyGly┐
│CCTAGAAGAGGTCCTAGACTGCTCATACATTACACATCTACATTACAGCCAGGCATCCCATCAAGGTTCGGTGGA 375
│     └━━━━━━━━CDR-2━━━━━━━━┘              └━━━━━FR-3━━━━
┌SerGlySerGlyLysLeuPheLeuArgHisGlnGlnLeuGluProGluTyrPheSerThrPheTyrCysProGln┐
│AGCGGGTCTGGAAAATTATTCCTTCGGCATCAGCAACTGGAGCCTGAATATTTTTCAACTTTTTATTGTCCACAG 450
┌TyrAspSerLeuLeuThrPheGlyGlyGlyThrLysLeuGluLeuLysArg┐
│TATGATAGTCTTCTCACGTTCGGTGGTGGGACCAAGCTGGAGCTGAAACGTAAGTACACTTTTCTCAACCTTTTT 525
│CDR-3━━━━━━━━━━━━━┘        └━━━━━━━━FR-4━━━━━━━━┘
TATGTGTAAGACACAGGTTTTCATATTAGGAGTTAAAGTCAGTTCAGAAAATCTTTTGAAAATAGGAGAGGTCTC 600
ATTATCAGTTGCCGTGGCATACAGTGTCAGATTTTCTGTTTATCACGCTAGTGAGGTTTGGGGCAAAAAGAGGCT 675
TTAGTTGAGAGGGAAACTAATTAATACTGTGGTCACCATCCAAGAGATTGGATCGGAGAATAAGCATGAGTAGTTA 750
TTGAGATCTGGGTCTGACTGCAG                                                    773
```

Fig. 3. Nucleotide sequences of the V_K gene and of the adjoining 3'- and
5'-regions. The amino acid sequence of the leader L, variable and
joining gene segments that corresponds to the nucleotide sequence
is given in the top part of the picture. Underlined are the regulatory pd and dc sequences, TATA and CAT boxes. The
conservative FR1-FR4 and hypervariable CDR1-CDR3 sites of the V
gene are shown in boxes.

At the same time, the 5'-region (Figure 3) bordering the gene is
AT-rich (A/T, 65%) and does not contain a GC motif.

We have established the structure of the enhancer zone for the same
κ-gene earlier and found a core similar to that of H chains and papova
viruses: TGTGGCTAA... 11 b.p....TGTGGTAA.

Therefore the expression of lymphocyte κ-genes is regulated by TATA
and CAT boxes common of many eukaryotic genes, but also a number of specific sequences located in the 5'-region and in the intron between the J and
C exons. It is noteworthy that the regulatory sequences involved in the

256

expression and translocation of κ-genes contain some elements in common: a TGCA present in pd and dc, a TGTG present in pd and at the 3'-end of dc, a signal sequence CACTGTG which takes part in the translocation of immunoglobulin gene (Sakano et al., 1979; Seidman et al., 1979). The above motifs exist in many regulatory sequences of genes and may play a universal role in the processes of recombination and transcription.

REFERENCES

Bartek, J., Viklicky, V., Franek, F., Angelisova, P., Draber, P., Jarosikova, T., Nemec, M., Verlova, H., 1982, Immunol.Lett. 4:231-236.

Bothwell, A. L. M., Paskind, M., Keth, M., Jamanishikavi, K., Baltimore, D., 1981, Cell. 24:p.625-637.

Deyev, S. M., Stepchenko, A. G., Polyanovsky, O. L., 1985, J.Mol.Biol., (in Russian) 19:209-217.

Dynan, W. S., Tjian, R., 1985, Cell., 35:79-87.

Falkner, F. G., Zachau, H. G., 1984, 310:71-74.

Grews, S., Griffin, J., Huang, H., Calame, K., Hood, L., 1981, Cell. 25:59-66.

Kabat, E. A., Wu, T. T., Bilofsky, H., Reid-Miller, M., Perry, H., 1983, Sequences of Proteins of Immunological Interest. USA. Bethesda, U.S. Depart. of Health and Human Services, p. 323.

Kearney, J. F., Radbrusha., Liesegang, B., Rayewsky, K., 1979, J.Immunology. 123:1548-1550.

Kohler, G., Milstein, G., 1976, Eur.J.Immunology., 6:511-519.

Sakano, H., Hüppi, K., Heinrich, G., Tonegawa, S., 1979, Nature, 280:288-294.

Seidman, J. G., Max, E. E., Leder, P., 1979, Nature. 280:370-375.

Serfling, E., Yasin, M., Schaffner, W., 1985, Trends Genetics, 8:224-230.

Singh, H., Sen, R., Baltimore, D., Sharp, 1986, Nature. 319:154-158.

Stepchenko, A. G., Deyev, S. M., Polyanovsky, O. L., Kovař, J., Franěk, F, 1986, Folia Biologica (Praha) 32:161-166.

Yamawaki-Kataoka, Y., Sato, K., Shimisu, A., Kataoka, T., Mano, J., Ono, M., Kawakami, M., Honjo, T, 1979, Biochemistry, 18:490-494.

Zuniga, M. C., D'Rustachio, P., Ruddle, N. H, 1982, Proc.Nat.Acad.Sci. U.S.A. 79:3015-3019.

REGULATION OF CYTOCHROME P-450 BIOSYNTHESIS IN

ALKANE ASSIMILATING YEAST

B. Wiedmann, M. Wiedmann, W. H. Schunck,
S. Mauersberger, E. Kargel and H.G. Müller

Central Institute of Molecular Biology
Academy of Sciences of the GDR, DDR 1115 Berlin-Buch
Robert-Rossle Str. 10

SUMMARY

In the yeast Candida maltosa both the substrate (n-alkane) and oxygen mediated induction of the alkane hydroxylating P-450 were shown to be regulated transcriptionally. That was concluded from good correlations of P-450 protein content (spectral determination of intact holo-enzyme and radioimmunological quantification of apo-enzyme) and relative amount of P-450 mRNA (in vitro translation with subsequent immunoprecipitation).

P-450 was taken as a model to investigate the biosynthesis of membrane proteins in homologous and heterologous cell-free translation/translocation systems. P-450 as well as alpha-factor could be translocated into mammalian and yeast membranes.

METHODS

Cultivation of C. maltosa (from the Institute of Biotechnology, Leipzig, GDR) on different carbon sources and at variable oxygen supply, the radioimmunological quantification and spectral determination of P-450 were performed as described by Schunck et al.(1987a, b). mRNA was isolated with phenol/chloroform and subsequent affinity chromatography on poly (U) sepharose (Wiedmann et al., 1986).

Translation in the wheat germ system and immunoprecipitation of in vitro synthesized P-450 was as published by Wiedmann et al. (1986).

To establish a C. maltosa derived cell-free protein synthesizing system complemented with yeast membranes, the method of Waters and Blobel (1986) was used, which had been elaborated originally for Saccharomyces cerevisiae.

Mammalian membranes and SRP were prepared from dog pancreas according to Walter and Blobel (1983a,b).

Abbreviations used: P-450, cytochrome P-450; SRP, signal recognition partical; RIA, radioimmunoassay; C. maltosa, Candida maltose.

INTRODUCTION

Like several other yeasts C. maltosa is able to use besides various sugar substrates n-alkanes as the only carbon source. An enzyme system consisting of P-450 and NADPH-P-450 reductase catalyses the first step of alkane degradation, e.g. the hydroxylation of alkane to the corresponding fatty alcohol (Riege et al., 1981). P-450 is induced by its own substrate (Mauersberger et al., 1981) and repressed by glucose.

Besides this substrate mediated induction we observed that oxygen-limitation during growth on n-alkanes caused an additional up to 6-fold increase of the cellular P-450 content. Similar results have been reported for other alkane-assimilating yeasts (Käppeli 1986).

The aim of the present paper was to get some insight into the under-lying mechanisms of regulation by means of comparing the content of intact holo-P-450 (spectral determination), apoprotein (RIA), and P-450 mRNA (in vitro translation).

Moreover, P-450 was taken as a model to get further insight into the first steps of biosynthesis of yeast membrane proteins, e.g. the incor-poration into rough endoplasmic membranes.

RESULTS AND DISCUSSION

Substrate Mediated Induction of P-450

A more than 100-fold increase of the alkane-hydroxylating P-450 and its specific mRNA was found when C. maltosa cells were transfered from glucose medium to alkane medium (see Table 1). Cycloheximide prevented this induction process.

Table 1. Comparison of P-450 level estimated by CO-difference spectroscopy and by RIA with P-450 mRNA content in dependence on growth substrate.

Growth substrate	Spectral determined $\dfrac{\text{nmol P-450}}{\text{mg microsomal protein}}$	RIA $\dfrac{\text{nmol P-450}}{\text{mg microsomal protein}}$	In vitro transl. % P-450 mRNA of total mRNA
glucose	0	0.003	0.002
alkane	0.47	0.410	1.000

The good correlation between all three determinations indicates a regulation by variable amounts of translatable mRNA. Formation of intact holo-enzyme from pre-existing apo-enzyme and heme was rather unlikely because nearly the same values were estimated with the RIA (which quan-tifies apo-enzyme too) as with CO-difference spectra.

Effect of Oxygen-limitation

As reported by Schunck et al. (1987a,b) oxygenlimitation ($pO_2 = 5\%$) caused an up to 6-fold increase of P-450 content during cultivation on alkanes. In a slightly different experimental set-up C. maltosa was culti-vated on hexadecane under the following conditions: 3 h at oxygen satu-ration (pO_2 70%), 5 h at oxygenlimitation (pO_2 about 5%) followed by 4 h at oxygen saturation. Cell samples were taken in each phase to isolate mRNA.

As shown in Figure 1 the variations in the cellular P-450 content (determined spectrophotometrically) were accompanied by corresponding alterations in the relative amount of P-450 mRNA. Oxygen limitation caused a reversible about 3-fold increase of both values. This process could be mimicked by selective inhibition of P-450 activity through binding of carbon monoxide.

A growing culture was streamed with low amounts of CO so that the oxygenconcenstration in the medium was only reduced to 80% of saturation. The reversible increase of P-450 was again accompanied by a greater value of translatable P-450 mRNA (Figure 2).

In a third experiment (Figure 3) the yeasts were cultivated on a non-repressive growth substrate - hexadecanol. A low level of P-450 was detectable. Streaming with CO caused no changes in growth rate, P-450 and P-450 mRNA contents.

Only the addition of the inducer hexadecane led to an increase of P-450 and its mRNA up to the values typical for oxygen limitation and CO-inhibition conditions.

Summarizing these results we conclude that the enhanced formation of P-450 during oxygen limited growth on n-alkanes is mainly regulated at the mRNA level.

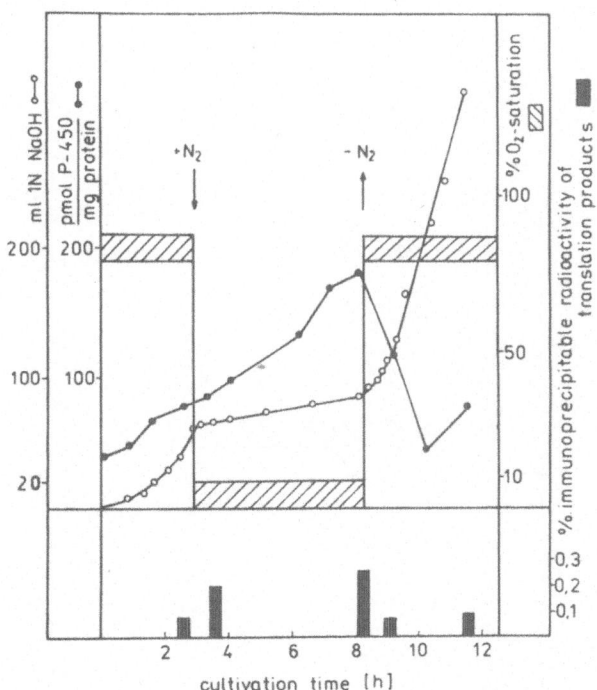

Fig. 1. Dependency of growth (NaOH-curve corresponds directly to culture
growth as shown in Schunck et al., 1983a) relative P-450 content
and P-450 mRNA on the oxygen concentration in the culture medium.
Candida maltosa cells were grown on hexadekane in a 1,5
l-fermentor. The oxygen concentration was varied by additionally
N₂-streaming. P-450 mRNA was quantified by in vitro translation
of total mRNA and immunoprecipitation of P-450 with specific IgG.

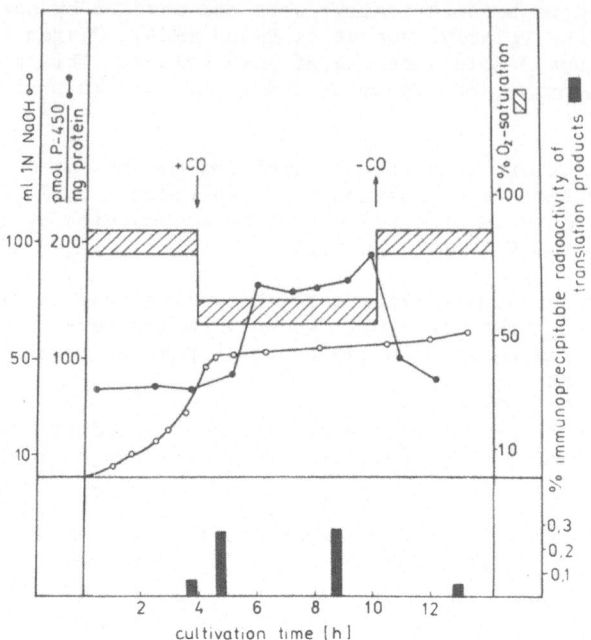

Fig. 2. See figure 1 with the exception that CO was used instead of N₂.

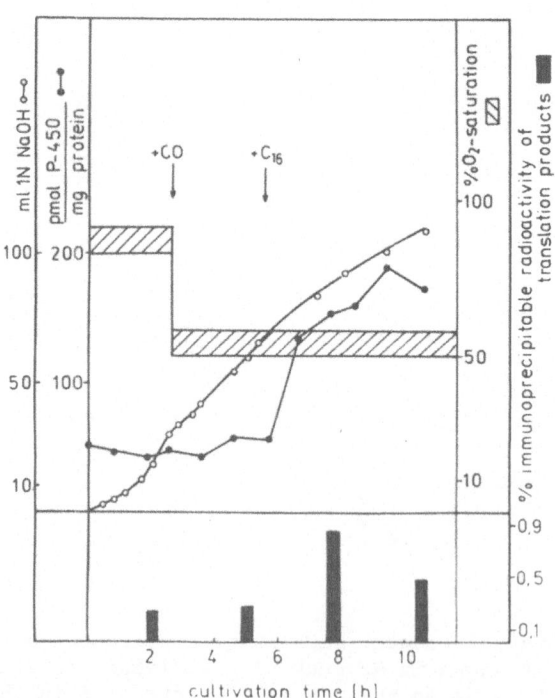

Fig. 3. See figure 1 with the exception that the cells were cultivated on hexadecanol and only after 2.5 h CO streaming hexadecan was fed additionally.

Obviously, neither a low oxygen concentration itself nor unknown factors arising from growth limitation could be the inducing signal. We suggest that the reduced P-450 activity (due to oxygen deficiency or CO-inhibition) leads to a change of the rate-limiting step of alkane degradation from uptake to hydroxylation and to an intracellular alkane accumulation. In this way oxygen limitation seems to enhance the substrate mediated induction.

In Vitro Translation/Translocation of P-450

According to early versions of the signal hypothesis proposed by Blobel and Dobberstein (1975) secretory and membrane proteins should be translocated into the membranes cotranslationally. Recently three groups (Rothblatt and Blobel 1986) found posttranslational transport of yeast alpha-factor (a secretory protein) in the Saccharomyces cerevisiae translation/translocation system. Only Hansen et al., (1986) detected little posttranslational translocation of alpha-factor in a wheat germ translation system supplemented with dog membranes.

To investigate the behaviour of another protein we translated the yeast membrane protein P-450 in the heterologous wheat germ/dog membrane system and in a newly developed homologous C. maltosa system.

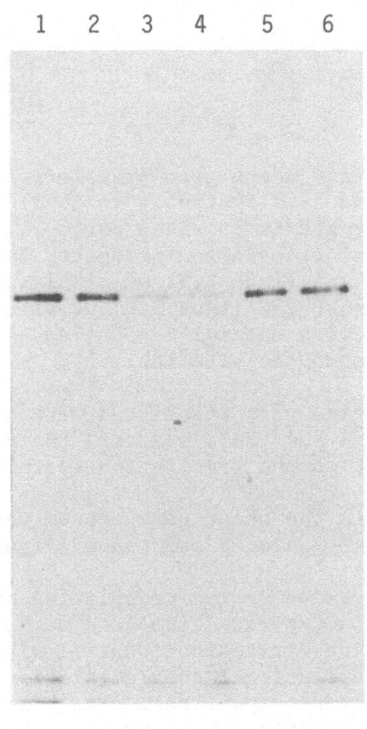

	1	2	3	4	5	6	
SRP	-	2.5	10	40	40	10	
K-RM	-	-	-	-	40	10	

Fig. 4.   P-450 immunoprecipitates from in vitro synthesized proteins in dependence of SRP and dog membranes. Yeast total mRNA was translated in a wheat germ cell-free system in the presence of different amounts of SRP and membranes and subsequent P-450 was immunoprecipitated.

263

Extract	C.m.	C.m.	C.m.	C.m.	C.m.	wg	wg	wg	wg
Nuclease	–	x	x	x	x	–	–	–	–
m RNA	–	–	x	x	x	x	x	x	–
IgG	–	–	–	x	x	x	x	–	–
cyt. P-450	–	–	–	x	–	–	x	–	–

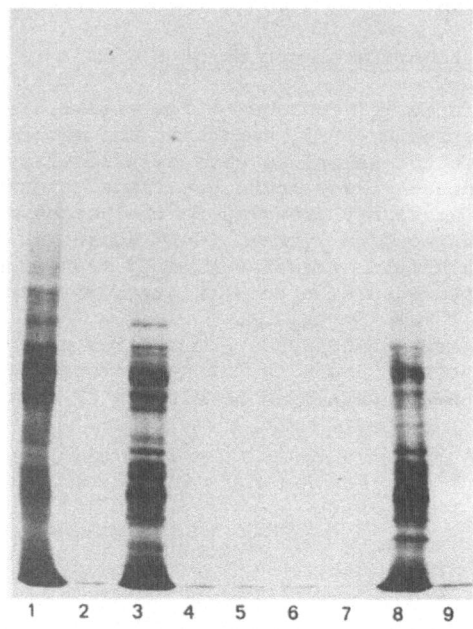

1 2 3 4 5 6 7 8 9

Fig. 5. Comparison of yeast and wheat germ translation systems regarding
P-450 synthesis. Lane 1-5: synthesis in Candida maltosa cell-free
system; lane 6-9: synthesis in wheat germ cell-free system; lane
1-3,8,9: 2.5 μl total translation products; lane 4-7: immunoprecipitates from 50 μl (lane 4,5) or 35 μl (lane 6,7) translation
product with anti P-450 IgG (lane 5,6) or with anti P-450 IgG
which was saturated with natural P-450 (lane 4,7); lane 4,7: show
the specifity of immunoprecipitation.

Though several mRNA species were translated with different
efficiencies in the wheat germ and yeast systems, the relative amount and
size of P-450 was independent on the translation system used (Figure 4).

The synthesis of P-450 in the wheat germ system was arrested by SRP
and was restored by adding SRP-depleted membranes (Figure 5).

If the membranes were present during translation nearly 64% of in
vitro synthesized P-450 was incorporated into the membranes, measured as
resistance of P-450 to alkaline extraction (Control without SRP or without
membranes: 5 - 9%). Posttranslational supplementation of membranes
yielded 16% P-450 in the membranes (publication in preparation). First
results with a homologous translation/translocation system indicate that
P-450 as well as alphafactor will be translocated with higher efficiency
posttranslationally than in the heterologous system. In every case the
cotranslational translocation was more efficient than the posttranslational
one. We have never found differences in size of synthesized P-450 in
SDS-gels, whether the cell-free translation was performed in the presence
of membranes or not.

Our datas indicate i) that the yeast P-450 was incorporated into dog
membranes SRP-dependent like membrane proteins of higher eukaryotes, and

ii) that a membrane protein besides the secretory protein alpha-factor can be incorporated into yeast membranes after finishing translation but with lower efficiency as during their synthesis. The phenomenon demands for some supplements to the original signal hypothesis.

REFERENCES

1. G. Blobel, B. Dobberstein, J.Cell Biol. 67; 835-851 (1975).
2. W. Hansen, P.D. Garcia, P. Walter, Cell 45; 397-406 (1986).
3. O. Käppeli, Microbiol. Rev. 50; 244-258 (1986).
4. E. Kärgel, H.G. Schmidt, W.-H. Schunck, P. Riege, S. Mauersberger, H.-G. Muller, Analyt. Letters 17; (B18). 2011-2024 (1984).
5. S. Mauersberger, W.-H. Schunck, H.-G. Müller, Z. Allg. Mikrobiol. 21; 313-321 (1981).
6. P. Riege, W.-H. Schunck, H. Honeck, H.-G. Müller, BBRC 98; 527-534 (1981)
7. J.A. Rothblatt, D.I. Meyerm Embo J. 5; 1030-1036 (1986).
8. W.-H. Schunck, S. Mauersberger³; J. Huth, P. Riege, H.-G. Müller, Arch. Microbiol 147; 240-220-244 (1987a).
9. W.-H. Schunck, S. Mauersberger, E. Kärgel, J. Huth, H.-G. Müller, Arch. Microbiol. 147; 245-248 (1987b).
10. P. Walter, G. Blobel, Meth. Enzymol. 96; 84-93 (1983b).
11. P. Walter, G. Blobel, Meth. Enzymol. 96; 682-691.
12. M.G. Waters, G. Blobel, J. Cell Biol.,102; 1543-1550 (1986)
13. B. Wiedmann, M. Wiedmann, E. Kärgel, W.-H., Schunck, H.-G. Müller, BBRC 136; 1148-1154 (1986)

ON THE MECHANISM OF REUTILIZATION OF NUCLEAR RNA DEGRADATION

PRODUCTS: A STUDY IN VIVO

Dobri D. Genchev

Bulgarian Academy of Sciences
Institute of Genetics
1113 Sofia, Bulgaria

More than twenty years ago it was established that a considerable proportion of the newly-synthesized RNA in animal cell nucleus is rapidly degraded (for review see Harris, 1963). Later it became clear that this phenomenon is due to the fact that the non-conserved segments, representing at least one-half of the precursor molecule to rRNA (45 S pre-rRNA) and considerably greater parts of the precursors to mRNA (hnRNA), are degraded (Hadjiolov and Nikolaev, 1976; Perry, 1976; Georgiev, 1980). The derived mononucleotides are reutilized in RNA synthesis (Harris, 1963). The exact mechanism of this reutilization remains unclear because the nature of mononucleotides resulting from RNA degradation is not recognized with certainty. These may be nucleoside 3'-phosphates or 5'-mono-nucleotides: in mammalian cells a variety of nucleases are found which produce both kinds of nucleotides (Siera-kowska and Shugar, 1977). The ways by which these may be reincorporated into RNA are quite different (Figure 1). Hence the knowledge of the nature of RNA degradation products is essential for the kinetic studies aimed at the determination of turnover parameters of RNA and its precursors.

The only experimental approach used up to now consists in incubation of nuclei (isolated from cells pre-labelled with radioactive RNA pre-curosors) in isotonic media and analysis of the acid-soluble products released. This kind of studies is subject to criticism as it is possible that redistribution, inhibition or loss of enzymes from different cell structures occur during cell fractionation. In fact, degradation of nuclear RNA to nucleoside 5'-disphosphates, 5'-monophosphates, nucleoside 3'-phosphates and nucleosides has been observed, the relative proportion of the aforementioned compounds varying greatly in different reports, leading consequently to quite controversial conclusions (Harris, 1963; Harris et al., 1963; Hurlbert et al., 1973, 1974; Kelley and Perry, 1971; Speers and Tigerstrom, 1975). Although highly desirable, studies on intact cells have never been carried out for a reason discussed below (see RESULTS AND DISCUSSION section).

In the present investigation the mononucleotides resulting from the degradation of nuclear RNA in intact Ehrlich ascites tumor cells were studied. The rationale of the study is given in the beginning of the same section.

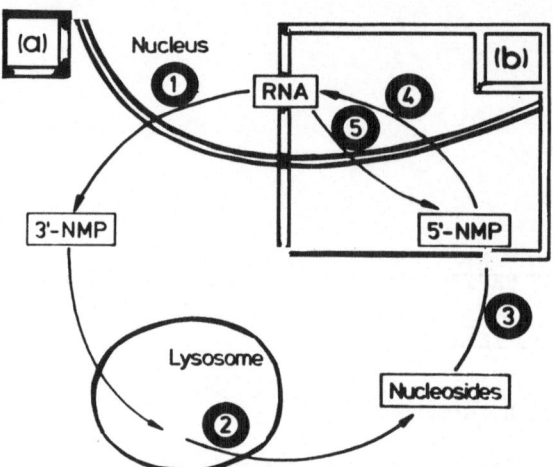

Fig. 1. Two main possibilities for reutilization of RNA degradation products: (a) consecutive realization of steps (1), (2), (3) and (4);
(<u>b</u>) consecutive realization of steps (5) and (4). These steps are
catalyzed by corresponding enzymes: (1) RNase(s) producing 3'-nu-
cleoside monophosphates, (2) unspecific (acid) phosphatase, (3)
nucleoside kinase(s), (4) nucleotide kinases followed by RNA polymerase, (5) RNase(s) producing 5'-mononucleotides.

METHODS

Ehrlich ascites tumor cells were obtained, labelled with (^{14}C)uridine
and their free nucleotides extracted with cold perchloric acid as described
previously (Genchev, 1980). To the extract carrier CMP, UMP and 2'(3')-
UMP* were added and the nucleotides were fractionated on a Dowex 1 column
with ("Method I") or without ("Method II") preliminary hot-acid hydrolysis
(Figure 2). After determination of the absorbance and the radioactivity of
all fractions, those corresponding to 3'-UMP were pooled, concentrated
through charcoal treatment and the nucleotide was further purified by TLC
on silica-gel plates developed with the solvent of Plessner as described by
Maruyama and Mizuno (1966). When "Method II" was used the following was
done in addition (expr. denoted (b), see the Table): the radioactive material remaining on the column after the appearance of the peak of 3'-UMP
was eluted <u>in toto</u> with 2 M HCOONH$_4$-6.5 M HCOOH, hydrolysed as above and
then subjected exactly to the same two-step purification procedure, namely
ion-exchange followed by thin-layer chromatography (Figure 2).

The total amount of radioactivity incorporated into a particular
nucleotide was calculated from the known quantity of the carrier added and
the specific radioactivity of the isolated nucleotide (determined through
UV-absorbance measurement and liquid scintillation counting).

---

* 2'(3')-UMP denotes the commercially supplied mixture of uridine 2'-phos-
phate and 3'-phosphate. As the chromatographic behavior of these two
isomers is absolutely the same in both chromatographic separation systems
employed, only the term 3'-UMP will be used throughout for brevity.

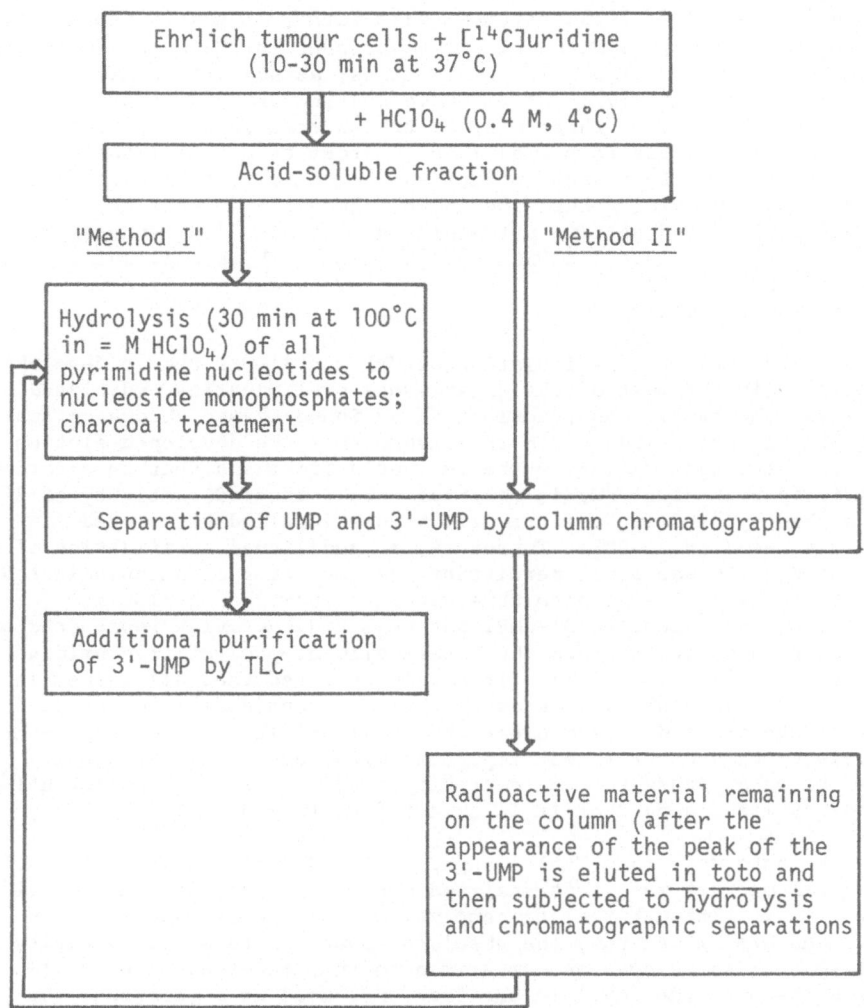

Fig. 2. Flow-diagram of the experiments.

Control experiments, in which highly-purified [^{14}C]CTP and [^{14}C]UTP were used, proved that: (1) the acid hydrolysis employed (cf. Genchev, 1980) brought about a quantitative conversion of pyrimidine nucleoside di- and triphosphates into CMP and UMP; (2) neither this hydrolysis nor the charcoal treatment caused any production of 3'-mononucleotides from CTP and UTP.

RESULTS AND DISCUSSION

As mentioned, it is desirable to study the nature of nuclear RNA degradation products in intact cells labelled with a suitable radioactive RNA precursor. Such studies require that the ratio between the radio-activity of the cellular mononucleotides due to direct incorporation of the exogeneous label and the radioactivity resulting from the degradation of newly-synthesized RNA be known. Evidently the lack of knowledge in this respect has been an obstacle to studies in vivo. However, our experimental data obtained previously (Genchev, 1980) permit to calculate that after

10-30 min incubation of Ehrlich tumor cells with [^{14}C]uridine about 50% or even more of the labelled mononucleotides present in the cell are in fact products of RNA degradation. If this process, as suggested (Maruyama and Mizuno, 1966; Bernardi, 1971; Sierakowska and Shugar, 1971), produces nucleoside 3'-phosphates, these must be transported into lysosomes and therein dephosphorylated to nucleosides in order to become liable to an enzymatic conversion to 5'-nucleotides and hence to reutilization (Figure 1). It follows, providing that the above suggestion is correct, that a detectable pool of radioactive pyrimidine nucleoside 3'-phosphates should exist in cells pulse-labelled with [^{14}C]uridine. Therefore, our task was to study if any significant portion of this precursor was incorporated into 3'-UMP in labelled cells.

As a first step to the isolation of UMP and 3'-UMP the acid-soluble cell extract (in the most of the experiments preliminarily hydrolysed, see Figure 2 and the Table) was fractionated by ion-exchange chromatography (Figure 3). It can be seen from the figure that the developed elution scheme separates satisfactorily the two nucleotides and that no apparent radioactivity peak of 3'-UMP is observed. Yet, the radioactivity of 3'-UMP may be masked by "tailing" of the very highly-labelled UMP eluted immediately ahead of 3'-UMP. Therefore, an additional purification of the latter through TLC was done, permitting the isolation of uncontaminated 3'-UMP (Figure 4). The results thus obtained show that a distinct radioactivity is present in 3'-UMP, but this radioactivity represents only 1-2% of that incorporated into the total cellular pool of free uridine nucleotides (Table, expr. 1-3). It should be noted that all values in the last column of the table are determined with a considerable error just because of the great difference between the labelling of 5'- and 3'-UMP. Nevertheless, this error is not important with respect to our final conclusion: in any case the radioactivity in 3'-UMP is negligible, all the more that at least a part of it is an artifact (see below).

Experiments were run (Table, expr. 4(a) and 5(a)) in which the acid-soluble cell extract was first analyzed without hydrolysis ("Method II", see Figure 2). Although the per cent values for the radioactivity in 3'-UMP found are higher now, the absolute amount of this radioactivity is in fact several-fold lower in comparison to that recorded in expr. 1-3. This is so because the label in the free UMP, used as a reference for calculation of the relative radioactivity in 3'-UMP in these experiments, comprises only 10-15% of the total label in uridine mononucleotides (Genchev, 1980). The results in the table (expr. 4(b) and 5(b)) confirm that a considerable part of the radioactive 3'-UMP recorded is not a product of enzymatic degradation of RNA. In experiments 4(b) and 5(b) the material eluted from the column after 3'-UMP was subjected to hydrolysis and then analyzed in the standard way: again some radioactive 3'-UMP was found. The only explanation of the finding is the following. The material retained on the column after the elution of 3'-UMP contains in addition to nucleoside di- and triphosphates some minor quantity of oligoribonucleotides. These are soluble in cold acids (Cleaver and Boyer, 1972), therefore extractable together with the free nucleotides, and yield upon hot-acid hydrolysis pyrimidine nucleoside 2'- and 3'-phosphates (Witzel, 1963). The presence of oligoribonucleotides in the cell should be expected as intermediates in the synthesis, degradation and maturation of RNA.

In addition, another source of formation of 3'-UMP may be some degradation of RNA to acid-soluble products by the cold perchloric acid. We observed (results not illustrated) that when highly-labelled RNA was incubated in cold 0.4 M HClO$_4$ a release, although very slow, of acid-soluble radioactivity did occur.

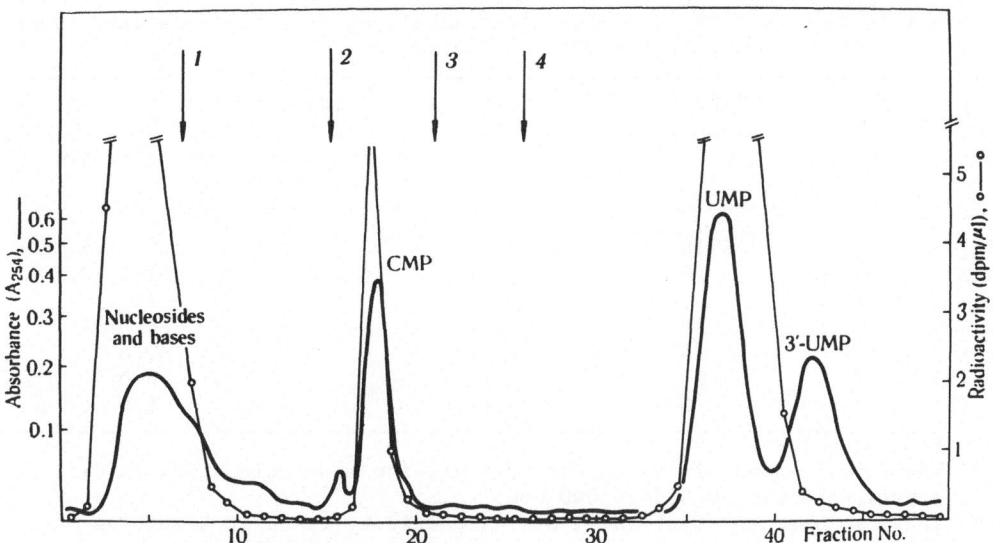

Fig. 3. Ion-exchange chromatoghraphy of a hydrolysed acid-soluble extract from ascites tumour cells labelled with [^{14}C]uridine (3 kBq/ml). The column Dowex 1 (X8 format) 11.6 cm x 0.75 cm was eluted at 0.8-0.9 ml/min and fractions of about 4 ml were collected. The numbered arrows indicate the eluent: 1, H_2O; 2, 0.15 M HCOOH; 3, H_2O; 4, linear gradient from H_2O to 1 M HCOOH$_4$ (mixer and reservoir each contained in the beginning 75 ml of H_2O and 1 M HCOONH$_4$ respectively).

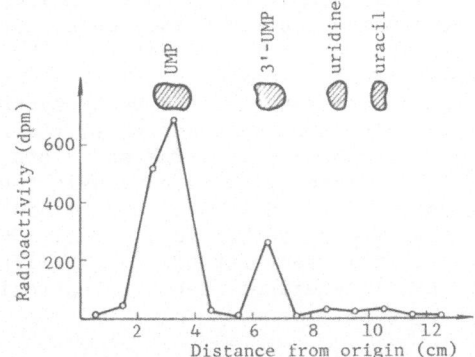

Fig. 4. Thin-layer chromatography of the concentrated through charcoal material obtained from the pooled fractions of the peak of 3'-UMP (see Figure 3). The radioactivity of 1 cm-wide strips scraped from the chromatogram was measured. The spots in the upper part of the figure show the position of the UV-markers added to the sample.

It must be underlined that after the short labelling of Ehrilich ascites tumor cells with radioactive uridine it becomes incorporated only into nuclear RNA species (predominantly hnRNA and pre-rRNA) but not into the cytoplasmic ribonucleic acids (Genchev et al, 1980; Georgiev, 1980). On this ground the results obtained above should be considered as related only to the nuclear RNA metabolism. It is quite possible that the cyto-

Table 1. Radioactivity in 3'-UMP after labelling of Ehrlich ascites tumor cells in suspension with [^{14}C]uridine

Experiment No.	Time of labelling (min)	Method of analysis*	Radioactivity in 3'-UMF (% of the total** radiactivity in acid soluble uridine nucleotides)
1,2,3	10	I	2.0, 1.5, 1.9
2,3	20	I	1.5, 0.7
1,	30	I	1.3
4 (a)	10	II	4.5***
4 (b)	10	II	0.6
5 (a)	20	II	6.0***
5 (b)	20	II	1.4

* See under "Methods"...

** The total radioactivity in the free uridine mononucleotides in different experiments was 20,000-45,000 dpm.

*** These values represent the % of radioactivity found in 3'-UMP as compared to the sum of the radioactivities in 3'-UMP and in UMP (as present in the unhydrolysed cell extract).

plasmic ribonucleic acids are degraded in a different way producing 3'-mononucleotides.

CONCLUSIONS

Under the experimental conditions used a considerable part (e.g. 50%) of the radioactive mononucleotides in intact cells labelled with [^{14}C] uridine is derived from nuclear RNA degradation. These nucleotides are practically only nucleoside 5'-phosphates. This indicates that the hydrolysis of the non-conserved segments of the newly-synthesized RNA is catalyzed by enzyme(s) with 5'-exonuclease activity, the presence of which in animal cell nuclei has been proven (Lazarus and Sporn, 1967; Sporn et al, 1969). Such a way of RNA degradation is obviously the most expedient with respect to the cell energetics and assures a very high rate of reutilization of the degradation products (Genchev, 1980). The model proposed (ibid.) for the mathematical treatment of the labelling kinetics of the free nucleotides and RNA is therefore the best approximation to the real situation in the living cell.

REFERENCES

Bernardi, G., 1971, In: "The Enzymes," Vol. 4:271-278.
Cleaver, J., Boyer, H., 1972, Biochim.Biophys.Acta, 262:116-124.
Genchev, D. D., 1980, Biochem J. 188:75-83.
Genchev, D. D., Kermekchiev, M. B., Hadjiolov, A. A., 1980, Biochem J. 188:85-90.
Georgiev, G. P., 1980, Biol.Revs. 1:215-255.
Hadjiolov, A. A., Nikolaev, N., 1976, Progr.Biophys.Mol.Biol., 31:95-144.
Harris, H., 1963, Progr.Nucl.Acid Res. 2:19-59.
Harris, H., Fisher, H.W., Rogers, A., Spencer, T., Watts, J. W., 1963, Proc.Roy.Soc. (London) 157:(SerB) 177-198.
Hurlbert, R. B., Hardy, D., Stankovich, M., Coleman, R. D., 1973, Adv.Enz. Regul. 11:323-341.
Hurlbert, R. B., Maxey, G. S., Coleman, R. D., 1974, Proc.Am.Assoc.Cancer Res. 15:147.

Kelley, D., Perry, R., 1971, <u>Biochim.Biophys.Acta</u>. 238:357-362.

Lazarus, H. M., Sporn, M. B., 1967, <u>Proc.Nat.Acad.Sci</u>. U.S. 57:1386-1390.

Maruyama, H., Mizuno, D., 1966, <u>Biochim.Biophys.Acta</u> 123:510-522.

Perry, R. P., 1976, <u>Annu.Rev.Biochem</u>. 45:605-629.

Sierakowska, Shugar, D., 1977, <u>Progr.Nucl.Acid Res</u>. 20:59-130.

Speers, E., TIgerstrom, R., 1975, <u>Canad.J.Biochem</u>. 53:79-90.

Sporn, M. B., Lazarus, H. M., SMith, J.M., Henderson, W. R., 1969,
    <u>Biochemistry</u> 8:1698-1706.

Witzel, H., 1963, <u>Progr.Nucl.Acid Res</u>. 2:221-258.

IMMUNOLOGICAL CHARACTERIZATION OF DNA-DEPENDENT RNA

POLYMERASE(S) OF SPINACH CHLOROPLASTS

E. Brautigam and S. Lerbs

Institute of Plant Biochemistry
Academy of Sciences of the GDR
4050 Halle (Saale), DDR

Chloroplasts from spinach are easily isolated in an intact form, free of cytoplasmic and nuclear contamination. The chloroplast genomes of tobacco and Marchantia have been sequenced (Shinozaki et al. 1986, Ohyama et al. 1986). Sequences homologous to the genes coding for the β, β' and α subunits of RNA polymerase of E. coli have been found. Two apparently independent transcriptional activities have been isolated. One, the soluble RNA poly-merase (sRNAP), is present within the chloroplast stroma. The second, the transcriptionally active DNA-protein complex (TAC), can be isolated from the thylakoid membrane fraction. The TAC contains chloroplast DNA, tightly bound chloroplast DNA-dependent RNA polymerase, and other proteins. According to Hallick and his co-workers the transcriptionally active chromosome of Euglena specifically transcribes the rRNA operon (Rushlow et al. 1980, Gruissem et al. 1983a, Greenberg et al. 1984, Narita et al. 1985), whereas the sRNAP is selective for transcription of tRNA genes from spinach (Gruissem et al. 1983b, Gruissem, 1984) and Euglena (Gruissem et al. 1983a, Greenberg et al. 1984). However, mustard chloroplast TAC is capable of transcribing plastid rRNA genes as well as protein coding genes (Reiss and Link, 1985).

The transcription complex from pea chloroplasts does not selectively transcribe certain regions of the DNA, only in short-term incubation, there are some regions of DNA that are relatively more transcribed than others (Tewari and Goel, 1983).

The TAC and the sRNAP are biochemically distinct. The TAC activity is resistant to heparin, while the soluble extract is quite sensitive to heparin. They respond differently to KCl and Mg^{2+}.

Differences observed in DNA binding, biochemical properties, sub-chloroplast location, gene selectivity suggest that there are at least two distinct chloroplast polymerase activities. At present it is unknown if the two RNA polymerase activities are two different and independent enzymes or if they use the same core RNA polymerase and differ in the factors used to transcribe different genes.

In this report we compare the spinach chloroplast TAC and the purified soluble RNA polymerase (sRNAP) by means of immunological methods. The membrane-bound RNA polymerase was isolated either as TAC as described by Briat et al. (1979), as high salt TAC (hsTAC) according to Narita et al.

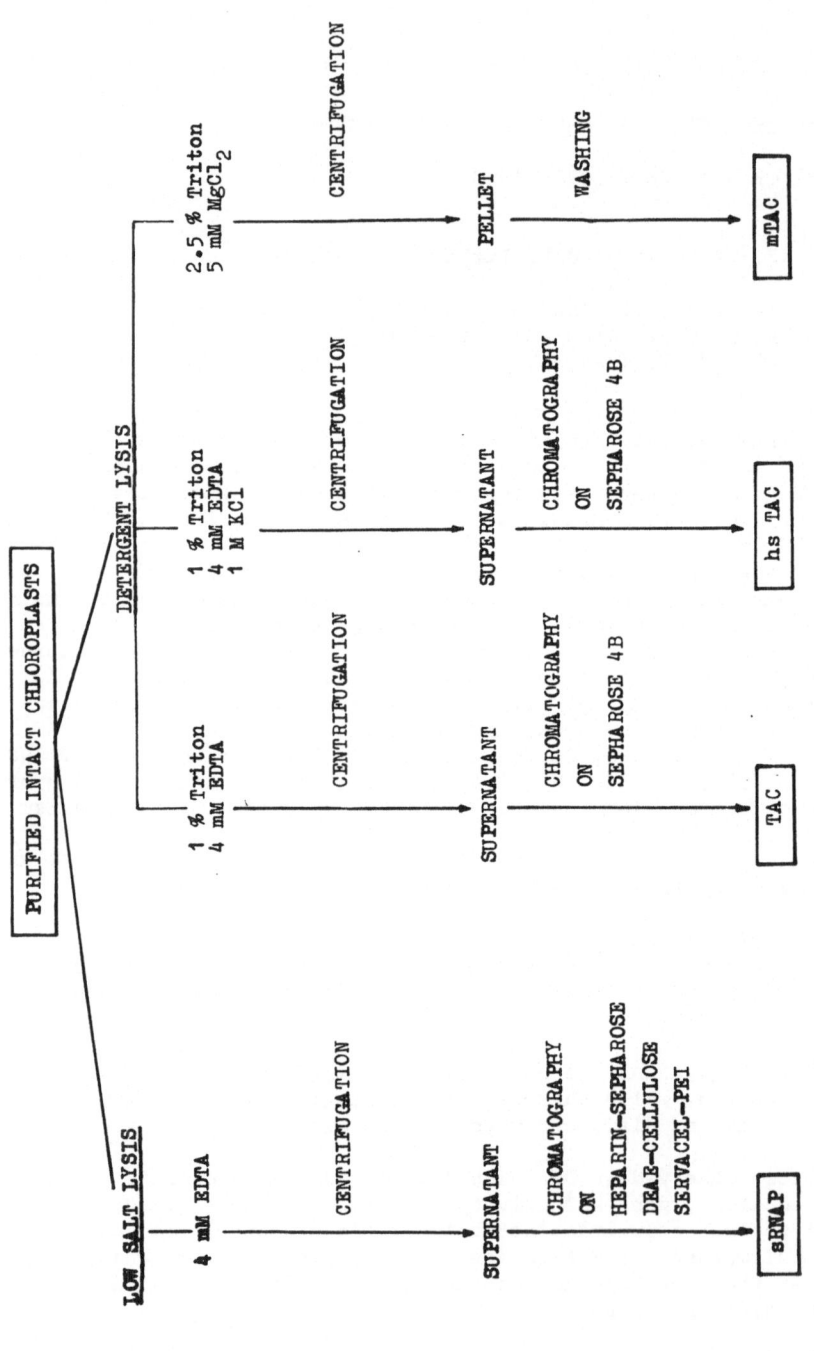

Fig. 1. Flow chart for preparation of soluble RNA polymerase (sRNAP) and transcriptionally active DNA-protein complex (TAC).

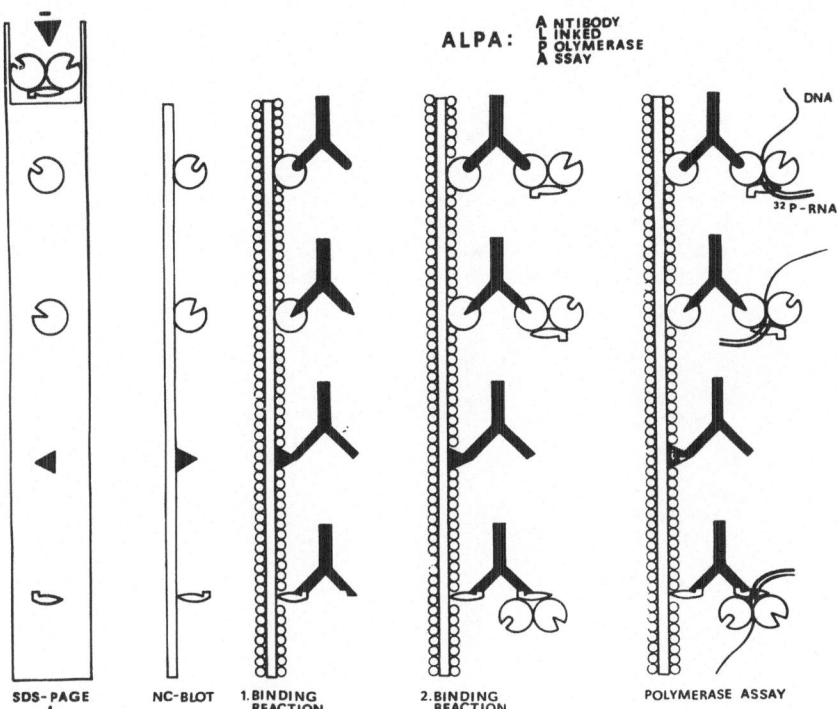

ALPA: **A** NTIBODY
L INKED
P OLYMERASE
A SSAY

SDS-PAGE +   NC-BLOT   1.BINDING REACTION   2.BINDING REACTION   POLYMERASE ASSAY

Fig. 2. Schematic representation of ALPA. Partially purified RNA polymerase is subjected to SDS-PAGE and blotted onto nitrocellulose.
The nitrocellulose was incubated subsequently. 1. with an excess
of antiserum (1. binding reaction); 2. with native RNA polymerase
(2. binding reaction); 3. with a complete transcription mixture
containing [^{32}P] ATP; 4. with trichloroacetic acid to precipitate
the nascent ^{32}P-labeled RNA in situ. Autoradiography reveals
^{32}-P-labeled RNA at the position of the subunits of RNA
polymerase.

(1985), or as TAC containing membrane fractions (mTAC) following the first
steps of a procedure established by Tewari and Goel (1983). The sRNAP was
isolated and purified from the soluble extract of spinach chloroplasts as
described by Lerbs et al. (1983, 1985) (Figure 1).

We have analysed the polypeptide composition of the isolated RNA
polymerase activities using the antibody linked polymerase assay (ALPA);
Lerbs et al. 1985, Van Der Meer et al. (1983). ALPA is a method for
correlating polymerase activity with a particular polypeptide band after
SDS polyacrylamide gel electrophoresis (SDS-PAGE) (Figure 2). Both non-
specific anti-polymerase serum and partially purified enzyme preparation
can be used for identification of the subunits of the polymerase, and
immunological cross-reactions with the subunits of E. coli polymerase can
be tested.

We prepared an antiserum against purified spinach chloroplast RNA
polymerase. Experiments with ^{125}J-labeled protein A indicate that our
poly-specific antiserum contains antibodies against all polypeptides of the
RNA polymerase preparation (Figure 3, lanes A and B). We found that the
150, 145, 110, 102, 80, 75 and the 38 kd polypeptides are labeled by ALPA

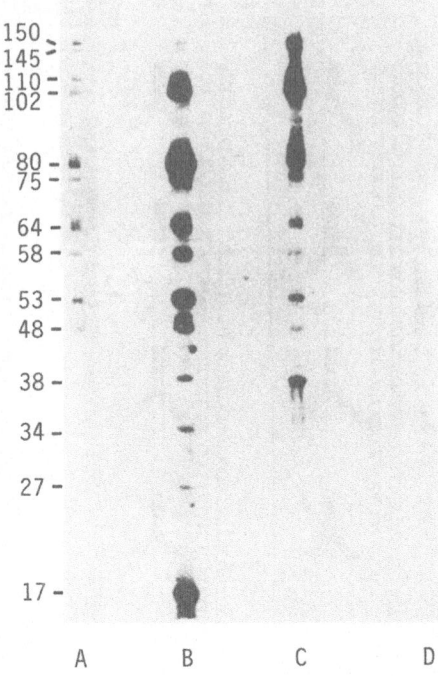

Fig. 3. Determination of "ALPA-reactive" polypeptides of purified chloroplast RNA polymerase. Nitrocellulose blots of purified RNA polymerase were treated as follows: lane A: staining of transferred
polypeptides with Amido-black; lane B: incubation with antiserum
raised against purified spinach chloroplast RNAP and [125J]-labeled
protein A; lane C: subsequent incubation with the same antiserum
and partially purified chloroplast RNA polymerase followed by the
transcription assay of antibody-linked RNA polymerase activity
(ALPA); lane D: control experiments (Lerbs et al., 1985).

(Figure 3, lane C), i.e. they are immunologically related to subunits (or
their degradation products) of the RNA polymerase. Some other proteins
ranging from 40-70 are also labeled by ALPA but in contrast to the seven
above mentioned polypeptides their amounts varies from one preparation to
another, for that reason these "minor" polypeptides were not considered as
subunits of the RNA polymerase (Lerbs et al. 1983, 1985). The 150, 145, 38
kd polypeptides are similar in size to the 165, 155, 39 kd subunits of E.
coli RNA polymerase. In order to find immunological cross-reactions of E.
coli RNAP and chloroplast RNAP nitrocellulose filters charged with E. coli
polymerase subunits were incubated with antiserum raised against sRNAP from
spinach and afterwards with [125J] protein A and nitrocellulose filters
charged with the sRNAP subunits from spinach were probed with native E.
coli RNAP in the ALPA (Figure 4). We found that the 150 and 145 kd polypeptides of the chloroplast polymerase are immunologically related to the
β and/or β' subunits of E. coli polymerase but also the 80 kd polypeptide
and some of the "minor" polypeptides (48, 53, 58, 64 kd), which may result
probably from proteolytic degradation of the 150 and 145 kd proteins during
purification. No cross-reactions with the 110, 102 and 38 kd polypeptides
could be detected.

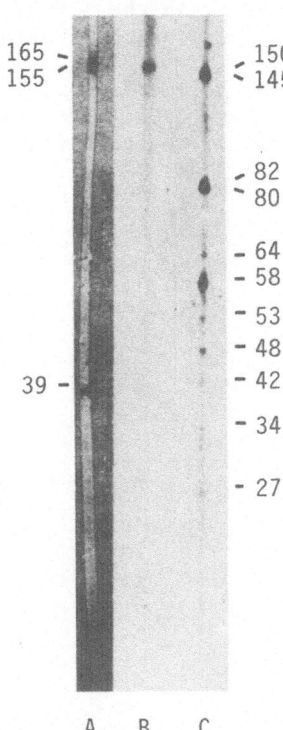

165 ⟩
155 ⟩   ⟨ 150
             ⟨ 145

⟨ 82
⟨ 80

– 64
– 58

– 53

– 48

39 –   – 42

– 34

– 27

A B C

Fig. 4.   Demonstration of immunochemical similarities between E. coli and
spinach chloroplast RNA polymerase subunits. Nitrocellulose blots
of E. coli RNA polymerase core enzyme (lanes A and B) and purified
chloroplast RNA polymerase (lane C) were treated as follows: lane
A: staining with Amido-black; lane B: incubation with antiserum
against chloroplast RNA polymerase holoenzyme, and [^{125}J] protein
A; lane C: incubation with the homologous antiserum and subsequent
binding of native E. coli polymerase followed by ALPA assay (Lerbs
et al., 1985).

Data obtained from immunological cross-reactions of E. coli enzyme and
spinach enzyme together with the results (now shown), that α-like and
δ-like polypeptides exist (Lerbs and Mache 1985) in the crude extract of
the chloroplasts, and that these polypeptides are separated by chromatography on Servacel-PEJ suggest that there is an E. coli-like RANP in
chloroplasts. It is tempting to speculate that the 110 and 102 kd
polypeptides belong to another RNA polymerase.

Fig. 5 compares the ALPA-reactive polypeptides of the soluble RNA
polymerase and of transcriptionally active DNA-protein complexes (TAC and
mTAC). The polypeptide composition of the TAC resembles that of the
soluble enzyme, however the 150 and 145 kd polypeptides are hardly labeled
(TAC) or not detectable (mTAC).

Kept at 4°C for 10 h the mTAC loses half of its activity, and the
degradation of those polypeptides which are immunological related to E.
coli RNA polymerase is nearly complete. In contrast the 110 and 102 kd
polypeptides remain stable.

The TAC was extensively purified using high salt concentrations to
dissociate loosely bound proteins. No cross-reactions of the polypeptides
of hsTAC with antiserum raised against E. coli RNAP was found. Spinach

serum did, however, cross-react with the 110 kd polypeptide of hs-TAC, here the antibody binding as visualized using [125]J-labeled protein A.

At last the different RNA polymerase preparations were probed with anti-sigma serum of E. coli (Figure 6). The sRNAP contains a 43 and a 90 kd polypeptide which cross-react, whereas neither the TAC nor the mTAC seem to contain a sigma-like polypeptide.

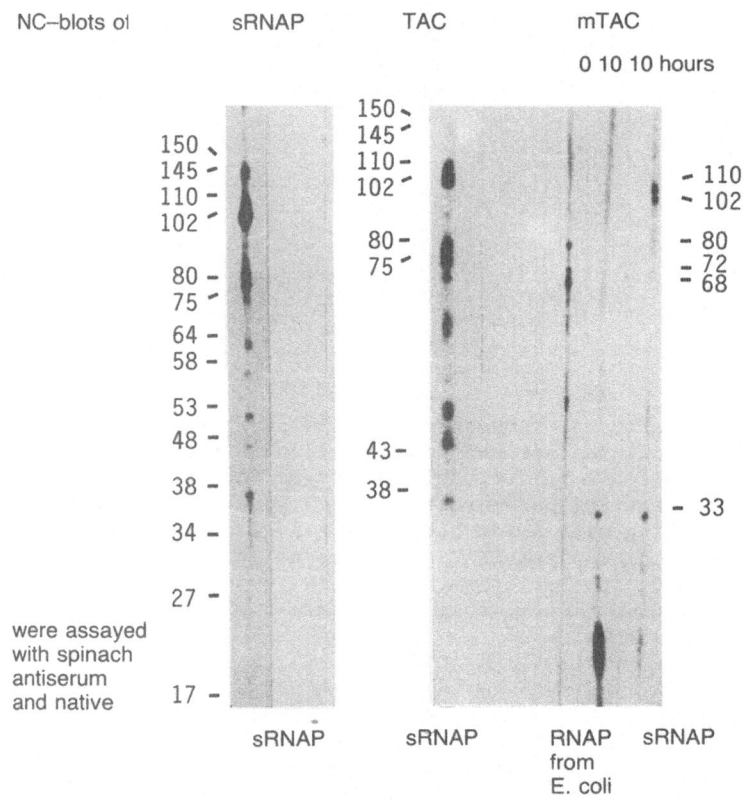

Fig. 5. Comparison of ALPA-reactive polypeptides of sRNAP and TAC.

Results presented here support the hypothesis that in chloroplasts at least two RNA polymerase activities exist. The soluble RNA polymerase and the RNA polymerase of the transcriptionally active DNA-protein complex have a similar polypeptide pattern, but differ in content of a sigma-like initiation factor used for specific transcription. Both the sRNAP and the TAC seem to contain two independent enzymes, one, the E.coli-like RNAP, which is composed of ββ'-like and α-like polypeptides (and sigma like protein in the case of sRNAP), and the other enzyme that has a 110 kd functional component.

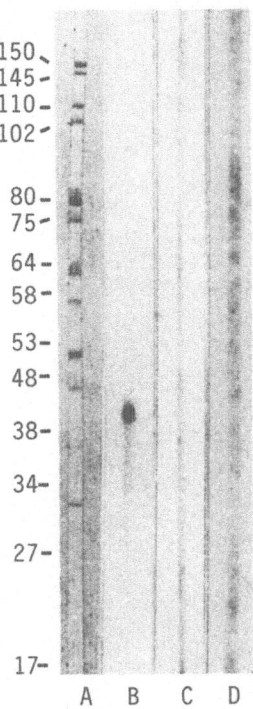

Fig. 6. Demonstration of immunological similarities between the sigma
subunit of E. coli polymerase with a 43 kd polypeptide of spinach
chloroplast sRNAP. A: Purified sRNAP, Coomassie staining after
SDS-PAGE; B-D: Incubation of NC-blots of sRNAP (B), TAC (C) and
mTAC (D) with anti-sigma serum of E. coli RNAP and subsequent
incubation with ^{125}J protein A.

REFERENCES

Briat, J.-F., Laulhere, J.P. and Mache, R. 1979, Eur. Biochem. 98; 285-292.
Greenberg, B.M., Narita, J.O., De Luca-Flaherty, C., Gruissem, W., Rushlow,
    K.A. and Hallick, R.B. 1984, J. Biol. Chem., 259; 14880-14887.
Gruissem, W., Narita, J.O., Greenberg, B.M., Prescott, D.M. and Hallick,
    R.B. 1983a, J. Cell. Biochem., 22; 31-46.
Gruissem, W., Greenberg, B.M., Zurawski, G., Prescott, D.M. and Hallick,
    R.B. 1983b, Cell, 35; 815-828.
Gruissem, W. 1984, Plant Molecular Reporter, 2; 15-23.
Lerbs, S., Briat, J.F. and Mache, R. 1983, Plant Mol. Biol., 2; 67-74.
Lerbs, S., Brautigam, E. and Parthier, B. 1985, Embo J., 7; 1661-1666.
Lerbs, S. and Mache, R. 1985, Chem. Hoppe-Seyler, 366; 819.
Meer, Van Der, J., Dorssers, L. and Zabel, P. 1983, Embo J., 2; 233-237.
Ohyama, K., Fukuzawa, H., Kohchi, T., Shirai, H., Sano, T., Sano, S.,
    Umesono, K., Shiki, Y., Takeuchi, M., Chang, Z., Acta, S., Inokuchi,
    H. and Ozeki, H. 1986, Nature, 322; 257-275.
Rushlow, K.E., Orozco, E.M., Jr., Lipper, C., Hallick, R.B. 1980, J. Biol.
    Chem., 255; 3786-3792.
Reiss, T. and Link, G. 1985, Eur. J. Biochem., 148; 207-212.
Shinozaki, K., Ohme, M., Tanaka, M., Wakasugi, T., Hayashida, N.,
    Matsubayashi, T., Zaita, N., Chunwongse,J., Obokata, J., Yamaguchi-
    Shinozaki, K., Ohto, C., Torazawa, K., Meng, B.Y., Sugita, M., Deno,
    H., Kamogashira, T., Yamada, K., Kusuda, J., Takaiwa, F., Kato, A.,
    Tohdoh, N., Shimada, H. and Sugiura, M. 1986, Embo J., 5; 2043-2049.
Tewari, K.K. and Goel, A. 1983, Biochemistry (Wash.), 22; 2142-2148.

DNA CROSSLINKED BY UV LIGHT TO THE NUCLEAR LAMINA IS

ENRICHED IN PULSE-LABELLED SEQUENCES

Z.I. Galcheva-Gargova

Institute of Molecular Biology
Bulgarian Academy of Sciences
1113 Sofia, Bulgaria

INTRODUCTION

The site of DNA replication in eukaryotes has been the subject of controversy. It is becoming increasingly accepted that the replication site is fixed and that DNA is synthesized as it moves through a polymerization complex attached to part of nuclear substructure, called matrix or cage (for review see Berezney, 1985). All recent evidence supporting this view is derived from studies in which cells or nuclei have been extracted in hypertonic salt solutions. The controversy centers on whether the resulting nuclear structures have counterparts in vivo or are simply artefacts and whether DNA is ever associated with the matrix (Jackson and Cook, 1986).

In the present study we used a recently developed method for isolating DNA fragments attached to the nuclear lamina (NL) by UV irradiation of Ehrlich Ascite Tumor (EAT) cells in vivo (Galcheva-Gargova and Dessev, 1987). The nuclear lamina represents a polymeric protein meshwork that lines the inner surface of the nuclear envelope (Gerace, 1986), which can be visualized in situ in certain eukaryotic cells (Fawcett, 1966) or in most cells after nuclear subfractionation (Krohne and Benavente, 1986). Crosslinking DNA to the NL proteins in vivo avoids the possible artefacts of redistribution during the isolation procedure, and besides provides an approach to study the role of the nuclear periphery in the process of DNA replication by pulse-labelling experiments.

METHODS

EAT cells were propagated in mice and DNA was long term-labelled with ^{14}C-thymidine (20 μCi per animal). All procedures were previously described (Galcheva-Gargova and Dessev, 1987). Briefly, for UV-irradiation the cell suspension (5.10^6 per ml) was irradiated for 30 min using a 15W germicidal lamp. Isolation of chromatin and NL was carried out using the standard procedure (Krachmarov et al, 1986). For the non-equilibrium metrizamide gradients (MA) one ml of 55% MA was overlayered with 5 ml linear 20-40% MA gradient and 3 ml 15% sucrose (all layers contained 2M NaCl, 10mM Tris.Cl, 1mM EDTA, pH 8.0). Samples of chromatin digested with DN-aseII and RNase dissociated in 2M NaCl were overlayered and centrifuged for 60 min at 10,000 rpm in a SW40 Beckman rotor. Fractions were collected from the meniscus.

RESULTS AND DISCUSSION

DNA of EAT cells was long term-labelled by intraperitoneal injection of ^{14}C-thymidine (16–18 hours). Pulse-labelling was performed for 2 minutes with ^{2}H-thymidine after UV-irradiation of the cell suspension. For the separation of DNA fragments, crosslinked to the nuclear lamina from the bulk of DNA and proteins, non-equilibrium metrizamide gradients were used. The density barrier created in such gradients allows only structure large enough to sediment (Galcheva-Gargova and Dessev, 1878). In the case of unirradiated controls, virtually no DNA cosediments with the nuclear shells (Figure 1A). The structures found at the bottom of such MA gradients represent nuclear lamina of very high purity, which were previously isolated and characterized (Krachmarov et al, 1986). In the irradiated samples (Figure 1B) about 2% of the total input of ^{14}C-thymidine (long-term label) is found covalently linked to the nuclear lamina. This result is in agreement with our previous data for DNA digested to an average length of 200–250 bp (GaLcheva-Gargova and Dessev, 1987). Meanwhile, the amount of the pulse-label (^{3}H-thymidine) in the pellet was three to four times higher. This enrichment of the crosslinked fragments with pulse labeled sequences is reflected also in the ratio of ^{3}H/^{14}C radioactivity of the NL structures, estimated 3 to 4 times higher in comparison to the ratio of the total ^{3}H/^{14}C radioactivity of DNA. It is likely that in vivo, at the periphery of the nucleus, very closely to the points of contact of DNA with the NL replication takes place. Further evidence in favor of this suggestion was obtained in experiments with aphidicolin, strong inhibitor of DNA polymerase in vivo (Oguro et al, 1979). In the conditions of our experiments, incubation of EAT cells for 15 minutes in the presence of 6 µ/ml aphidicolin, inhibited incorporation of labeled thymidine to more than 75% (results not shown). The dashed line in Figure 1B shows the distribution of the ^{3}H-DNA after incubation of irradiated EAT cells with the inhibitor, prior pulse labelling. Virtually no nascent DNA was found associated with the NL when replication was not allowed to proceed. On the other hand these results indicate that the enrichment of pulse-labelled DNA in the fragments crosslinked to the nuclear shells, could hardly be due to a reparation process. Experiments confirming this suggestion were next performed. Pulse labelling of DNA was done before UV-irradiation of the cells. The EAT suspension was incubated for 2 minutes in the presence of ^{3}H-thymidine, washed and irradiated with UV light for 30 minutes. In this case the distribution of ^{3}H-radioactivity in the MA gradients was as in Figure 1B, except that the total amount was higher, probably because of the incomplete washing of the isotope.

The above results indicate that the enrichment of pulse labelled sequences in the DNA fragments covalently linked to the NL, most probably is due to the process of replication.

DNA isolated from the bottom and from the top fractions of the MA gradients had an average length of 200–250 bp, as estimated from the electrophoresis (Figure 2, lane A). Experiments with lower level of DNA fragmentation, shown on the same figure, lane B, were next performed. The results are summarized in Figure 3. The amount of nascent DNA associated with the nuclear periphery correlates with the average size of the DNA fragments. This fact probably reflects the dynamic of the process of enrichment with pulse-label. This suggestion was further confirmed in a series of experiments. The level of DNA fragmentation was constant (about 450 bp) but the time of pulse-labelling was prolonged (Figure 4). The percentage of nascent DNA crosslinked to the nuclear lamina also increased.

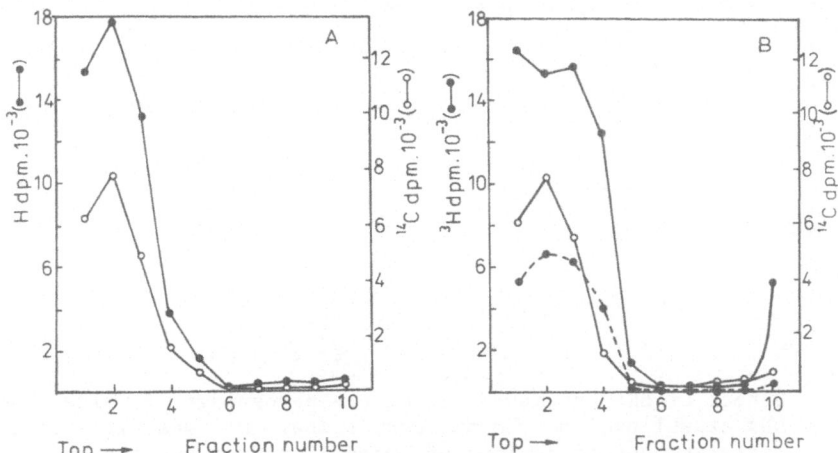

Fig. 1. Distribution of long term-labelled DNA (^{14}C-thymidine o——o) and pulse labelled DNA (^{3}H-thymidine, ●——●) in MA gradients. (A) control cells; (B) UV-irradiated cells for 30 min. ^{3}H-DNA in the pellet-8.6%, ^{14}C-DNA- 1.9%. For the total radioactivity – 2.3 ^{3}H/^{14}C ratio, for the bottom fraction – 12.5. The dashed line represents the distribution of the pulse-label after treatment of the cells with aphidicolin prior ^{3}H-labeling.

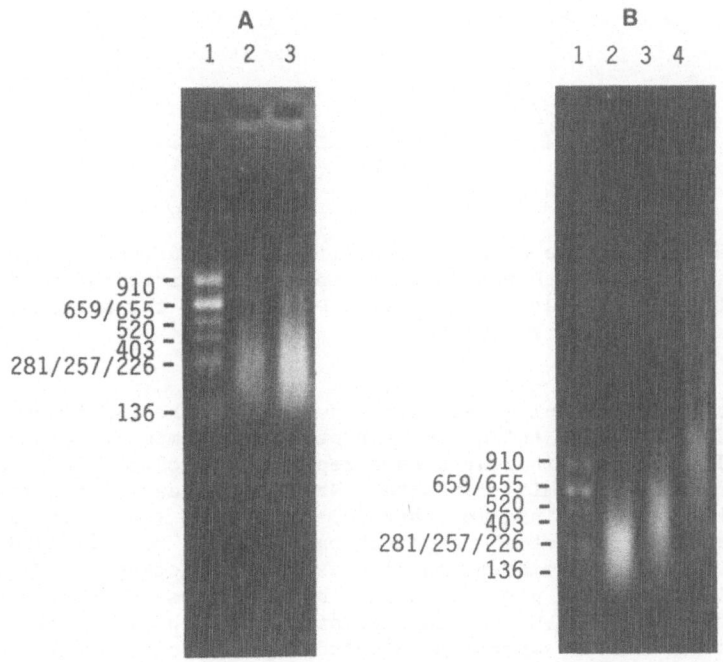

Fig. 2. The 1% agarose gel electrophoresis of DNA. Lane A: 1-DNA-size marker (pBR 322 digested with Alu 1). 2-DNA isolated from the bottom fraction of the gradient, after digestion with 2 units DNase II per A_{260}. 3-DNA isolated from the top of the same gradient. Lane B: 1-DNA size marker (pBR 322 digested with Alu 1). 2-DNA digested with 2 units DNase II per A_{260}. 3-DNA digested with 1 unit DNase II per A_{260}. 4-DNA digested with 0.5 units DNase II per A_{260}.

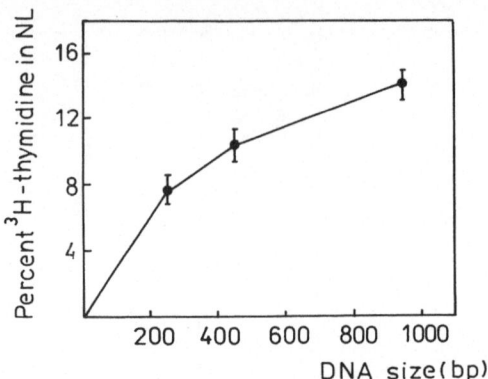

Fig. 3.   Effect of DNA fragmentation on the percentage of pulse-labelled
           DNA crosslinked to the nuclear lamina. DNA was digested with
           different concentrations of DNase II to the average length
           illustrated in Figure 2, lane B.

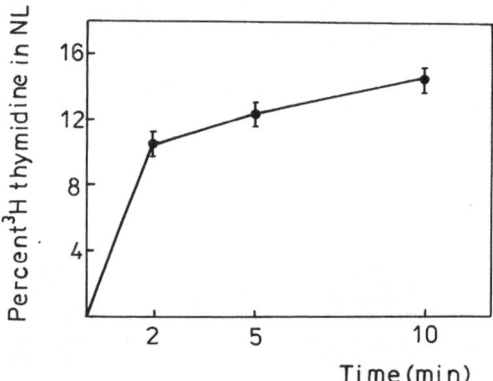

Fig. 4.   Effect of time of pulse labelling on the percentage of DNA
           associated with the nuclear lamina after irradiation with UV
           light. DNA was digested with DNase II to the average length
           shown in Figure 2, lane B-2.

      The enrichment of pulse-labelled sequences in the DNA fragments coval-
ently linked by UV light to the nuclear periphery demonstrated in the
present study, probably indicates that replication takes place near the
points of DNA-nuclear lamina contacts. Results suggesting the role of the
nuclear lamina were obtained in other investigations (Wanka et al, 1982;
Vogelstein et al, 1982) of the replication process. But all evidence
presented from the pulse-chase experiments, autoradiography and so on, was
obtained using isolation procedures with hypertonic salt solutions. The
method of "freezing" DNA-NL contacts existing in vivo, applied in the
present work provides an approach for studying the role of the nuclear
periphery in the process of replication in more details.

REFERENCES

Berezney, R., 1985, "Nuclear Structure and RNA Maturation," Alan R. Liss,
     (ed.), New York, 99-117.
Fawcett, D, 1966, Am.J.Anat. 119:129-146.
Galcheva-Gargova, Z. I., and Dessev, G., 1987, in press.

Gerace, L., 1986, TIBS II:443-446.

Jackson, D. A., and Cook, P. R., 1986, EMBO J. 5:1403-1410.

Krachmarov, Ch., Iovcheva, Ch., Hancock, R., and Dessev, G, 1986, J.Cell. Biol. 31:59-47.

Krohne, G., and Benavente, R., 1986, Exp.Cell Res. 162:1-10.

Oguro, M., Suzuki-Hori, Ch., Nagano, H., Mano, Y., and Ikegami, S., 1979, Eur.J.Biochem. 97:603-607.

Vogelstein, B., Nelkin, B., Pardoll, D., and Hunt, B. F., 1982, "The Nuclear Envelope and the Nuclear Matrix", Alan R. Liss, (ed.), New York, 169-181.

Wanka, F., Pieck, A. C. M., Bekers, Ad., G. M., and Mullenders, L. H. F., 1982, "The Nuclear Envelope and the Nuclear Matrix", Alan R. Liss (ed.), New York, 199-211.

THE MECHANISM OF ARABINOFURANOSYL-CYTOSINE INHIBITION IN

UNDIFFERENTIATED HUMAN B LYMPHOCYTES

T. Spasokukotskaja, J. Taljanidisz, M. Benczur* and M. Staub

1st Institute of Biochemistry Semmelweis University Medical
School, H-1444 Budapest 8, PO Box 260

*National Institute of Hematology and Bloodtransfusion
H-1502 Budapest, PO Box 44

INTRODUCTION

Main event of lymphocyte differentiation is the rearrangement of their
genetic material. Rearrangement of the chromatin is accompanied by the
replication of DNA. The supply of this DNA replication by nucleotides was
supposed to occur in the same way, as in normal, mitotic DNA replication.

Two ways are known to produce deoxynucleotides for the DNA synthesis:
DE NOVO biosynthesis of nucleotides from metabolites and the SALVAGE of
nucleosides. The salvage of "ready-made" purines and pyrimidines are of
higher importance than it has been thought some years ago, when de novo
biosynthesis received more attention. The salvage pathways of nucleotides
must have especially a striking importance in the immune system. Several
inherited defects of the purine nucleotide salvage enzymes are associated
with severe immunodefficiences, characterized as disorders in lymphocyte
differentiation without affecting the function of other tissues (reviewed
by Van Laarhoven and De Bruyn, 1983). The salvage of purine nucleosides
has been widely investigated during normal lymphocyte differentiation.
Different enzyme levels have been observed in the mature and immature
lymphocytes, and also in the different malignant states of the lymphoid
leukemia.

The salvage of pyrimidine nucleosides is measured very often, first of
all the salvage of thymidine is used as a universal marker for DNA synthesis in cells. However little is known about the connection of pyrimidine salvage and lymphocyte maturation. Autoradiographic studies have
revealed, that not only the utilization of purine nucleotides but also that
of deoxypirimidine nucleotides are related to normal lymphocyte differentiation. Differences have been found between ^{3}H-CdR and ^{3}H-TdR labelling
pattern of lymphocytes according to the localization of lymphocytes in the
lymphoid organ. Lymphocytes in the thymic cortex and germinal centers of
lymph nodes are more heavily labelled by ^{3}H-CdR than by ^{3}H-TdR, and opposite distribution of the labelling has been found outside the germinal
centers. The results of Hamatani and Amano (1980) point to the differences
of deoxypriimidine salvages in connection with the function of the cells.

In addition the salvage pathways of nucleosides are important not only for the activation of the natural nucleosides, but also for the activation of their analogues used as cytostatic agents in cancer therapy (e.g. araC).

In this work the salvage of deoxycytidine and thymidine was analyzed in an activated lymphocyte population isolated from infant's tonsils, along with the effect of araC on these processes.

RESULTS AND DISCUSSION

Lymphocytes isolated from tonsils of 3-6 years old children were separated by 25% BSA according to their density into a "light" so called "S-phase" and into a "heavy" so called "G-phase" cell population. We have compared the phosphorylation of CdR and TdR, their incorporation into DNA (Table 1) and the level of the enzymes engaged in their salvage (Table II) of the two lymphocyte populations, differed in their buoyant density.

It can be seen in Table 1, that the "S-phase" lymphocyte population incorporated 5 times more ^{3}H-TdR and 4-7 times more 5-^{3}H-CdR than "G-phase" lymphocytes. "S-phase" cells also contained 3-10 times more radioactivity in ethanol soluble thymidine nucleotide pools and 6-14 times more radioactivity in the pool of deoxycytidine nucleotides. It can also be seen, that the soluble pool of thymidine nucleotides if much lower than that of deoxycytidine nucleotides in both "S" and "G-phase" lymphocytes.

To give further evidences for the more intense DNA synthesis in "S-phase" tonsillar lymphocytes, the corresponding enzyme activities were determined (Table 2). Specific activity of DNA polymerase α was 2-5 times, of thymidine kinase was about 7 times and of deoxycytidine kinase was 2-3 times higher in the "S-phase" cells, than in the "G-phase" cells.

The "S-phase" lymphocyte subpopulation presented about 5-15% of the total cell population and contained about 20-30% more DNA per cell than the "G-phase" lymphocytes (date not present). The two lymphocyte subpopulations differed also in cell surface phenotypes characterized by mono-clonal-antibody reagents in the Institute of Hematology (Table 3). The main part of the "S-phase" cells is bearing early B cell markers and surface immunglobulins. The high rate of CD_5 and CD_{10} markers indicate to their proliferative state (Hsu and Jaffe, 1984), in contrast to "G-phase" cells, which are bearing mainly T cell markers (CD_3), and are poor in markers characteristic for cell proliferation.

Thus, infant's tonsils contain lymphocytes which occur mainly immature B lymphocytes and are active in DNA synthesis. This cell population was used further and signed as "S-phase" cells.

DNA synthesis of B lymphocytes is very sensitive to araC, a widely used drug for the treatment of acute leukemia and solid human tumors. When araC enters the cell it is first phosphorylated by deoxycitidine kinase and then by other kinases, giving rise to araCTP, the active form of the drug (Kassel and al, 1967). AraCTP can compete with dCTP at the level of the DNA polymerase (Graham and Whitmore, 1970), or it incorporates into the DNA blocking the elongation of the latter (Momparler, 1972).

We can summarize, araC has to be metabolized by the salvage enzymes of CdR, to exert its cytostatic effect on the mammalian cells. It is known that the sensitivity of mammalian tumor cells against araC is very different. Lymphocyte cell lines differ also very much in this sensitivity that can also be dependent on the actual stage of the nucleoside salvage of the

Table 1. Phosphorylation and Incorporation of Deoxypyrimidine Nucleoside into DNA by Tonsillar lymphocyte Subpopulations.

Cell	Experiment	Phosphorylation (ethanol soluble)		Incorporation (acid insoluble)	
		^{3}H–TdR	5–^{3}HCdR	^{3}H–TdR	5–^{3}H–CdR
"S-phase"	1	1628	3780	12963	3241
	2	276	805	7218	1293
"G-phase"	1	560	587	2590	450
	2	26	58	1432	304
"S-"/"G"	1	2.9	6.4	5.0	7.2
	2	10.6	13.9	5.0	4.3

cpm/10^6 cells

Human tonsillar lymphocyte suspensions were prepared from infant's tonsils (Staub et al, 1976). ^{3}H-TdR and 5-^{3}H-CdR phosphorylation and incorporation into DNA was measured as described by Taljanidisz et al, (1986). Lymphocyte separation was performed as follows: 1 ml lymphocyte suspension (10^8 cell/ml) was layered on 1 ml of 25% BSA (in Eagle's MEM) in 10 mm tubes and centrifuged for 30 min at 1,000 g. Lymphoblasts ("S-phase" cells) were concentrated in the interphase, small lymphocytes ("G-phase" cells) were pelleted at the bottom. Cells were washed twice with Eagle's MEM and were immediately used.

Table 2. Enzyme Activities Related to the Deoxypyrimidine Nucleoside Metabolism

Cell	Experiment	DNA polymerase	Thymidine kinase (TK)	Deoxycytidine kinase (dCK)
"S-phase"	1	109.5	1581	262
	2	106.0	2274	352
"G-phase"	1	21.3	225	123
	2	48.4	328	120
"S"/"G"	1	5.1	7.0	2.1
	2	2.2	6.9	2.9

Enzyme activity pmol/10^6cell/hour

Enzyme activities were measured as described by Staub et al, (1983), with small modification: in all cases the reaction mixtures contained 20 mM DTT, 20 mM NaF (TK) or 10 mM NaF (DNA-polymerase and dCK) and the concentration of labelled substrates was 50 μM.

Table 3. Surface Markers on the Tonsillar Lymphocyte Subpopulations

Cells	CD_1	CD_3	CD_5	CD_{10}	CD_{19}	CD_{20}	CD_{21}	Surface IgG,A,M
"S-phase"	1	8	59	68	82	61	72	82
"G-phase"	1	51	3	24	–	11	8	31

Surface markers (% of cell number)

Monoclonal antibodies used for surface marker determinations were as follows: CD_1, CD_3, CD_5 from Ortho, CD_{10} from Coulter, CD_{19}, CD_{20}, CD_{21} from Beckton - Dickinson and SIg G, A, M from Hyland.

cells. This prompted us to investigate the sensitivity against araC in B lymphocyte population characterized by surface markers as undifferentiated.

Figure 1.Shows the effect of araC on [14]C- and 5-^{3}H-CdR and ^{3}H-TdR incorporation into the "S-phase" lymphocytes. As it can be seen the incorporation of TdR is more sensitive to araC than that of CdR-s, but there is also a big difference in the sensitivity of the two CdR-s labelled in a different way. Under the same conditions 1 μM araC caused 56% inhibition on ^{14}C-CdR and only 13% inhibition on 5-^{3}H-CdR incorporation into DNA. To explain the differences between incorporation of the differently labelled CdR-s, the possible pathways of this nucleoside are shown in Figure 2.

One part of CdR finds its direct way into DNA, while another part is deaminated and converted to dUMP (or UdR). The dUMP can be methylated by thymidilate synthase to dTMP and than incorporated into DNA via the dTTP pool. However if the CdR has ^{3}H label in the position 5 of the cytosine ring, the ^{3}H will be released as [3]H$_2$O during this interconversion and labelling in DNA originates only from CdR built in directly into DNA. In the case of ^{14}C-CdR, not only cytosine, but also thymine in DNA can become radioactive, since a part of ^{14}C-dCMP can be converted to ^{14}C-dTMP, and be used for DNA synthesis. It seems probable that the capacity for conversion of dCMP to TMP and utilizing it for DNA synthesis is one of the important factors.

Table 4 shows the metabolism of 5-^{3}H-CdR in "S-phase" tonsillar lymphocytes as a function of CdR concentration. As can be seen from the extent of ^{3}H-release, at all concentrations about 70% of CdR was converted to dTMP, 13-16% of radioactivity was incorporated into DNA as ^{3}H-dCMP, 3-8%

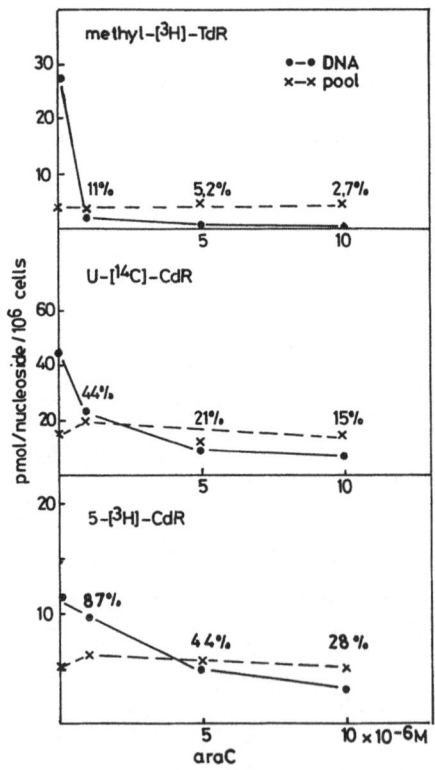

Fig. 1.   Effect of araC on ^{3}H-TdR, ^{14}C-CdR and 5-^{3}H-CdR incorporation into "S-phase" lymphocytes

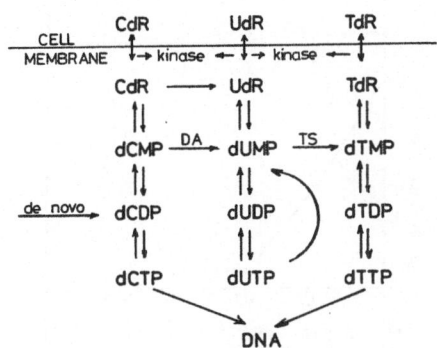

Fig. 2. Major pathways of pyrimidine deoxynucleotide metabolism.

Table 4. Metabolism of 5-³H-CdR In "S-Phase" Tonsillar Lymphocytes

5-³H-CdR, μM acitivity	0.2		0.4		0.8 pmol/10⁶ cells/hour (%)		1.6		3.2	
³H-release (conversion to dTMP)	2.5	(69.5)	4.8	(65.8)	8.5	(71.4)	11.9	(69.4)	16.3	(68.2)
Phosphory- lation (et- hanol soluble, DEAE-cell- ulose bound)	0.3	(8.3)	0.5	(6.8)	0.6	(5.0)	0.8	(4.6)	0.8	(3.4)
Ethanol sol- uble (not bound to DEAE- cellulose)	0.3	(8.3)	1.0	(13.7)	1.2	(10.1)	1.6	(9.4)	3.3	(13.8)
Incorporation into DNA	0.5	(13.9)	1.0	(13.7)	1.6	(13.5)	2.8	(16.4)	3.5	(14.6)
Total radio- activity in Cells	3.6	(100)	7.3	(100)	11,9	(100)	17,1	(100)	23.9	(100)

5-³H-CdR phosphyrylation and incorporation into DNA was measured as in
Table 1. ³H-release was detected in cell supernatants as radioactivity not
bound to Norit-A.

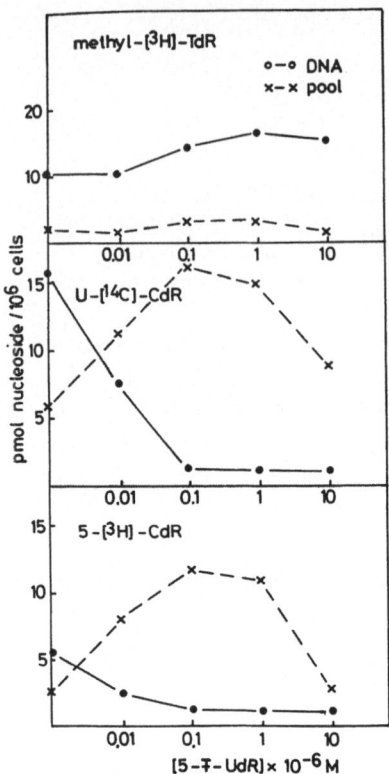

Fig. 3. The effect of 5–F–UdR on the incorporation of ³H–TdR, ¹⁴C–CdR and 5–³H–CdR into the DNA in "S-phase" tonsillar lymphocytes. The concentration of labelled deoxynucleosides was 3.3 μM.

was present in the soluble 5–³H–CdR pool in the form of nucleotides, and 8–14% of the total radioactivity taken up by cells was incorporated into an unknown substance. This substance presented however 1/3 of the radio-activity not converted to dTMP and contained the labelled deoxycytidine.

As we have seen the incorporation of ¹⁴C–CdR was bout twice more sensitive against araC than the incorporation of 5–³H–CdR. One can explain this effect, if a definite part of araC is also deaminated to araUMP, in the same way as CdR is. In this case araC would be converted to araTTP, competing with the low dTTP pool of DNA synthesis. This suggestion could be answered directly by measuring the fate of labelled araC. At this moment we have only indirect experiments supporting our suggestion.

A crucial point of our suggestion is the role of thymidilate synthase, which should be responsible for the methylation not only for dUMP but also for araUMP. To switch off this step in the cells, the well known inhibitor of thymidilate synthase 5–Fluoro–d–Uridine (5–F–UdR) was used. The effect of 5–F–UdR on the incorporation of ³H–TdR, ¹⁴C–CdR and of 5–³H–CdR is shown in Figure 3. 5–F–UdR inhibited only the incorporation of CdR and could be interrupted by thymidine (date not presented).

Following experiment (Figure 4) shows the inhibitory effect of araC on the incorporation of ³H–thymidine in the presence of 1μM FdU, totally inhibiting thymidilate synthase i.e. the interconversion of araC into araTTP. No synergistic but rather "extinguishing" effect was observed. The inhibition by araC was lower in the presence of 5–F–UdR, than in the absence of it.

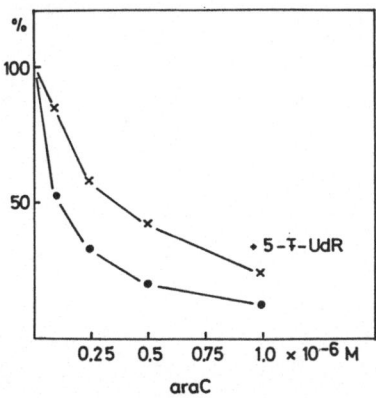

Fig. 4. The effect of araC on ³H–TdR Incorporation in the presence of
5–F–UdR. The concentration of ³H–TdR was 0.16 µM. The 100%
incorporation was 5.76 pmol ³H–dTMP in the presence and 4.08 pmol
in the absence of 5–F–UdR.

The phosphorylated nucleotides were analyzed on PEI–cellulose sheets
after 5–³H–CdR labelling. As can be seen in Figure 5, under the experimental conditions, the amount of ³H–dCDP and ³H–DCTP did not change.
There is a substantial increase in the concentration of the monophosphates,
originating from ³H–dUMP (data not presented). We can conclude from this
experiment that the decrease of araC inhibition in the presence of 5–F–UdR
is not a consequence of the increase of the dCTP pool.

Data presented here support our suggestions that the interconversion
of araC into araTTP may be a very important condition for the cytostatic
effect of the drug. The sensitivity of lymphocytes to araC depends on the
normal salvage pathway of deoxypyrimidines, which changes during differentiation, and can be characterized by the changes of the ratio of dC-
deaminase to dCMP–deaminase. Transformed cells are cells stopped at the
different levels of differentiation, containing different amounts of salvage enzymes. The function of the changes of the salvage pathways during
the normal differentiation process are still not known.

SUMMARY

An activated lymphocyte population was isolated from tonsils of 3–6
year old children by density gradient centrifugation. The isolated "S-
phase" cells were bearing early B lymphocyte markers and were 5–6 times
more active in DNA synthesis than the "G–phase" lymphocytes.

It was found that about 70% of CdR was deaminated and converted into
dTMP. The cells were very sensitive to araC, but the incorporation of
¹⁴C–CdR was twice more sensitive to araC than that of 5–³H–CdR. This
effect can be explained by the interconversion of araC into araT nucleotide
via the CdR interconversion pathway. This suggestion was also supported by
the effect of 5–FUdR, which decreased the inhibition of DNA synthesis
caused by araC. 5–FUdR the inhibitor of thymidilate synthase, possibly
decreases the interconversion of araC into araTTP, and its action via a
smaller dTTP pool of the cells.

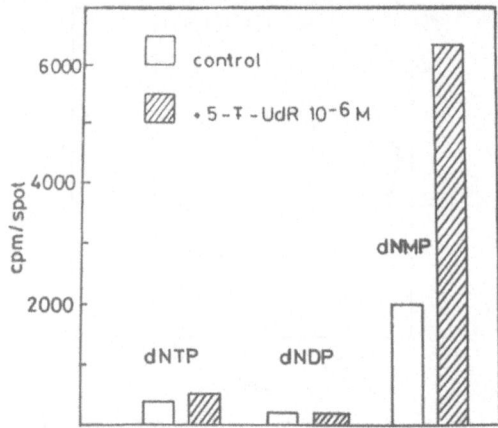

Fig. 5. The effect of 5-F-UdR on the phosphorylation of 5-³H-CdR. 3.3 μM
³H-CdR was used. Chromatography of the ethanol soluble fraction
was performed on PEI-cellulose plates in 0.75 M LiCl.

Our data suggest, that the sensitivity of cells to araC may depend on
the capacity of the CdR→→TdR interconversion pathway, which, however,
changes during the normal differentiation process of lymphocytes. Thus,
the sensitivity of different malignant cells to araC is determined by the
differentiation stage at which they were arrested during cell transformation.

REFERENCES

Graham, F. L. and Whitmore, G. F., 1970, Cancer Res. 30:2627-2635.
Hamatani, K. and Amano, M., 1980, Cell Tissue Kinet. 13:435-443.
Hsu, S. M. and Jaffe, E. S., 1984, Am.J.Pathol. 114:387-395.
Kassel, D., Hall, T., and Woolinsky, I., 1967, Science (Wash.DC)
      156:1240-1241.
Van Laarchoven, J. P. R. M. and De Bruyn, Ch. H. M. M., 1983, Leuchemia
      Res. 7:451-480.
Momparler, R. L., 1972, Mol.Pharmacol. 8:362-370.
Staub, M., Antoni, F., and Sellyei, M., 1976, Biochem.Med., 15:246-253.
Staub, M., Spasokukotskaja, T., Taljanidisz, J., Sasvari-Szekely, M.,
      Antoni, F., 1983, Immunol.Lett., 6:137-142.
Taljanidisz, J., Sasvari-Szekely, M., Spasokukotskaja, T., Antoni, F.,
Staub, M., 1986, Biochem.Biophys.Acta 885:266-271.

CLOSING OF THE SYMPOSIUM BY J. ZELINKA, CORRESPONDING MEMBER OF
THE CZECHOSLOVAK AND SLOVAK ACADEMY OF SCIENCES, DIRECTOR OF
THE INSTITUTE OF MOLECULAR BIOLOGY, SLOVAK ACADEMY OF SCIENCES

Ladies and gentlemen, dear friends,

Everything in life has its end and the same applies also for this friendly and useful meeting.

Let me thank you for your attendance, for the presented papers of very high quality and for your valuable comments in friendly discussions as well as for your helpful advice in personal meetings and discussions.

At the opening of this Symposium I have stated that all our former meetings had a positive effect on the scientific work of our Institute. The successive shift in the scopes and topics of our work was very fittingly shown in the introductory slide of professor Follmann which has pleasantly surprised us.

I sincerely hope that you have had a nice time here, that you will remember us and that we will again meet here in 1990.

I am also using this opportunity to express my sincere thanks to all my co-workers who have helped to create the pleasant atmosphere we have enjoyed here and especially to Dr. Balan and to Mrs. Sadloňová as well as to all employees of this beautiful castle.

I would also like to wish you a pleasant journey home to your families and friends. Good-bye at the seventh Symposium.

Thank you for your attention.

PARTICIPANTS

Bakalova, A., Institute of Molecular Biology, Bulgarian Academy of
Sciences, Sofia 1113, Bulgaria.

Balan, J., Institute of Molecular Biology, Slovak Academy of Sciences, 842
51 Bratislava, Czechoslovakia.

Balanová J., Institute of Molecular Biology, Slovak Academy of Sciences,
842 51 Bratislava, Czechoslovakia.

Barák, I., Institute of Molecular Biology, Slovak Academy of Sciences, 842
51 Bratislava, Czechoslovakia.

Bräutigam, E., Institute of Plant Biochemistry, Academy of Sciences of the
GDR, 4020 Halle (Saale), GDR.

Butkus, V., ESP Fermentas, Fermentu 8, Vilnius 232028, Lithuanian SSR,
USSR.

Deyev, S., Institute of Molecular Biology, USSR Academy of Sciences,
Vavilov 32, Moscow 117984 GSP-1, USSR.

Dobias, J., Institute of Molecular Biology, Slovak Academy of Sciences, 842
51 Bratislava, Czechoslovakia.

Fábry, M., Institute of Molecular Genetics, Czechoslovak Academy of
Sciences, Flemingovo nám 2, 166 37 Prague 6, Czechoslovakia.

Farkašovský, M., Institute of Molecular Biology, Slovak Academy of
Sciences, 842 51 Bratislava, Czechoslovakia.

Follmann, H., Fachbereich Chemie, Abteilung Biochemie, Philipps Universität, D-3550 Marburg, Federal Republic of Germany.

Galcheva-Gargova, Z., Institute of Molecular Biology, Bulgarian Academy of
Sciences, Sofia 1113, Bulgaria.

Gálová, M., Institute of Molecular Biology, Slovak Academy of Sciences,
842 51 Bratislava, Czechoslovakia.

Gašperík, J., Institute of Molecular Biology, Slovak Academy of Sciences,
842 51 Bratislava, Czechoslovakia.

Gavrilova, L., Institute of Protein Research, Academy of Sciences of the
USSR, 142292 Pushchino, Moscow region, USSR.

Genchev, D., Institute of Genetics, Bulgarian Academy of Sciences, POB 96,
113 Sofia, Bulgaria.

Godány, A., Institute of Molecular Biology, Slovak Academy of Sciences, 842
51 Bratislava, Czechoslovakia.

Harišová, M., Institute of Molecular Biology, Slovak Academy of Sciences,
842 51 Bratislava, Czechoslovakia.

Hill, C., Chemistry Department, University of York, Heslington, York, YO1
5DD, United Kingdom.

Hogenkamp, H., Department of Biochemistry, University of Minnesota,
Minneapolis 55455 USA.

Homerová, D., Institute of Molecular Biology, Slovak Academy of Sciences,
842 51 Bratislava, Czechoslovakia.

Janulaitis, A., Institute of Applied Enzymology, Fermanty, 8, Vilnius
232028, Lithuanian SSR, USSR.

Jonák, J., Institute of Molecular Genetics, Czechoslovak Academy of
Sciences, Flemingovo nám. 2, 166 37 Prague 6, Czechoslovakia.

Jucovič, M., Institute of Molecular Biology, Slovak Academy of Sciences, 842 51 Bratislava, Czechoslovakia.

Kliment, J., Institute of Molecular Biology, Slovak Academy of Sciences, 842 51 Bratislava, Czechoslovakia.

Kollárová, M., Department of Biochemistry, Faculty of Science, Comenius University, 842 15 Bratislava, Czechoslovakia.

Kormanec, J., Institute of Molecular Biology, Slovak Academy of Sciences, 842 51 Bratislava, Czechoslovakia.

Küntzel, H., Max Planck Institute for Experimental Medicine, Herman Rein Str. 3., D-3400 Göttingen, Federal Republic of Germany.

Kutejová, E., Institute of Molecular Biology, Slovak Academy of Sciences, 842 51 Bratislava, Czechoslovakia.

Limborska, S., Institute of Molecular Genetics, Academy of Sciences of the USSR, Kurchatov Sq. 46, Moscow 123128, USSR.

Muchová, J., Institute of Molecular Biology, Slovak Academy of Sciences, 842 51 Bratislava, Czechoslovakia.

Muchová K., Institute of Molecular Biology, Slovak Academy of Sciences, 842 51 Bratislava, Czechoslovakia.

Nazarov, V., Institute of Molecular Biology, Slovak Academy of Sciences, 842 51 Bratislava, Czechoslovakia.

Nikolov, E., Institute of Molecular Biology, Bulgarian Academy of Sciences, Sofia 1113, Bulgaria.

Nosikov, V., Institute of Genetics and Selection of Industrial Micro-organisms, Moscow 113545, USSR.

Pačes, V., Institute of Molecular Genetics, Czechoslovak Academy of Sciences, Flemingovo nám. 2, 166 37 Prague, Czechoslovakia.

Parmeggiani, A., Laboratoire de Biochimie, Ecole Polytechnique, F-91128, Palaiseau Cedex, France.

Perečko, D., Institute of Molecular Biology, Slovak Academy of Sciences, 842 51 Bratislava, Czechoslovakia.

Polyakov, K., Institute of Crystallography, Academy of Sciences of the USSR, Leninski pr. 59, Moscow 117333, USSR.

Popovňáková, Institute of Molecular Biology, Slovak Academy of Sciences, 842 51 Bratislava, Czechoslovakia.

Pristaš, P., Institute of Molecular Biology, Slovak Academy of Sciences, 842 51 Bratislava, Czechoslovakia.

Réti, M., Institute of Molecular Biology, Slovak Academy of Sciences, 842 51 Bratislava, Czechoslovakia.

Ritchie, D., Department of Genetics, University of Liverpool, Liverpool L69 3BX, United Kingdom.

Rüterjans, H., Institute of Biophysical Chemistry, J.W. Goethe University, Haus 75A, Theodor Stern Kai 7, D-6000, Frankfurt/Main 70, Bundersrepublik Deutschland.

Rybajlák, I., Institute of Molecular Biology, Slovak Academy of Sciences, 842 51 Bratislava, Czechoslovakia.

Rychlík, I., Institute of Molecular Genetics, Czechoslovak Academy of Sciences, Flemingovo nám. 2, 166 37 Prague 6, Czechoslovakia.

Sawicki, M., Biotechnology Research and Development Centre, Starościńska 5. 02-516, Warsaw, Poland.

Sedláček, J., Institute of Molecular Genetics, Czechoslovak Academy of Sciences, Flemingovo nám. 166 37 Prague Czechoslovakia.

Ševčík, J., Institute of Molecular Biology, Slovak Academy of Sciences, 842 41 Bratislava, Czechoslovakia.

Šimbochová, G., Institute of Molecular Biology, Slovak Academy of Sciences, 842 51 Bratislava, Czechoslovakia.

Šišová, M., Institute of Molecular Biology, Slovak Academy of Sciences, 842 51 Bratislava, Czechoslovakia.

Spaskokukotskaja, T., First Institute of Biochemistry, Semmelweis University Medical School, POB 260, 1444 Budapest 8, Hungary.

Steinborn, G., Central Institute for Genetics and Plant Breeding, Academy of Sciences of the GDR, Gatersleben 4325, GDR.

Syrovatka, M., Boehringer Mannheim GmbH. Wien, Pasettistr. 64, 1020 Wien, Austria.
Timko, J., Institute of Molecular Biology, Slovak Academy of Sciences, 842 51 Bratislava, Czechoslovakia.
Turňa, J., Department of Biochemistry, Faculty of Science, Comenius University, 842 15 Bratislava, Czechoslovakia.
Vaško, M., Institute of Molecular Biology, Slovak Academy of Sciences, 842 51 Bratislava.
Vassileva, R., Institute of Molecular Biology, Bulgarian Academy of Sciences, Sofia 113, Bulgaria.
Waltschewa, L., Institute of Molecular Biology, Bulgarian Academy of Sciences, Sofia 1113, Bulgaria.
Wasternack, C., Division of Plant Biochemistry, Section of Biosciences, University of Halle/Saale, 402 Halle, Neuwerk 1, GDR.
Wiedmann, B., Central Institute of Molecular Biology, Academy of Sciences of the GDR, Röbert Rossle Str. 10, 1115 Berlin-Buch, GDR.
Wilkinson, A., Chemistry Department, University of York, Heslington, York YO1 5DD, United Kingdom.
Witzel, H., Institute of Biochemistry, University of Muenster, Wilhelm Klemm Str. 2, D-4400 Muenster, Federal Republic of Germany.
Zelinka, J., Institute of Molecular Biology, Slovak Academy of Sciences, 842 51 Bratislava, Czechoslovakia.
Zelinková, E., Institute of Molecular Biology, Slovak Academy of Sciences, 842 51 Bratislava, Czechoslovakia.

AUTHOR INDEX

Calf chymosin, activity, 123
Calf thymus, 2
Candida maltosa, cytochrome P-450
        synthesis, 259-264
N-Carbamyl-β-alanine (NCBA), 143-147
Carbamyl phosphate synthetase
        (CPSase), 143
Carrots, phosphotransferase, 59
CAT boxes, 255, 256
Catharanthus roseus, 143
Cell division cycle (cdc genes), 127
Cells, intact, RNA degradation
        studies, 267-272
Cheese, by recombinant technology,
        126
Chicken c-myb gene, 183-189
  exon-intron structure, 184
Chlamydomonas reinhardii,
  DNA methyltransferase, 2
Chlorella sp., 21
Chloroplasts
  'ALPA-reactive' polypeptides, 278
  spinach, immunology, RNA
        polymerases, 275-281
  sRNAp, 275, 281
  transcriptionally active complex
        (TAC), 275-281
Chromotofocusing, 246, 249
Chromatography
  affinity, 234
  Concanavaline-Sepharose, 234
  DNA methyltransferase, 2
  ECTEOLA-cellulose, 234
  HPLC reverse phase, 4
  ion exchange, ascites tumor cells,
        271
    hydroxylamine reaction, 56
  ribosomal proteins, rye, 246-250
  thermal hydroxylapatite, 4
  thioredoxin separation, 38-40
  TLC, ascites tumor cells, 270-271
Chymosin, prochymosin, cDNA, 123-126
Citrate synthase (CS), 145-146
Clam, ribonucleotide reductase, 22
Codon-specific translation, GTP
        expense, 101-107
Corynebacterium nephridii,
        thioredoxin,
  activity, 30
  analysis, 32, 33
  assay, 28-29
  cloning, 28
  DNA sequencing, 28
  molecular weight, 34
  purification, 29, 34-35
  restriction marks, 31
Coryneforms, ribonucleotide
        reductases, 19-20
COSY spectroscopy, 43
Crotalus see Venom exonuclease
Cruciferae, 6

Cucurbitaceae, 4
  mitochondrial DNA, 131
Culture methods, fermentation, 125
Cyanobacteria, ribonucleotide
        reductases, 20
Cyclic AMP
  control of yeast cell cycle START,
        127-129
  -dependent protein kinase, 127
  receptor protein complex, 161
Cycloheximide, 260
Cytochrome p-450 synthesis, 259-264
Cytosine, methylation sites, 1-2
  N4- and C5-methylation of DNA,
        73-78
  residues, plants, 1-6

Deoxyadenosylcobalamin
  as cofactor, 19
  -requiring nucleotide reductase,
        20
  tritium exchange reaction, 21
Deoxypyrimidine nucleoside, 289-291
  pathways, 293
Deoxyribonucleotides
  formation from ribonucleotides,
        37-43
DEP (diethyl pyrocarbonate), 236,
        238, 241, 242
Dihydro-orotic acid hydrolase
        (DHOase), 143
Dihydropyrimidinase, 144
Dinucleoside oligophosphate
        phosphorylase, 149-153
Dithioerythritol, 40-41
DNA
  4mC, 5mC methylations, 74-78
  AraCTP incorporation, 290
  C5-methylation of cytosine, 73-78
  cDNA,
    brain cells, Alu-carrying
        clones, 191-196
    chicken, c-myb, 183-189
    cloning and characterization,
        175-181
    cross-linkage, nuclear lamina,
        283-286
  DNase II, 285-286
  electrophoresis, 285
  gyrase, active site, 63
  ID sequences, 177-178
  library, recombinant clones, 197
  methylation, in plants, 1-6
  methyltransferase, 1-6
  mitochondrial, restriction
        analysis, 131-135
  N4-methylation, cytosine, 73-78
  nuclear lamina 'freezing', 286
  organelle transfer, in plants, 135
  plant characteristics, 1
  plastid DNA, tomato, 132

306